Andreas Martin (Ed.)

Zeolite Catalysis

MDPI

This book is a reprint of the Special Issue that appeared in the online, open access journal, *Catalysts* (ISSN 2073-4344) from 2015–2016, available at:

http://www.mdpi.com/journal/catalysts/special_issues/zeolite-catalysis

Guest Editor
Andreas Martin
Leibniz-Institute for Catalysis
Germany

Editorial Office
MDPI AG
St. Alban-Anlage 66
Basel, Switzerland

Publisher
Shu-Kun Lin

Managing Editor
Zu Qiu

1. Edition 2016

MDPI • Basel • Beijing • Wuhan • Barcelona • Belgrade

ISBN 978-3-03842-264-8 (Hbk)
ISBN 978-3-03842-265-5 (PDF)

Table of Contents

List of Contributors

Mattia Ardizzi Dipartimento di Chimica Industriale "Toso Montanari", Viale Risorgimento 4, Università di Bologna, 40136 Bologna, Italy.

Udo Armbruster Leibniz-Institut für Katalyse e.V. an der Universität Rostock, Albert-Einstein-Straße 29a, 18059 Rostock, Germany.

Francesca Bortolani Dipartimento di Chimica Industriale "Toso Montanari", Viale Risorgimento 4, Università di Bologna, 40136 Bologna, Italy.

Fabrizio Cavani Dipartimento di Chimica Industriale "Toso Montanari", Viale Risorgimento 4, Università di Bologna, 40136 Bologna, Italy.

Tung Thanh Dang Vietnam Petroleum Institute, 167 Trung Kinh, Yen Hoa, Cau Giay, Hanoi 10000, Vietnam.

Radostina Dragomirova Leibniz Institute for Catalysis, University of Rostock (LIKAT Rostock), Albert-Einstein-Str. 29a, 18059 Rostock, Germany.

Aijun Duan State Key Laboratory of Heavy Oil Processing, China University of Petroleum, Beijing 102249, China.

Wolf-Dietrich Einicke Institute of Chemical Technology, Universität Leipzig, Linnéstraße 3, 04103 Leipzig, Germany.

Dongyu Fan School of Science, Beijing University of Posts and Telecommunications, No. 10, Xitucheng Road, Haidian District, Beijing 100876, China.

Weiwei Fu College of Chemistry, Experimental Center of Shenyang Normal University, Shenyang 110034, China | State Key Laboratory of Inorganic Synthesis and Preparative Chemistry, College of Chemistry, Jilin University, 2699 Qianjin Street, Changchun 130012, China.

Roger Gläser Institute of Chemical Technology, Universität Leipzig, Linnéstraße 3, 04103 Leipzig, Germany.

Gherardo Gliozzi Dipartimento di Chimica Industriale "Toso Montanari", Viale Risorgimento 4, Università di Bologna, 40136 Bologna, Italy.

Jinhua Guo Key Laboratory for Green Chemical Technology of the Ministry of Education, School of Chemical Engineering and Technology, Collaborative Innovation Center of Chemical Science and Engineering (Tianjin), Tianjin University, Tianjin 300072, China.

Ceri Hammond Cardiff Catalysis Institute, School of Chemistry, Cardiff University, Cardiff CF10 3AT, UK.

Jing Han State Key Laboratory of Heavy Oil Processing, China University of Petroleum, Beijing 102249, China.

Shanlei Han State Key Laboratory of Heavy Oil Processing, China University of Petroleum, Beijing 102249, China.

Guiyuan Jiang State Key Laboratory of Heavy Oil Processing, China University of Petroleum, Beijing 102249, China.

Ruifeng Li Institute of Special Chemicals, School of Chemistry and Chemical Engineering, Taiyuan University of Technology, Taiyuan 030024, China.

Guozhu Liu Key Laboratory for Green Chemical Technology of the Ministry of Education, School of Chemical Engineering and Technology, Collaborative Innovation Center of Chemical Science and Engineering (Tianjin), Tianjin University, Tianjin 300072, China.

Jia Liu State Key Laboratory of Heavy Oil Processing, China University of Petroleum, Beijing 102249, China.

Jian Liu State Key Laboratory of Heavy Oil Processing, China University of Petroleum, Beijing 102249, China.

Lijia Liu State Key Laboratory of Inorganic Synthesis and Preparative Chemistry, College of Chemistry, Jilin University, 2699 Qianjin Street, Changchun 130012, China.

Yeming Liu State Key Laboratory of Heavy Oil Processing, China University of Petroleum, Beijing 102249, China.

Benoit Louis Institute of Chemistry, UMR 7177, University of Strasbourg, 1 rue Blaise Pascal, 67000 Strasbourg Cedex, France.

Jinghong Ma Institute of Special Chemicals, School of Chemistry and Chemical Engineering, Taiyuan University of Technology, Taiyuan 030024, China.

Patrizia Mangifesta Dipartimento di Chimica Industriale "Toso Montanari", Viale Risorgimento 4, Università di Bologna, 40136 Bologna, Italy.

Andreas Martin Leibniz-Institute for Catalysis, Albert-Einstein-Str. 29a, 18059 Rostock, Germany.

Leandro S. M. Miranda Instituto de Química, Universidade Federal do Rio de Janeiro, Av. Athos da Silveira Ramos 149, CT Bloco A, Cidade Universitária, 21941-909 Rio de Janeiro, Brazil.

Seitaro Namba Chemical Resources Laboratory, Tokyo Institute of Technology, 4259 Nagatsuta, Midori-ku, Yokohama 226-8503, Japan.

Márcio Nele Universidade Federal do Rio de Janeiro, Escola de Química Av. Athos da Silveira Ramos 149, CT Bloco E, Cidade Universitária, 21941-909 Rio de Janeiro, Brazil.

Duc Anh Nguyen Vietnam Petroleum Institute, 167 Trung Kinh, Yen Hoa, Cau Giay, Hanoi 10000, Vietnam.

Sura Nguyen Vietnam Petroleum Institute, 167 Trung Kinh, Yen Hoa, Cau Giay, Hanoi 10000, Vietnam.

Sauro Passeri Dipartimento di Chimica Industriale "Toso Montanari", Viale Risorgimento 4, Università di Bologna, 40136 Bologna, Italy.

Marcelo M. Pereira Instituto de Química, Universidade Federal do Rio de Janeiro, Av. Athos da Silveira Ramos 149, CT Bloco A, Cidade Universitária, 21941-909 Rio de Janeiro, Brazil.

Binh Minh Quoc Phan Vietnam Petroleum Institute, 167 Trung Kinh, Yen Hoa, Cau Giay, Hanoi 10000, Vietnam.

Alessandra V. Silva Instituto de Química, Universidade Federal do Rio de Janeiro, Av. Athos da Silveira Ramos 149, CT Bloco A, Cidade Universitária, 21941-909 Rio de Janeiro, Brazil.

Aixia Song Institute of Special Chemicals, School of Chemistry and Chemical Engineering, Taiyuan University of Technology, Taiyuan 030024, China.

Qingli Sun State Key Laboratory of Inorganic Synthesis and Preparative Chemistry, College of Chemistry, Jilin University, 2699 Qianjin Street, Changchun 130012, China.

Giulia Tarantino Cardiff Catalysis Institute, School of Chemistry, Cardiff University, Cardiff CF10 3AT, UK.

Takashi Tatsumi Chemical Resources Laboratory, Tokyo Institute of Technology, 4259 Nagatsuta, Midori-ku, Yokohama 226-8503, Japan.

Akram Tawari Institute of Chemical Technology, Universität Leipzig, Linnéstraβe 3, 04103 Leipzig, Germany.

Xuan Hoan Vu Vietnam Petroleum Institute, 167 Trung Kinh, Yen Hoa, Cau Giay, Hanoi 10000, Vietnam.

Ruipu Wang State Key Laboratory of Heavy Oil Processing, China University of Petroleum, Beijing 102249, China.

Runwei Wang State Key Laboratory of Inorganic Synthesis and Preparative Chemistry, College of Chemistry, Jilin University, 2699 Qianjin Street, Changchun 130012, China.

Yajun Wang State Key Laboratory of Heavy Oil Processing, China University of Petroleum, Beijing 102249, China.

Yong Wang Chemical Resources Laboratory, Tokyo Institute of Technology, 4259 Nagatsuta, Midori-ku, Yokohama 226-8503, Japan.

Yuechang Wei State Key Laboratory of Heavy Oil Processing, China University of Petroleum, Beijing 102249, China.

Sebastian Wohlrab Leibniz Institute for Catalysis, University of Rostock (LIKAT Rostock), Albert-Einstein-Str. 29a, D-18059 Rostock, Germany.

Chunming Xu State Key Laboratory of Heavy Oil Processing, China University of Petroleum, Beijing 102249, China.

Duo Xu Institute of Special Chemicals, School of Chemistry and Chemical Engineering, Taiyuan University of Technology, Taiyuan 030024, China.

Toshiyuki Yokoi Chemical Resources Laboratory, Tokyo Institute of Technology, 4259 Nagatsuta, Midori-ku, Yokohama 226-8503, Japan.

Yaoyuan Zhang State Key Laboratory of Heavy Oil Processing, China University of Petroleum, Beijing 102249, China.

Yunxia Zhao Key Laboratory for Green Chemical Technology of the Ministry of Education, School of Chemical Engineering and Technology, Collaborative Innovation Center of Chemical Science and Engineering (Tianjin), Tianjin University, Tianjin 300072, China.

Zhen Zhao State Key Laboratory of Heavy Oil Processing, China University of Petroleum, Beijing 102249, China.

Houbing Zou State Key Laboratory of Inorganic Synthesis and Preparative Chemistry, College of Chemistry, Jilin University, 2699 Qianjin Street, Changchun 130012, China | School of Science, Beijing University of Posts and Telecommunications, No. 10, Xitucheng Road, Haidian District, Beijing 100876, China.

About the Guest Editor

Andreas Martin was born in Wilkau-Haßlau, East-Germany (1955) and received his M.Sc. degree from Dresden University of Technology (TU Dresden), Germany (1980) and his Ph.D. (1986) from the "Academy of Sciences", Berlin, Germany. He received postdoctoral qualification (2005) and "venia legendi" from the University of Jena, Germany. Since 1980 he has been working as a researcher at the Academy of Sciences and since 1993 as project leader at the „Institute of Applied Chemistry" (ACA), Berlin. He had several research stays at the German Universities of Darmstadt, Karlsruhe and Bremen. He was Head of the "Catalytic Processes" Department at ACA from 2002–2005. From 2006 to date he is the Head of "Heterogeneously Catalyzed Processes" Department at Leibniz-Institute for Catalysis (LIKAT), Rostock, Germany. At present he is also an associate professor at the Leibniz-Institute for Catalysis (LIKAT) and University of Rostock. He is the author and co-author of over 250 peer-reviewed publications and holds over 40 German and international patents.

Preface to "Zeolite Catalysis"

The term zeolite is based on the Greek words for "to boil" and "stone" and has been known for more than 250 years. At that time, the Swedish mineralogist, A.F. Cronstedt (1722–1765), observed the formation of a large amount of steam when heating the material Stilbite, pointing to its porous character and adsorption capacity. At present, over 200 different zeolite frameworks have been identified. In general, zeolites are crystalline aluminosilicates with a defined micropore structure. Within zeolites, a good number of elements can be isomorphously incorporated and many more elements or their oxides can be hosted by zeolites. In addition, zeolites display a large variety of pore-mouths sizes, channels, crossings, etc. which has led to their designation as molecular sieves and uses in membrane applications. Nowadays, various hierarchical and composite materials are available offering further interesting properties, e.g., by introduction of mesopores or generation of fibers. Zeolites reveal Brønsted and Lewis acidic properties that can be highly varied as well, thereby justifying the name "solid acids". Zeolites are immensely important in diverse industrial applications such as catalysts and adsorbents, for example in the refinery industry, chemical industry, detergent sector or for solar thermal collectors and adsorption refrigeration.

This Special Issue collection focusses on new developments and recent progress with respect to zeolite-catalyzed chemical reactions, adsorption applications and membrane uses, as well as improved synthesis strategies and characterization techniques. It brought together the recent research of well-known research teams from all over the world.

The editor thanks MDPI and Keith Hohn as Editor-in-Chief of *Catalysts*—the open access catalysis journal—and Mary Fan, as the Senior Assistant Editor, for the opportunity to organize this Special Issue on "Zeolite Catalysis" and for their immense support, and significant encouragement and patience. I would also like to thank the contributing authors and colleagues acting as peer reviewers for their efforts in preparing high quality manuscripts, and in further improving the manuscripts several times following comments and suggestions of the reviewers.

Andreas Martin
Guest Editor

Zeolite Catalysis

Andreas Martin

Reprinted from *Catalysts*. Cite as: Martin, A. Zeolite Catalysis. *Catalysts* **2016**, *6*, 118.

1. Background

The Special Issue "Zeolite Catalysis" published in the online journal *Catalysts* was recently successfully completed. A good number of peer-reviewed publications were published reflecting the broadness of the zeolite syntheses, characterizations and various application fields. This issue brought together recent research of well-known research teams from all over the world.

The term "zeolite" is based on the Greek words for "to boil" and "stone" and it has been known since 250 years ago. At that time, the Swedish mineralogist, A.F. Cronstedt (1722–1765), observed the formation of a large amount of steam when heating the material Stilbite which indicated its porous character and adsorption capacity. At present, over 200 different zeolite frameworks have been identified. In general, zeolites are crystalline aluminosilicates with a defined micropore structure. Within zeolites, a good number of elements can be isomorphously incorporated and much more elements or their oxides can be hosted by zeolites. In addition, zeolites comprise a large variety in size of pore mouths, channels, crossings, etc. leading also to their designation as molecular sieves and use in membrane applications. Nowadays, various hierarchical and composite materials are designed offering further interesting properties, e.g., by the introduction of mesopores or generation of fibers. Zeolites reveal Brønsted and Lewis acidic properties that can be varied in wide limits as well. Thus, they deserve the name "solid acids". Zeolites have an immense importance in diverse industrial applications such as catalysts and adsorbents, for example in the refinery industry, chemical industry, detergent sector or for solar thermal collectors and adsorption refrigeration.

2. This Special Issue

The aim of the Special Issue was directed to new developments and recent progress with respect to zeolite-catalyzed chemical reactions, adsorption applications and membrane uses, as well as improved syntheses strategies and characterization techniques. Xuan Hoan Vu and colleagues [1] reported on the synthesis of novel ZSM-5 containing hierarchical composites and their use in catalytic cracking of triglyceride-rich biomass to lower olefins. It could be proven that the yield for propene and butenes can be increased using such composites. Aixia Song et al. [2] studied the adsorption and diffusion properties of zeolite Beta using three xylene isomers. Adsorption isotherms from microporous and mesoporous zeolite

Beta were recorded showing the impact of mesopores on adsorption properties. Houbing Zou and co-workers [3] reported on facile synthesis of yolk/core-shell structured TS1@mesosilica composites, catalytic properties were checked in the challenging hydroxylation of phenol. The catalyst characterizations showed a high surface area of 560–700 m^2/g and a hierarchical pore structure with mesochannels and micropores. In comparison to well-known TS-1, the synthesized solids reveal enhanced activity at comparable selectivity. The research group of Fabrizio Cavani [4] contributed to the Special Issue with an article on the use of zeolite catalysts for phenol benzoylation with benzoic acid. The aim of this work was the synthesis of hydroxybenzophenons which are important intermediates in the chemical industry. H-Beta zeolites offer superior performance compared to H-Y solid. The studies were supported by various mechanistic insights. Radostina Dragomirova and Sebastian Wohlrab [5] extensively summarized the application of zeolite membranes in catalysis. The detailed review is backed by ca. 300 references on zeolite membrane preparation, separation principles as well as basic considerations on membrane reactors. The given classification according to membrane location considers: (i) membranes spatially decoupled from the reaction zone; (ii) packed bed membrane reactors; (iii) catalytic membrane reactors; and (iv) zeolite capsuled catalyst particles. Ceri Hammond and Giulia Tarantino [6] reported on post-synthesis modifications of TS-1 to suppress undesirable H_2O_2 decomposition in hydroxylations. Ti site speciation changes were observed by in-situ spectroscopic techniques. Takashi Tatsumi and colleagues [7] in their contribution described effects of dealumination and desilication of Beta zeolite and the consequences for their catalytic performance in n-hexane cracking to propene. Dealumination was carried out by HNO_3 treatment; desilication was obtained by alkali treatment. The propene selectivity at high n-hexane conversions was increased after alkali treatment followed by acid treatment. This is due to: (i) the decrease in number of acidic sites; and (ii) by an increase in number of mesopores which are beneficial to the diffusion of coke precursor compounds. Jing Han et al. [8] reported on the manufacture of Ga_2O_3/ZSM-5 hollow fibers for use as efficient dehydrogenation catalysts for n-butane conversion. Light olefin yields could be increased significantly compared to Ga_2O_3, ZSM-5 fibers and GaO_3 supported on ZSM-5. Guozhu Liu and coworkers [9] in their article showed the catalytic properties of Pt/H-ZSM-5 in the conversion of lignin-based phenols into xylene isomers. The addition of methanol to the reaction mixtures leads to increased xylene yields. The impact of MeOH addition is attributed to the combined action in both the reaction pathways: methylation of m-cresol into xylenols followed by hydrodeoxygenation to form p-/m-xylene, and hydrodeoxygenation of m-cresol into toluene followed by methylation into p-/m-xylene. Alessandra Silva et al. [10] reported on the synthesis of ZSM-5 zeolites using biomass such as sugar cane bagasse as structure directing agent. MFI crystals with different morphologies were obtained that were different from the pristine zeolite formed in the absence of biomass. The research team of Roger Gläser [11] contributed to the Special Issue with a report on photocatalytic oxidation of NO over TiO_2/ZSM-5 composites. Various composites were synthesized using different TiO_2 sources. The highest NO conversion of ca. 40% was obtained with a catalyst from sol–gel synthesis with equal amounts of the two components after calcination at 250 °C.

This short survey proves the potential of zeolites and zeolite-based materials in modern catalysis and related research areas. I have no doubt that further articles on the above mentioned topics will be published in *Catalysts* soon.

Conflicts of Interest: The author declares no conflict of interest.

References

1. Vu, X.H.; Nguyen, S.; Dang, T.T.; Phan, B.M.Q.; Nguyen, D.A.; Armbruster, U.; Martin, A. Catalytic Cracking of Triglyceride-Rich Biomass toward Lower Olefins over a Nano-ZSM-5/SBA-15 Analog Composite. *Catalysts* **2015**, *5*, 1692–1703.
2. Song, A.; Ma, J.; Xu, D.; Li, R. Adsorption and Diffusion of Xylene Isomers on Mesoporous Beta Zeolite. *Catalysts* **2015**, *5*, 2098–2114.
3. Zou, H.; Sun, Q.; Fan, D.; Fu, W.; Liu, L.; Wang, R. Facile Synthesis of Yolk/Core-Shell Structured TS-1@Mesosilica Composites for Enhanced Hydroxylation of Phenol. *Catalysts* **2015**, *5*, 2134–2146.
4. Gliozzi, G.; Passeri, S.; Bortolani, F.; Ardizzi, M.; Mangifesta, P.; Cavani, F. Zeolite Catalysts for Phenol Benzoylation with Benzoic Acid: Exploring the Synthesis of Hydroxybenzophenones. *Catalysts* **2015**, *5*, 2223–2243.
5. Dragomirova, R.; Wohlrab, S. Zeolite Membranes in Catalysis—From Separate Units to Particle Coatings. *Catalysts* **2015**, *5*, 2161–2222.
6. Hammond, C.; Tarantino, G. Switching off H_2O_2 Decomposition during TS-1 Catalysed Epoxidation via Post-Synthetic Active Site Modification. *Catalysts* **2015**, *5*, 2309–2323.
7. Wang, Y.; Yokoi, T.; Namba, S.; Tatsumi, T. Effects of Dealumination and Desilication of Beta Zeolite on Catalytic Performance in n-Hexane Cracking. *Catalysts* **2016**, *6*, 8.
8. Han, J.; Jiang, G.; Han, S.; Liu, J.; Zhang, Y.; Liu, Y.; Wang, R.; Zhao, Z.; Xu, C.; Wang, Y.; et al. The Fabrication of Ga_2O_3/ZSM-5 Hollow Fibers for Efficient Catalytic Conversion of n-Butane into Light Olefins and Aromatics. *Catalysts* **2016**, *6*, 13.
9. Liu, G.; Zhao, Y.; Guo, J. High Selectively Catalytic Conversion of Lignin-Based Phenols into para-/m-Xylene over Pt/HZSM-5. *Catalysts* **2016**, *6*, 19.
10. Silva, A.V.; Miranda, L.S.M.; Nele, M.; Louis, B.; Pereira, M.M. Insights to Achieve a Better Control of Silicon-Aluminum Ratio and ZSM-5 Zeolite Crystal Morphology through the Assistance of Biomass. *Catalysts* **2016**, *6*, 30.
11. Tawari, A.; Einicke, W.-D.; Gläser, R. Photocatalytic Oxidation of NO over Composites of Titanium Dioxide and Zeolite ZSM-5. *Catalysts* **2016**, *6*, 31.

Zeolite Membranes in Catalysis—From Separate Units to Particle Coatings

Radostina Dragomirova and Sebastian Wohlrab

Abstract: Literature on zeolite membranes in catalytic reactions is reviewed and categorized according to membrane location. From this perspective, the classification is as follows: (i) membranes spatially decoupled from the reaction zone; (ii) packed bed membrane reactors; (iii) catalytic membrane reactors and (iv) zeolite capsuled catalyst particles. Each of the resulting four chapters is subdivided by the kind of reactions performed. Over the whole sum of references, the advantage of zeolite membranes in catalytic reactions in terms of conversion, selectivity or yield is evident. Furthermore, zeolite membrane preparation, separation principles as well as basic considerations on membrane reactors are discussed.

Reprinted from *Catalysts*. Cite as: Dragomirova, R.; Wohlrab, S. Zeolite Membranes in Catalysis—From Separate Units to Particle Coatings. *Catalysts* **2015**, *5*, 2161–2222.

1. Introduction

It is not possible to imagine industrial catalysis without zeolites. Over the years, zeolites gained that importance owing to their outstanding properties which are (i) high surface area; (ii) pore sizes in the molecular range; (iii) adsorption capacity; (iv) controllable adsorption properties; (v) inherent active sites; (vi) shape selectivity and (vii) stability [1,2]. Certainly, due to these unique features the application of zeolites is not only restricted to catalysis. Their potential to serve as highly selective sorption materials make zeolites also indispensable for industrial separation tasks [3]. Highlighting, separation and purification of gases [4], especially swing adsorption techniques [5–7], as well as water and waste water treatment [8,9] are also not imaginable without zeolitic materials.

To run separation processes continuously and without recurring regeneration steps zeolite membranes have been developed and studied over the last decades [10–20]. With this development, the utilization of zeolite membranes was progressively investigated for catalytic reactions [21–32]. The special interest in membranes for catalysis science lies in the possibilities of equilibrium shifts, improved yields and selectivities as well as more compact operations compared to conventional processes. Nevertheless, no industrial commercialization of zeolite membrane reactors has been occurring until now. However, recent pioneering developments towards sub-μm membranes [33–38] or the reduction of defects [39–41] promise a revival of zeolite membrane applications. Furthermore, novel cost reduction concepts [42] will make zeolite membranes more attractive for industry.

However, a first application of zeolite membranes in industry already exists. It is the use of hydrophilic zeolite membranes in the dehydration of organic solvents. In detail, NaA membranes have been used in a large-scale pervaporation plant mainly for alcohols by BNRI (Mitsui Holding) [43], and further progress on LTA membranes in the pervaporation separation of water was achieved over the years [44–47]. An important consideration on the way for industrial application might be the availability of membranes characterized by suitable performances at reasonable prices. On the one hand, a 10,000 m^2/year production line for LTA membranes was established [48]. In this context, the further up-scaling also of other membrane types will be of significant importance whether zeolite membranes will be applied in industrial catalysis or not. On the other hand, over the last years, membrane coatings on catalyst particles have been successfully developed and applied as micro membrane reactors [24]. Perhaps, these zeolite capsuled catalysts will be the final breakthrough for zeolite membrane reactors in industry?

A review of the state of the art of permselective zeolite membranes in reaction processing is given in the following for anyone who is already engaged with the matter and for those who want to start with. In particular, the application of zeolite membranes for a bunch of different possible reactions is presented. In the first part, a short overview on zeolite membrane synthesis will be given, followed by discussion on the transport mechanisms in a zeolite membrane. Afterwards, the application of zeolite membrane reactors in diverse configurations will be presented and discussed for the different reaction types. Starting with zeolite membranes apart from the reaction zone over packed bed membrane reactors and catalytic membrane reactors, zeolite coatings on catalyst particles are finally covered.

2. Separation by Zeolite Membranes

2.1. Synthesis of Zeolite Membranes

Over the last decades a great research effort is allocated in the preparation of zeolite membranes applying different synthesis techniques. Generally, for the synthesis of zeolite membranes two procedure routes are followed. On the one hand, one-step techniques referred to as direct *in situ* crystallization are applied. Thereby, the surface of the untreated either tubular or disc support is brought in a direct contact with an aluminosilicate precursor solution and the membrane crystallization is performed under hydrothermal conditions as shown exemplarily in the following references [49–55]. Alternatively, Caro *et al.* proposed a seeding-free synthesis strategy for the preparation of dense and phase-pure zeolite LTA and FAU membranes using 3-aminopropyltriethoxysilane (APTES) as covalent linker between the zeolite layer and the alumina support [56,57]. Moreover, considerable effort has been directed towards controlled crystal orientation of MFI by *in situ* crystallization [58–61].

On the other hand, two-step syntheses referred to as secondary growth are applied for zeolite membrane preparation. Therefore, a seeding layer is deposited on the membrane support at first and the membrane layer is grown in a second step via hydrothermal synthesis [62–67]. Seeds can be prepared either by bottom-up approaches [15] or by top-down techniques *ex situ*. The latter include template assisted nanoscale zeolite syntheses [68,69], exfoliation of zeolite sheets [34,70,71] or simply crushing. The nature of the obtained zeolite membrane depends strongly on the seeding and the subsequent secondary growth technique applied. Figure 1 displays two examples of MFI membranes obtained by the two-step hydrothermal preparation procedure. Figure 1a illustrates the outcome of a recent development resulting in the most probable thinnest membrane ever reported [34]. In detail, MFI-nanosheets were prepared by exfoliation of multilamellar MFI. After purification, these sheets were supported onto either Stöber silica supports or novel highflux, high-strength porous silica fibre supports. Secondary intergrowth of the film was performed using the so called gel-less growth technique first described by Pham *et al.* [35]. This approach could be a more powerful strategy for the preparation of sub-µm membranes compared to the alternative Langmuir through assembly [36,72].

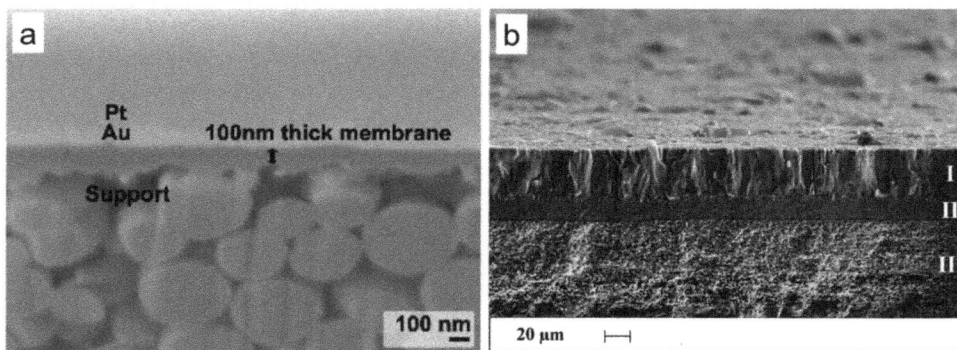

Figure 1. Scanning electron micrographs of (**a**) MFI film prepared by gel-less growth of MFI-nanosheets on a Stöber silica support (Reprinted with permission from [34]. Copyright (2015) John Wiley and Sons.); and (**b**) MFI membrane from ball milled MFI seeds and subsequent secondary growth (I) MFI membrane; (II) MFI seed layer; and (III) alumina microfiltration layer on a macroporous alumina support. (Reprinted from [67]. Copyright (2011) Elsevier).

Figure 1b shows a MFI membrane obtained by seeding of ball milled silicalite used for the functionalization of an inert alumina microfiltration layer (denoted as III in Figure 1b) located on a macroporous Al_2O_3 support. Thus, the resulting MFI seed layer (denoted as II in Figure 1b) acts both as heterogeneous nucleation site and as flexible distance holder between support and MFI membrane, suppressing defect formations during thermal template removal. Onto the MFI seed layer the MFI membrane layer (denoted as I in Figure 1b) was formed by hydrothermal synthesis [67]. Of course, both archetypes provide different properties. The thin membrane from Figure 1a offers high flux whereas the membrane in Figure 1b possesses high pressure stability.

Parasitic twin crystals were identified as defect-forming during the thermal removal of the structure directing agent (SDA) and should be prevented. By using a low SDA/Si reactant ratio (≤ 0.05) it was found, the formation of twin crystals can be reduced [40]. Alternatively, SDA-free zeolite syntheses are available [39,41], and a template-free secondary growth synthesis of MFI type zeolite membranes [73] is known for nearly 15 years.

On the search of the most efficient synthesis technique, several more methods were reported. A secondary growth method with vacuum seeding for the preparation of A-type zeolite membranes [74] and diverse techniques for crystal orientation in zeolite membranes [75–82] were proposed. A vapor-phase transport (VPT) method was first reported by Xu *et al.* as an alternative approach to the hydrothermal synthesis for the preparation of ZSM-5 zeolite [83]. The VPT method was extensively studied by Matsukata *et al.* for the synthesis of defect-free zeolite-alumina composite membranes [84], MFI-type zeolitic membrane [85] as well as preferentially oriented MFI layers [86], compact ferrierite (FER)-alumina composite layer [87,88], where a dry aluminosilicate gel layer was deposited on the support and then further crystallized under vapors of amines and water. Besides, the achieved progress in microwave assisted syntheses of zeolite membrane has been outlined by Li *et al.* [89]. Gascon and coworkers debated on the limitations of the existing preparation techniques and evaluated future perspectives of zeolite and zeolite-type materials for membrane production [19].

For further reading on the progress in the seeding and secondary growth techniques used for the preparation of zeolite membranes two recent reviews are recommended [42,48].

2.2. Permeation in Zeolite Membranes

According to Weisz, classical catalytic reactors should work in a certain "window of reality" in order to run efficiently. Consequently, the optimal space time yield of a conventional membrane reactor should be centered around STY = 10^{-6}–10^{-5} mol·cm^{-3}·s^{-1} [90]. Weisz explained this window with time limitations at the lower border and issues on mass flow, diffusion and heat transfer at the upper limit. Years later, in 1997, Boudart suggested membrane reactors should be classified analogously [91]. When an industrial reactor should have its STY around 10^{-6}–10^{-5} mol·cm^{-3}·s^{-1}, a membrane reactor should be located in the same window—defined by its areal time yield (ATY). Boudart referred to a Pd/Al$_2$O$_3$ membrane and its ability to permeate hydrogen at a permeability P of at least 10^{-5} mol·cm^{-2}·s^{-1} [92]. By assuming a cylindrical reactor of diameter d, ATY can be calculated from the STY by multiplying with the surface to volume ratio, $d/4$. As an example, a reactor tube of 40 cm in diameter would fit in a "window of reality" of a membrane reactor matching P, or ATY $(40/4 \times 10^{-6}$ mol·cm^{-2}·s$^{-1})$. By simply adjusting the diameter of the membrane tube, the "window of reality" can be reached in the case of high permeation rates. What about zeolite membranes?

Van de Graaf et al. compared the volume ratio of the catalytic reactor derived from the productivity per unit volume (defined as STY) to the permeation per membrane area (defined as permeation flux), or in other words the ATY [93,94]. By dividing STY through ATY, the area to volume ratio (A/V) of the catalytic membrane reactor is obtained as a simple measure of the industrial feasibility of membrane reactors. However, the authors calculated A/V values between 20 and 5000 m^{-1} for porous inorganic membranes, whereas the example referred by Boudart (Pd/Al$_2$O$_3$ [92]) shows a much better performance of $A/V = 10$. It is clear, the smaller the A/V ratio the more realistic becomes an industrial transfer.

Deeper insights between catalytic reaction and permeation can be obtained by comparison of catalytic performance and permeation rate—which are the two limiting factors of a membrane reactor. The catalytic performance can be understood as ratio between reaction rate and convective transport rate of the feed, given as Damkohler number (Da). The ratio of convective transport to permeation rate through the membrane is the so called Peclet number (Pe). The product of both numbers defines the efficiency of a given membrane reactor [95–97]. Hence, a catalytic membrane reactor can be optimized either by catalyst activity adjustment or by manipulating the permeability of the membrane. For industrial applications the focus should be on the latter: (i) diameters of membrane supports can be reduced up to a certain value (e.g., as hollow support fibres); and (ii) permeability can be increased. Recent developments in the fabrication of ultrathin membranes (see Section 2.1.) are promising enough to overcome barriers. Since permeation is inversely proportional to membrane thickness [98] a novel generation of fast permeating zeolite membranes can be directed towards industrial applications.

Generally, the transport of molecules through zeolite membranes depends strongly on the membrane pore size and the interaction of the permeating species with the zeolite structure and can be modelled mainly as combined effect between adsorption and diffusion. This surface diffusion of adsorbed species from multi-component mixtures can be described by the aid of the Maxwell-Stefan model as Krishna and co-workers impressively have been demonstrated over the past years [99–103]. For further reading on modelling the permeation through zeolite membranes a recent review by Rangnekar *et al.* is recommended [42].

The permeation through the zeolite membrane is controlled by either shape selectivity, diffusion or adsorption properties [12]. Considering shape selectivity, the separation ability of a membrane is based on retaining components larger in size than the zeolite membrane pores and permeation of only the smaller components [65,104–106]. For mixtures having components with similar adsorption properties the gas transport is determined by the mobility of molecules inside the zeolite pores. Exemplarily, the diffusion controlled permeation was demonstrated for *n*-butane/*i*-butane mixtures in MFI membranes [107–109]. In detail, both components show strong adsorption in the zeolite so that the permeation is mainly governed by the diffusion mobilities of the components. For mixtures comprising components with different adsorption and diffusion properties the selective gas transport is predominantly controlled by adsorption. So for example, for hydrogen/*n*-butane and methane/*n*-butane mixtures the permeation fluxes of the less adsorbing components, in these cases hydrogen and methane, respectively, are significantly suppressed by the strongly adsorbing component *n*-butane. Thereby, the arising higher occupancy of *n*-butane leads to a higher driving force for its diffusion [99,110]. Similarly, for mixtures of ethane and *i*-butane the passage of ethane through the membrane was retained by pore blocking effect caused by the stronger adsorption of *i*-butane on the MFI zeolite [111]. For sorption-driven separation processes the transport through porous single-crystal membranes [112] further adopted for zeolite membrane [113] was described by a five-step transport model, including: (1) molecule adsorption from the gas phase at the external surface of the zeolite; (2) transport from the external surface into the pores; (3) intracrystalline transport; (4) transport from the pores to the external surface and (5) desorption from the external surface to the gas phase. For mixtures comprising strongly and weakly adsorbing components, the membrane performance is significantly affected by the operating parameters—pressure and temperature. In this regard, by employing experimental configurations with varying operating conditions the significance of adsorption and diffusion [67,114], desorption [115] as well as condensation [116] was recently demonstrated by our group for the separation of methane/*n*-butane mixtures by MFI membranes.

As shown until now, zeolite membranes could offer diverse separation properties. According to the specific needs of a process, the type of membranes but also their localization to the reaction zone has to be well-chosen. In the following subchapter the advantages of implementing zeolite membranes in catalytic reactions are reviewed for spatial decoupled processes before we proceed with arrays of membranes and catalysts in close contact with each other in the following main chapters.

2.3. Zeolite Membrane Separation Spatially Decoupled from the Catalytic Unit

Zeolite membrane modules could be applied apart from the reaction zone as alternatives to the complex conventional separation processes. Figure 2 reveals two application possibilities of zeolite membranes where the separation process is either for (i) feed treatment (Figure 2A) or (ii) product treatment (Figure 2B) with optional retentate stream recycle. The current chapter briefly discusses only examples where the catalytic process is spatially decoupled from the membrane unit.

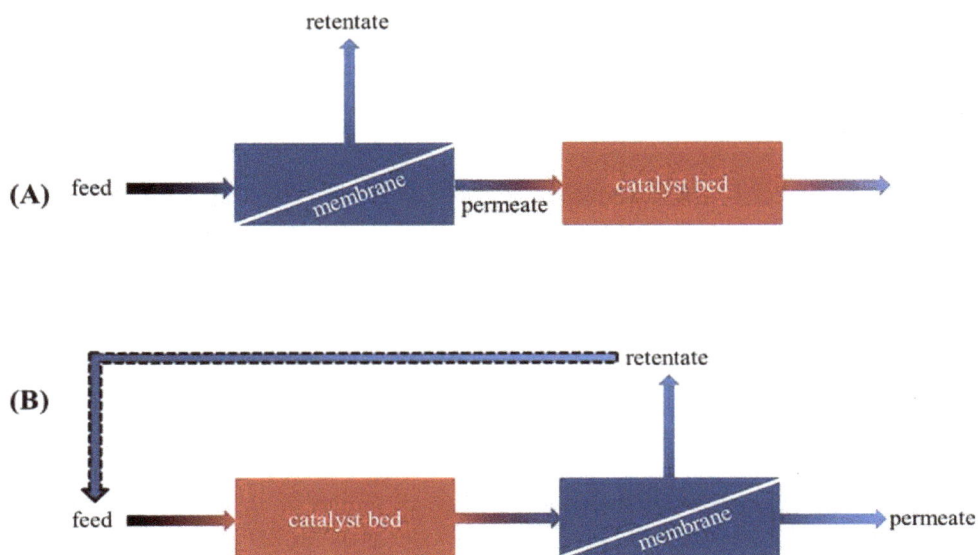

Figure 2. Schematic drawing of different set-ups of zeolite membranes apart from the reaction zone; (**A**) feed treatment; and (**B**) product treatment with optional retentate recycle.

Starting with the feed treatment configuration, the membrane module could be used to adjust the feed stream in order to intensify processes. Beside polymer membranes [117], MFI zeolite membranes can be applied for the conditioning of natural gas [118]. Simplified natural gas model mixtures comprising methane and n-butane were separated with high efficiency under permeate vacuum [115]. Well-pronounced loading gradients across the membrane, decreased coverages of the adsorbed n-butane molecules at the permeate side as well as decreased diffusion resistances were registered. The resulting permeate streams, highly enriched with n-butane, were converted with steam at low (450 °C) and high (750 °C) temperatures over 1 wt. % Rh/γ-Al$_2$O$_3$ for syngas production [119]. The positive influence of the membrane upstream on the steam reforming was demonstrated. For example, at a temperature of only 450 °C a H_2 volume content of 57 vol. % at a H_2/CO ratio of 3.6 could be obtained from a methane/n-butane mixture with φ_{C4H10} 70.4 vol. % underlining the potential for natural gas processing and conversion.

Considering product treatment, the membrane module is placed downstream to the catalytic reactor in order to purify the product stream, thus allowing membrane process conductions at operating conditions different from those of the catalytic unit. In this context, the applicability of zeolite membranes for the separation of hydrogen from reforming streams during syngas production has been recently reviewed [31,120]. Interestingly, during the evaluation of as-prepared and ion-exchanged zeolite LTA membranes for the separation of hydrogen from a simulated gas reformer mixtures, Cs-exchanged LTA membrane demonstrated stable H_2 permeance in the presence of water [121]. Nenoff et al. evaluated theoretically the hydrogen separation selectivities of silicalite and ETS-10 membranes [122] as well as of zeolite NaA and zinc phosphate molecular sieve $Na_3ZnO(PO_4)_3$ [123] using Grand Canonical Monte Carlo techniques. The same working group modified the internal surface of MFI-type zeolite membranes by silane precursors and obtained a H_2/CO_2 permselectivity of 141 combined with high hydrogen permeance [124].

On the other hand, the well-established capability of hydrophilic NaA zeolite membranes for selective water separation has entailed the first industrial implementation of zeolite membranes for de-watering of ethanol and i-propanol [43]. The high separation efficiency of this kind of membrane in pervaporation processes (see for instance reviews [13,125]) has motivated further research effort coupling water releasing reactions and water separations in equilibrium limited reactions. As an alternative to reactive distillation, different reactor configurations including pervaporation membranes could be applied in the esterification reactions [126], where the water permeation through the membrane is assisted by the mean of applied vacuum or sweep gas and then further condensed in a cold trap. In the last decades, membranes coupled to esterification processes have been investigated considering different zeolite structures, including zeolite A [127], sodalite (SOD) [128], chabazite CHA [129].

So for instance, Jafar et al. [127] investigated the homogeneously catalyzed esterification of lactic acid and ethanol by p-toluene sulphonic acid as catalyst, using a zeolite A membrane supported on carbon/zirconia tube for the separation of the produced water. Despite the superior performance of the zeolite A membrane in terms of water flux and selectivity [130,131], its low stability in acidic media [132] imposed the need of performing the process under vapor permeation conditions in order to avoid the direct contact of the membrane with the acidic reaction environment. Not surprisingly, high separation selectivities were achieved during the experiments resulting in significantly improved yields above the equilibrium limit. Similarly, Hasegawa et al. [129] adopted CHA-type zeolite membrane in the

11

vapor phase during the esterification of adipic acid with isopropyl alcohol using sulfuric acid as catalyst. Thus, the profit from the effective water removal by the membrane was revealed since the yield of diisopropyl adipate reached 98%. In comparison an equilibrium yield of 56% was obtained during the operation without membrane. After 10 runs the permeation fluxes of water were reduced, however the reaction maintained stable performance showing 99% esterification conversion and 97% yield. On the other hand, acid-resistant hydrophilic merlionite (MER), phillipsite (PHI) and chabazite (CHA) zeolite membranes [133] have been also applied in pervaporation-aided ester condensation reactions with alcohols in order to extract water and so shift the equilibrium position. The authors recorded more than 20% increased yield compared to the equilibrium conversion demonstrating the profit of using zeolite membranes for selective water removal.

Hydroxy sodalite (SOD) zeolite membranes combine excellent separation of water from various organic alcohol streams and good resistance to the acidic medium in esterification reactions as shown by Khajavi *et al.* [128,134]. For this reason, SOD zeolite membranes were applied in the pervaporation-aided esterification of acetic acid with either ethanol or 1-butanol. The membrane exhibited absolute water selectivity, thereby. Moreover, the membrane was able to permeate water in rates comparable to its formation rate giving rise to enhanced yield and almost complete conversion. Importantly, the membrane demonstrated stable water permeation under mild acidic conditions for long period of operation. This behavior was ascribed by the authors mainly to low aluminum leaching out of the zeolite structure due the narrow window openings of the SOD zeolite.

Very recently, kinetic modeling of pervaporation-aided esterification of propionic acid and ethanol with T-type zeolite membranes has been performed and compared with the experimental data obtained in order to provide further insights on the effect of the operating parameters temperature, ethanol/acid molar ration, as well as ratio of membrane area to amount of initial reaction liquid [135]. As a result, more effective water removal and thus improved esterification conversion could be ensured by the use of membranes with larger area, however to the disadvantages of higher equipment costs. Further example of process intensification by incorporation of NaA zeolite membrane tubes into a reactive distillation column for the etherification of tert-amyl alcohol with ethanol was given by Aiouache *et al.* [136]. The hybrid configuration was found to be effective for the removal of water and consequently for the surpassing of the thermodynamic limitations leading to increased tert-amyl ethyl ether yield.

However, the special implementation of zeolite membranes in membrane reactors, where catalysis and selective separation are combined in the same unit, could offer several advantages in comparison to the conventional reactor (CR) configuration and is considered a promising concept to overcome limitations in the performance of catalytic reactors [22,30,137–140].

3. Membrane Reactor Concepts

According to the IUPAC definition, a membrane reactor couples a chemical reaction and a membrane-based separation process in the same unit so as to intensify the whole process [141]. Generally, for this purpose permselective and non-permselective membranes can be employed in different membrane reactor configurations as show in Figure 3.

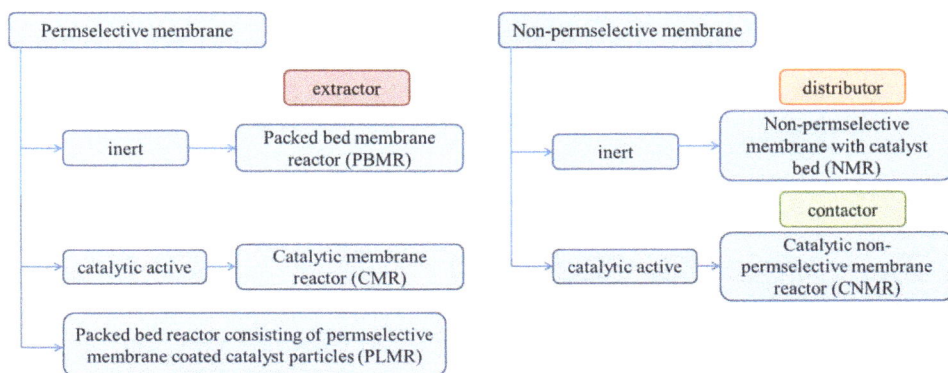

Figure 3. Classification of membrane reactor configurations according to membrane function, after conventional classifications [23,26,29,142,143].

On the one hand, depending on how permselective membranes are combined with catalysts in the reactor unit, membrane reactors can be divided generally into packed bed membrane reactors (PBMR) and catalytic membrane reactors (CMR). On the other hand, the membrane utilized in the membrane reactor can also be non-permselective so that further membrane reactor configurations are possible, namely non-permselective membrane reactors (NMR) with direct contact to a catalyst bed and catalytic (active) non-permselective membrane reactors (CNMR). Another membrane reactor configuration is represented by reactors packed with permselective membrane-coated catalyst particles (PLMR). These capsuled particles can be also understood as microscopic structured membrane reactors.

In the following, we want to discuss the classification of zeolite membranes on the basis of their function either as extractor, distributor or contactor.

3.1. Extractor Type Zeolite Membrane Reactors

As extractors either catalytic or inert permselective membranes can be applied so that here the CMR and PBMR configurations are included. The membrane has thereby the function to remove selectively one or more products from the reaction zone, thus improving the conversion/selectivity/yield in equilibrium limited reactions. Alternatively, the membrane could be employed to supply selective reactants from the feed mixture in order to enhance the selectivity. Possible applications of extractors with either catalytic active permselective membranes in CMR or inert permselective membranes in PBMR configurations are listed in Table 1 and will be discussed in further details in Sections 4 and 5.

Although the main focus of this review is laid on permselective membranes some examples considering the application of non-permselective membranes in membrane reactor should be given in the following paragraphs for completeness.

3.2. Distributor Type Zeolite Membrane Reactors

The distributor type membrane reactor configuration is characterized by controlled permeation/dosing of reactants via a non-permselective membrane. Especially for oxidation reactions, such a configuration could offer diverse benefits compared to conventional reactors, namely precise distribution of the reactant along the catalyst bed, thus minimizing the local appearance of dangerous conditions. Mota *et al.* [172] carried out the selective oxidation of butane to maleic anhydride in membrane reactors of the distributor type combining zeolite MFI membranes used to distribute oxygen and vanadium phosphorus mixed oxides-based catalysts known for their high selectivity and conversion of alkanes [173]. Even though, the obtained results were in fact quite similar with those of the conventional co-feed configuration, the authors pointed out, that the separated O_2 feeding could be beneficial to avoid flammability problems. Nevertheless, employing non-permselective mesoporous ceramic membranes to distribute the oxygen allows an operation with higher n-butane concentrations leading to higher maleic anhydride yields [174,175]. Moreover, Mallada *et al.* [174] reversed the butane flow in the inner volume of the membrane reactor in order to overcome the problem with observed heterogeneity of the oxidation state of the catalyst bed. As a result, maleic anhydride yields above those obtained in the conventional reactor mode were recorded. In a later work from the same working group, zeolite membranes were combined with cobalt-doped vanadium phosphorus mixed oxides-based catalysts in membrane reactors operating at high butane concentrations in the feed [176]. Due to the O_2 distribution, the studied membrane reactors were able to eliminate the formation of critical concentrations and allowed the operation at the flammability zone, giving rise to three times higher maleic anhydride productivity.

Table 1. Overview of the application of permselective inert or catalytic active zeolite membranes in membrane reactor configurations for different processes reported in the literature.

Reaction	Reactor Type	Feed	Operating Conditions	Catalyst	Membrane	X_{CR} (%) S_{CR} (%) Y_{CR} (%)	X_{MR} (%) S_{MR} (%) Y_{MR} (%)	References
Dehydrogenation of ethylbenzene to styrene	CMR	water, ethyl-benzene	600 °C sweep gas: nitrogen	Fe-MFI/α-Al₂O₃ tube		$X_{ethylbenzene}=45.1$ $S_{styrene}=92.8$ $Y_{styrene}=41.9$	$X_{ethylbenzene}=60.1$ $S_{styrene}=96.9$ $Y_{styrene}=58.6$	[144]
Dehydrogenation of i-butane	PBMR	i-butane, hydrogen, balance nitrogen	730 K $p=100$–170 kPa sweep gas: nitrogen	PtIn/MFI 0.8 wt. % In 0.5 wt. % Pt	MFI/α-Al₂O₃ tube	n.r.	n.r.	[145]
Dehydrogenation of i-butane	PBMR	pure i-butane	510 °C WHSV = 0.5–1.6 h⁻¹ sweep gas: nitrogen	Cr₂O₃/Al₂O₃	MFI/α-Al₂O₃ tube	$X_{i\text{-butane}}=29.1$ $S_{i\text{-butene}}=\sim90$	$X_{i\text{-butane}}=41.7$–48.6 $S_{i\text{-butene}}=96$	[146]
Dehydrogenation of i-butane	PBMR	pure i-butane	712–762 K $p_{feed}=101$ kPa sweep gas: nitrogen	Cr₂O₃/Al₂O₃	DD3R/α-Al₂O₃ tube	$Y_{i\text{-butene}}=0.28$ at 762 °C	$Y_{i\text{-butene}}=0.41$ at 762 °C	[147]
Dehydrogenation of cyclohexane	PBMR	cyclo-hexane diluted in argon	423–523 K $p=101.3$ kPa Sweep gas: argon	Pt/Al₂O₃ 1 wt. % Pt	FAU/α-Al₂O₃ tube	$X_{cyclohexane}=32.2$	$X_{cyclohexane}=72.1$	[148,149]
Dehydrogenation of ethylbenzene to styrene	PBMR	water and ethyl-benzene	580–640 °C $\Delta p = 0.8$ atm sweep gas: nitrogen	Fe₂O₃	silicalite-1/stainless steel tube	$X_{ethylbenzene}=67.5$ at 610 °C	$X_{ethylbenzene}=74.8$ at 610 °C	[150]
Dehydrogenation of ethane	PBMR	pure ethane	500–550 °C $p_{feed}=104$ kPa $p_{perm}=101.3$ kPa sweep gas: argon	Pt-Sn/Al₂O₃ 1 wt. % Pt, 0.3 wt. % Sn	natural mordenite disk	$X_{ethane}=9.7$ $S_{ethylene}=92.2$ $Y_{ethylene}=9$ at 550 °C	$X_{ethane}=10.5$ $S_{ethylene}=93.7$ $Y_{ethylene}=9.8$ at 550 °C	[151]
High-temperature water gas shift reaction	PBMR	carbon monoxide, water steam	400–550 °C $H_2O/CO=1.0$–3.5 sweep gas: nitrogen	Fe/Ce	MFI/α-Al₂O₃ tube	$X_{CO}=62.5$	$X_{CO}=81.7$	[152]
Low-temperature water gas shift reaction	PBMR	carbon monoxide and water steam diluted in nitrogen	220–290 °C $p=6$ bar GHSV = 1000–7500 L_N/kg_{cat} sweep gas: nitrogen	CuO-ZnO/Al₂O₃	MFI/α-Al₂O₃ disc	$X_{CO}=89.1$	$X_{CO}=95.4$	[153]

Table 1. Cont.

Reaction	Reactor Type	Feed	Operating Conditions	Catalyst	Membrane	X_{CR} (%) S_{CR} (%) Y_{CR} (%)	X_{MR} (%) S_{MR} (%) Y_{MR} (%)	References
High-temperature water gas shift reaction	PBMR	carbon monoxide, water steam and nitrogen	400–550 °C H_2O/CO = 1.0–3.5 WHSV = 7500–60,000 h^{-1} p = 2-6 atm sweep gas: nitrogen	Fe/Ce	MFI/α-Al$_2$O$_3$ disc	X_{CO} = ~90	X_{CO} > 95	[154,155]
Water gas shift reaction	PBMR	carbon monoxide, water steam	500 °C p = 5 atm H_2O/CO = 3.0 GHSV = 72,000 h^{-1} sweep gas: argon	Fe-Cr-Cu	ZSM-5/silicalite bilayer/α-Al$_2$O$_3$	n.r.	X_{CO} = 89.8	[156]
High-temperature water gas shift reaction	PBMR	carbon monoxide, hydrogen, preheated steam	300–450 °C p_{feed} = 0.1–0.15 MPa p_{perm} = 0.1 MPa sweep gas: steam	Fe$_2$O$_3$/Cr$_2$O$_3$/Al$_2$O$_3$	MFI/α-Al$_2$O$_3$ hollow fibre	X_{CO} = 63.4	X_{CO} = 73.6	[157]
Xylene isomerization	PBMR	m-xylene diluted in nitrogen	577 K sweep gas: nitrogen in counter-current mode	Pt on zeolite	MFI/α-Al$_2$O$_3$ tube	$S_{p\text{-xylene}}$ = 58 $Y_{p\text{-xylene}}$ = 21	$S_{perm.\ only}$ = 100 $S_{perm.+Ret.}$ = 65 $Y_{p\text{-xylene}}$ = 23	[158]
Xylene isomerization	CMR	pure m-xylene; carrier gas: nitrogen	300–400 °C sweep gas: nitrogen	H-ZSM-5/316L stainless steel disc		$X_{m\text{-xylene}}$ = 5.87 $S_{p\text{-xylene}}$ = 55.6 $S_{o\text{-xylene}}$ = 44.4	$X_{m\text{-xylene}}$ = 6.9 $S_{p\text{-xylene}}$ = 66.7 $S_{o\text{-xylene}}$ = 33.3	[159]
Xylene isomerization	CMR	m-xylene diluted in helium	370 °C sweep gas: nitrogen	Pt/H-ZSM-5/stainless steel tube		n.r.	$S_{p\text{-xylene}}$ = 67	[160]
Xylene isomerization	PBMR	mixture of m-, p- and o-xylene carrier gas: hydrogen	340–390 °C WHSV = 550 h^{-1}	Pt/H-ZSM-5	Ba-ZSM-5/ Stainless steel	$S_{p\text{-xylene}}$ = 52	$S_{p\text{-xylene}}$ = 69	[160]

Table 1. *Cont.*

Reaction	Reactor Type	Feed	Operating Conditions	Catalyst	Membrane	X_{CR} (%) S_{CR} (%) Y_{CR} (%)	X_{MR} (%) S_{MR} (%) Y_{MR} (%)	References
m-xylene isomerization	PBMR	*m*-xylene diluted in helium	270–390 °C sweep gas: helium diverse packing configurations	HZSM-5	silicalite-1/ α-Al$_2$O$_3$ disc	GHSV = 1574 h^{-1} $X_{m\text{-xylene}}$ = 51.9 $S_{p\text{-xylene}}$ = 35.7 GHSV = 4722 h^{-1} $X_{m\text{-xylene}}$ = 36.5 $S_{p\text{-xylene}}$ = 47.3	- $X_{m\text{-xylene}}$ = 47.8 $S_{p\text{-xylene}}$ = 44.6 - $X_{m\text{-xylene}}$ = 36.1 $S_{p\text{-xylene}}$ = 49.6	[161]
m-xylene isomerization	PBMR	*m*-xylene, carrier gas: nitrogen	473–573 K sweep gas: nitrogen	Pt-HZSM-5	MFI/α-Al$_2$O$_3$ tube	$S_{p\text{-xylene}}$ = 42 $Y_{p\text{-xylene}}$ = 27	$S_{p\text{-xylene}}$ = 49 $Y_{p\text{-xylene}}$ = 23	[162,163]
xylene isomerization	CMR	*m*-xylene diluted in hydrogen	355–450 °C p = 101 kPa sweep gas: nitrogen	acid-functionalized silicalite-1/ α-Al$_2$O$_3$ disc propylsulfonic and arenesulfonic acid sites		n.r.	$X_{m\text{-xylene}}$ = 52 $Y_{p\text{-xylene}}$ = 32 at 450 °C	[164]
m-xylene isomerization	CMR	*m*-xylene diluted in helium	270 °C sweep gas: helium	H-MFI/α-Al$_2$O$_3$ disc		n.r.	$X_{m\text{-xylene}}$ = 6.5 $S_{p\text{-xylene}}$ = 92.1	[165]
Double-bond isomerization of 1-butene	CMR	1-butene diluted in nitrogen	120–250 °C p = 1 bar sweep gas: nitrogen	[B]MFI/α-Al$_2$O$_3$ tube		n.r.	X = 44.5 ratio $trans/cis$ = 2.2 at 250 °C	[166]
Esterification of ethanol with acetic acid	CMR	ethanol, acetic acid	333–363 K Δp = 0–1 bar sweep gas: He	H-ZSM-5/α-Al$_2$O$_3$ or stainless steel tubes		X = 49.4	X = 63.1	[167]
Esterification of acetic acid with ethanol	PBMR	ethanol, acetic acid	358 K p_{ret} = 1.3 bar p_{perm} = 2 mbar	Amberlyst 15	mordenite/α-Al$_2$O$_3$ zeolite A/α-Al$_2$O$_3$	X = 66.9	X = ~90	[168]
Catalytic dehydration of methanol	PBMR	methanol	150–250 °C WHSV= 0.5–2.6 h^{-1} p_{feed} = 1–1.7 bar p_{perm}= 1 mbar	γ-alumina	NaA/stainless steel wire mesh	X_{CH3OH} = 61 at 230 °C	X_{CH3OH} = 85 at 230 °C	[169]
CO$_2$ hydrogenation into methanol	PBMR	carbon dioxide, hydrogen	200–263 °C p = 20–24 bar H$_2$/CO$_2$ = 3–7	Cu/ZnO/Al$_2$O$_3$	NaA/α-Al$_2$O$_3$ tube	X_{CO2} = 5 S_{CH3OH} = 48 Y_{CH3OH} = 2.4	X_{CO2} = 11.6 S_{CH3OH} = 75 Y_{CH3OH} = 8.7	[170]

17

Table 1. *Cont.*

Reaction	Reactor Type	Feed	Operating Conditions	Catalyst	Membrane	X_{CR} (%) S_{CR} (%) Y_{CR} (%)	X_{MR} (%) S_{MR} (%) Y_{MR} (%)	References
Metathesis of propene and geometrical isomerization of *cis*-2-butene	PBMR	pure propene	296 K sweep gas: helium	Re_2O_7/γ-Al_2O_3	silicalite-1/stainless steel disc	$X_{propene} = 33.4$ $X_{cis\text{-}2\text{-}butene} = 76.1$	$X_{propene} = 38.4$ $X_{cis\text{-}2\text{-}butene} = 79.4$ $Y_{trans\text{-}2\text{-}butene} = 79$	[93,94]
Hydro-isomerization of C_6	PBMR	*n*-hexane, 2-methyl-pentane (MP); carrier gas: helium	393 K WHSV = 0.21 $g_{HC}/(g_{cat}\,h)$ sweep gas: hydrogen	Pt-chlorinated alumina (AT-2G)	silicalite-1/TiO_2/stainless steel tube	n.r.	$X_{n\text{-hexane}} = 71.8$ at 393 K	[171]

18

Micro- and mesoporous zeolite membranes enclosing V-Mg-O catalyst beds have been employed to control the oxygen partial pressure in order to enhance the selectivity towards propene in the oxidative dehydrogenation of propane [177]. The microporous zeolite membrane evinced to be an effective gas barrier since only small amounts of the reactant propane diffused through the membrane and all the oxygen permeating was consumed in the dehydrogenation reaction causing remarkable increase in the propene yields at low C_3H_8/O_2 ratios under separate feeding configuration. Julbe *et al.* [178] studied MFI and vanadium-loaded MFI membranes for the same reaction using them in two configurations either as oxygen distributors or as flow-through contactors. The V-MFI membrane outperformed the MFI and in the flow-through configuration a propene yield of nearly 8% at selectivity of 40%–50% was obtained.

As an alternative to the conventional fixed-bed reactors, methanol oxidative dehydrogenation to yield formaldehyde was studied in a NMR with either methanol or oxygen as permeating species [179]. The non-permselective stainless steel membrane was packed with Fe-Mo oxide catalyst. The configuration with oxygen feeding though the membrane outperformed the membrane reactor with methanol as permeating component and that of the traditional reactor, assuring lower oxygen concentration and thus selectivity improvement and higher formaldehyde yields. Later on, the same authors optimized the oxygen feed distribution on the basis of productivity of the desired product formaldehyde [180].

Modified MFI membranes were utilized in membrane reactors of the distributor type to run butadiene hydrogenation in the vapor phase [181]. By the use of the zeolite membrane to distribute hydrogen in a controlled manner along the catalyst bed, it was possible to overcome the selectivity drawback, and to greatly improve the butadiene conversion.

3.3. Contactor Type Membrane Reactors

The last membrane reactor principle considering the membrane function is the contactor type, where catalytic active and non-permselective membranes can be employed. The use of contactors is usually announced when a reaction front should be set in order to intensify the contact between the catalyst and the reactants.

Torres *et al.* reported on *i*-butene oligomerization with catalytic BEA zeolite membranes in the temperature range from 373 to 423 K [182,183]. The non-permselective membrane was used as a contact medium demonstrating high activity towards *i*-butene dimers. Furthermore, Pt-ZSM-5 membranes prepared on tubular supports were applied for the combustion of volatile organic compounds (VOCs) at low concentrations [184]. The authors fed hexane as representative for VOC at the one side of the membrane and O_2 at the other side. Interestingly, a membrane with intermediate concentration of defects performed better due to improved contact between the reactants and the catalytic material leading to nearly complete combustion of *n*-hexane at 210 °C.

4. Applications in Packed Bed Membrane Reactors (PBMR)

The utilization of permselective zeolite membranes in a direct contact to the catalyst bed, as shown in Figure 4, could offer several advantages over alternative reactor concepts. On the one hand, zeolite membranes can remove products, catalyst poisons or inhibiting products from the reaction zone in order to enhance the conversion of a given reaction. On the other hand, zeolite membranes might provide selective reactant supply from the feed mixture as well as control of the reactant traffic or residence time, thus contributing to selectivity enhancement.

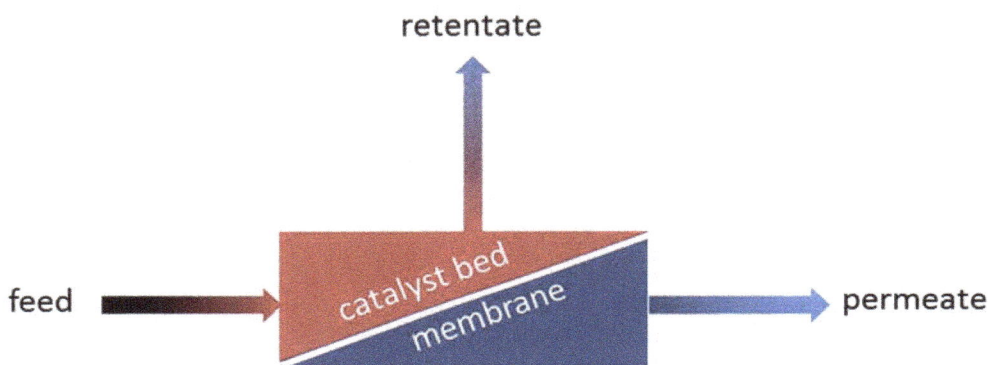

retentate

feed   catalyst bed   membrane   permeate

Figure 4. Schematic drawing of a packed bed membrane reactor (PBMR).

4.1. Product Removal: Enhanced Conversions by Shifting the Chemical Equilibrium

Zeolite membranes being able to preferentially transport one or more products from the reaction zone are intimately located to the catalyst bed. The so called extractor type membrane reactor unit is often utilized in equilibrium limited reactions in order to break this limitation and improve the yield of the desired products. However, sufficient permeation and sharp separation, as well as mechanical, thermal and chemical stabilities are considered as crucial membrane requirements for successful application in this membrane reactor mode. Contrarily, by employing membranes with poor selectivity, the reaction will suffer from significant reactant loss. Similarly, low permeation rates will demand increased membrane area that will lead to discrepancies in regard to catalyst volume and higher investment costs.

4.1.1. Equilibrium Shift by Water Removal

As already discussed in Section 2.2. esterification is an equilibrium limited reaction, where water separation is required to increase conversion. Tanaka *et al.* [185,186] proposed zeolite T membranes for the pervaporation-aided esterification of acetic acid or lactic acid with ethanol, catalyzed by an exchange resin (Amberlyst 15, Organo) in batch reactor at 343 K. The membrane displayed good stability even being submerged in the acidic media and was able to selectively remove the produced water, thus exceeding the equilibrium limit. In such a way, nearly 100% conversion within 8 h of operation was reached. In this context, de la Iglesia *et al.* [168] employed modernite and zeolite A membranes in a continuous membrane reactor packed with Amberlyst™ as catalyst and evaluated their joint performance during esterification of ethanol and acetic acid. The used membranes exceeded the equilibrium conversion towards the esterification products in less than 1 day of operation, however displayed different resistance to the acidic reaction conditions affecting their long term stability. The modernite membrane leveled at a conversion of nearly 90% for 5 days of experiment attributed to its stability under the conditions applied, whereas the instability of zeolite A caused tremendous conversion loss. Interestingly, in the case of the modernite membrane significant rise in the separation factors of H_2O/Ethanol and H_2O/acetic acid from 55 and 25, respectively, at the beginning of the experiment towards 200 and 95 at the third day was observed associated with the increasing production of water in the catalyst zone. On the contrary, the initial higher separation factors of H_2O/Ethanol (315) and H_2O/acetic acid (89) for the zeolite A membrane declined sharply to 32 and 18, respectively, leading to reactants loss and conversion decreasing from 72% at the beginning of the experiment to 53% after two days of operation.

Beside esterification, further water releasing reactions can be enhanced using zeolite membranes for equilibrium shift. For instance, modernite or NaA zeolite membranes together with Amberlyst™ were implemented in membrane reactors for the gas-phase synthesis of methyl-*tert*-butyl ether (MTBE) from *tert*-butanol (TBA) and methanol (MeOH) [187]. In the first step, the membranes were characterized by separation experiments of multicomponent mixtures containing water, MeOH, TBA, MTBE and *i*-butene (IB) displaying the following selectivity trend: $S_{H_2O/IB} > S_{H_2O/MTBE} > S_{H_2O/TBA} > S_{H_2O/MeOH}$. The selective permeation of water across the zeolite membranes was attributed to their high polarities resulting in preferential adsorption and pore blocking for the other components. Then, during the reaction experiments the water removal in the membrane reactor operation contributed to 67.5% MTBE yield showing 6.7% absolute increase in the MTBE yield compared to the conventional configuration without water removal. Very recently, NaA zeolite membranes were also used as water extractors in the methanol dehydration to dimethyl ether (DME) [169]. The γ-Al_2O_3 catalyst was located between two

21

disc membranes giving a ratio of membrane area to reactor volume of 200 m^{-1}, whereas 10–100 m^{-1} is said being reasonable for industrial application [94]. For comparison, the authors performed the same reaction in a conventional fixed-bed reactor operated under the same conditions. Water and DME were the only products of the catalytic experiments. Increased temperature in the range of 150–250 °C and decreased weight hourly space velocity (WHSV) from 2.6 to 0.5 h^{-1} evinced to be more beneficial for the membrane reactor configuration, resulting in more than 20% absolute improvement of the methanol conversion. However, the limiting factor in course of the experimental study turned out to be on the one hand the reactor design and on the other hand the membrane selectivity since the separation of methanol and water is mainly based on competitive adsorption being so dependent on the process parameter temperature, pressure and composition.

Apart from overcoming equilibrium limited reactions, hydrophilic zeolite membranes can be employed in reactions where the water removal will contribute to decreased catalyst deactivation. So, the feasibility of NaA zeolite membranes for water liquid-phase etherification of *n*-pentanol to di-*n*-pentyl ether (DNPE) catalyzed by ion-exchange sulfonated resins has been evaluated [188]. In this reaction, the formed water has a strong deactivating effect on the catalyst. However, due to the excellent dehydration performance of the membrane an enhancement of the *n*-pentanol conversion (64% conversion compared to 35% in the fixed bed-reactor) was possible. Moreover, the authors analyzed the economic aspects of the membrane-based reactor configuration for DNPE production and estimated by assuming a 35,000 tm DNPE/year production about 30% of the investment costs for the membrane reactor. According to analysis of operational costs (OPEX) for 1 L of DNPE it would be related to 1.6 US$ of unit costs.

Meanwhile, the carbon dioxide utilization is one of the most challenging tasks in chemistry [189,190]. Here, zeolite membranes might be used for the direct dimethyl ether synthesis from a mixture of syngas and carbon dioxide [191]. However, in the multistep reaction via methanol water is being released [192,193]. The advantageous performance of a packed bed zeolite membrane reactor was theoretically demonstrated for *in situ* water removal. However, based on mathematical models, the authors concluded that the dimethyl ether yield is highly dependent on the membrane water permselectivity. For instance, for membranes with low permselectivity, the dimethyl ether yield was nearly 50% lower due to reactant loss than that obtained in a conventional reactor (7.0% yield in the PBMR *vs.* 14.8% yield in the CR). Therefore, in further studies, the same authors analyzed and optimized theoretically the operating conditions leading to enhanced dimethyl ether yield and CO_2 recovery [194]. Thereby, the authors found an increase in CO_2 conversion up to 85% by using high sweep gas stream. Moreover, approximately 30% yield of DME could be obtained in the PBMR at high recirculation factors of the sweep gas stream due to reduction in the methanol loss across the membrane.

In addition, the application of hydrophilic membrane reactors for the selective *in situ* removal of water has been considered for Fischer–Tropsch synthesis in three directions: (i) improvement of the catalyst lifetime; (ii) rise of the reactor productivity; and (iii) displacement of the water gas shift equilibrium in favor of CO [195,196]. The use of hydrophilic zeolite membranes and their positive effect on the conversion in the Fischer-Tropsch process was earlier discussed by Espinoza *et al.* [197]. Later, Rohde *et al.* [196] evaluated thoroughly different membranes. The authors concluded that membranes with proper fluxes for the selective removal of water from mixtures comprising H_2, CO, CO_2 and hydrocarbons are still required for industrial application. For such applications H_2O permeances of over 1×10^{-7} mol/(s m^2 Pa) and ratios of H_2O permeance to that of respective reactants greater than 75 were defined. In this context, the literature study showed that zeolite membranes outperform both amorphous membranes and polymer membranes in regard to H_2O permeance (in the range of 1×10^{-7} and 10^{-6} mol/(s m^2 Pa)) and H_2O/H_2 permselectivity (>10), being however far from meeting the defined requirements for technical application. Hence, the authors suggested hydroxy sodalite zeolite membranes with a layer thickness of 2 μm as auspicious candidate for the *in situ* water removal due to their extraordinary separation and permeation performance demonstrated as well earlier by Khajavi *et al.* [198]. Recently, a novel reactor configuration denoted as fixed-bed membrane reactor followed by fluidized-bed membrane reactor (FMFMDR) has been proposed for high temperature Fischer-Tropsch synthesis, in particular gasoline production from syngas [199,200]. This configuration combines a fixed-bed water permselective membrane reactor equipped with H-SOD zeolite membrane coated on α-Al_2O_3 substrate and a fluidized-bed hydrogen permselective membrane reactor equipped with Pd-Ag membrane. By means of theoretical modeling, the authors stated the benefits of the studied reactor configuration in respect to improved gasoline yield and reduced CO_2 yield.

4.1.2. Hydrogen Permeation in Dehydrogenation Reactions

The dehydrogenation of alkanes to olefins is a strongly endothermic and thermodynamically equilibrium limited reaction [201]. By removing products from the reaction zone the reaction equilibrium will be displaced towards the product side, thus increasing the overall conversion. The performance of hydrogen selective zeolite membranes in terms of H_2 (product) removal and its effect on the reaction efficiency has been studied in dehydrogenation of alkanes by several groups. For instance, MFI zeolite membranes were applied in a PBMR configuration and extensively investigated in dehydrogenation of *i*-butane by the group of Dalmon *et al.* [145,202–204]. Either commercial Pt-Sn/γ-Al$_2$O$_3$ [202], Pt-In/silicalite [145,203] or MFI supported Pt-In-Ge [204] catalysts placed in the core of the membrane tube were used. The effectiveness of the membrane reactor to improve dehydrogenation yields compared to the conventional reactor was up to four times higher [145]. When comparing mesoporous membranes with microporous zeolite membranes, the authors found hydrogen selectivity only for the latter, while the usage of larger-pored membranes led just to gas mixing at both sides of the membrane [202]. Moreover, the authors pointed on the importance of the precise control/selection of the operating parameters such as feed flow and sweep gas flow, as well as sweep gas configuration and found correlations between membrane reactor performance and membrane permeability or catalyst activity depending on the applied conditions [145,203]. In this regard, the reactor performance was limited by the catalyst activity under counter-current sweep flow conditions so that the catalyst was not active enough to follow the high membrane permeability. On the other side, choosing the co-current sweep configuration, the reactor performance was controlled by the membrane permeation efficiency and insufficient selectivity resulted thereby in reactants (*i*-butane) loss. Van Dyk *et al.* [204] confirmed these observations conducting a comparative study with microporous MFI and dense Pd membranes in membrane reactors packed with Pt-In-Ge/MFI catalyst. Generally, better yields were obtained with the membrane reactor configurations than with the conventional reactor. Nevertheless, the two membranes reached almost equal yields despite their different separation efficiencies so that the authors concluded that the membrane reactor performance was limited by the catalyst activity. At this point the fundamental work of Gokhale *et al.* [205] has to be mentioned. The authors presented insights on the relationship between permeation rates as function of separation selectivity and residence time. They focused on the possible operating conditions under which reactant loss controls conversion in dehydrogenation reactions. However, Illgen *et al.* [146] commented a feed dilution effect should always be considered during product analysis. It was stated that despite the high separation efficiency of the MFI zeolite membranes, e.g. a H$_2$/*i*-butane mixture separation factor of 70 and a permeance of 1 m^3/m^2 h bar at the reaction temperature of 510 °C, the increased conversion up to 49% obtained in the PBMR at WHSV of 0.5 h^{-1}, where the conversion of the conventional reactor was 29.1% was due to a great extent to the dilution of the reactant feed by the sweep gas and less to the removal of H$_2$ from the reaction zone.

Consequently, van den Bergh *et al.* [147] revealed the benefit of using small-pore zeolite DD3R membranes coupled to Cr$_2$O$_3$/Al$_2$O$_3$ catalysts in PBMR for the dehydrogenation of *i*-butane. The DD3R membrane is believed to be quite attractive for the present application since it is able to separate H$_2$ and *i*-butane by molecular sieving effects. Accordingly, only H$_2$ could pass the membrane and *i*-butane will be retained due to its bigger size. In this context, the membrane exhibited outstanding H_2/i-butane ideal selectivity (based on the single gas permeation fluxes) of over 500 at 773 K. However, slightly lower mixture selectivity was recorded since the driving force for permeation of H_2 in the mixture was reduced by the increased partial pressure of H_2 in the permeate and its decreased partial pressure in the feed. About 50% increase in yield compared to the equilibrium value was recorded mainly due to the effective removal of H_2 from the reaction zone. On the other hand, the studies provided evidence of a minor decrease in the catalyst activity compared to the conventional reactor. Moreover, a slightly increased coke formation, however with selectivity towards coke being still low than 5% was observed attributed by the authors to the lower H_2 partial pressure at the reaction side. Despite the fact that the catalyst activity and the H_2 removal hold a good balance, the authors concluded that both parameters limit to some extent the overall performance. Such being the case, further improvements in the catalyst activity and stability as well as in the permeation fluxes of the membrane are necessitated for successful application in dehydrogenation reactions at industrial scale.

The performance of large-pore FAU type zeolite membranes prepared on porous α-Al$_2$O$_3$ support tubes in the catalytic dehydrogenation of cyclohexane conducted in membrane reactors was evaluated experimentally [148] and theoretically [149] by the group of Kusakabe. The membranes were able to simultaneously remove hydrogen and benzene from the reaction zone filled with a Pt/Al$_2$O$_3$ catalyst. Moreover, higher sweep or lower feed flow rates affected positively the cyclohexane conversion. The authors evaluated mathematically the trade-off effect of membrane permeance and separation factor on the conversion and concluded that in terms of industrial practice high membrane permeability accompanied with reasonable selectivity might be the more favorable option than a high selectivity at the expense of low permeability. Furthermore, it was highlighted that the H_2 addition to the cyclohexane feed compensates the reduced H_2 partial pressure on the reaction side, caused by its continuous extraction, hence preventing the catalyst from coking [149].

Defect-free silicalite-1 zeolite membranes were used for the catalytic dehydrogenation of ethylbenzene to styrene in membrane reactors packed with Fe$_2$O$_3$ exposing their advantages (74.8% at 610 °C) over a conventional reactor (67.5% at the same temperature) in regard to conversion due to the instant extraction of the produced H_2 across the membrane [150]. However, the benefit of the membrane reactor diminished with increasing space velocities ranging from 0.5 to 1.5 and approached the performance of the conventional reactor configuration. On the contrary, rising the ratios of sweep gas to reactant feed from 0.5 to 2 contributed to increase in the ethylbenzene conversion (from 70.5% to 74.8%) since higher ratios induce generally larger driving force for permeation resulting in higher H_2 permeation rates. Further rise in the sweep/feed ratio from 2 to 6 resulted in a nearly constant conversion pointing out that ratio of 2 was sufficient in order to remove the desorbing hydrogen from the permeate side of the membrane. The authors stressed once again the benefit of the zeolite membrane reactor operation in terms of reduced partial pressure of H_2 contributing to higher reaction rates.

Alternatively, small-pore size SOD membranes have been recently considered as an attractive candidate for the selective removal of H_2 in catalytic dehydrogenation of ethylbenzene to styrene [206]. The performance of a membrane reactor and a conventional plug flow reactor (PFR) was predicted, confirming the benefit of the membrane reactor with regard to an absolute ethylbenzene conversion increase of 3.45% and yield increase of 8.99% ascribed to the effective H_2 extraction from the reaction side. Due to the dynamic limitations, the PFR reached 80% conversion and 44.5% yield.

Very recent results promise the potential application of cost-effective natural mordenite membranes. Indeed, the mordenite membranes were fabricated by using rock material (Paradise Quarry Limited, Whangarei, New Zealand) processed by a diamond saw. In membrane reactors, packed with Pt/Al_2O_3 beads for the dehydrogenation reaction of ethane [151], the membrane was able to shift the reaction equilibrium at the studied temperature range of 500–550 °C due to its hydrogen-selective properties. Based on evaluation of the membrane reactor effectiveness in terms of permeation area to reactor volume ratio (A/V ratio), the authors demonstrated that increasing the ratio from 0.04 m^{-1} to 0.16 m^{-1} gives rise to additional reduction of the H_2 content in the reaction zone in relation with its formation rate and consequent increase in the reaction rate contributing to more enhanced ethane conversion compared to the PBMR with the smaller permeation area.

4.1.3. Hydrogen Permeation in Water Gas Shift Reaction

The water gas shift (WGS) reaction represents another equilibrium limited reaction for which the application of extractor type membrane reactors has been reported. The reversible and mildly exothermic WGS reaction is a subsequent step for the increased production of H_2 from initially produced CO gained from fossil fuel reforming. Generally, in order to overcome the thermodynamic and kinetic limitations, the WGS reaction is performed in two steps including high temperature shift favoring higher space time yields followed by low temperature shift to obtain high CO conversion [207]. However, coupling the reaction with a H_2-selective membrane can break the equilibrium constrains and facilitate the CO conversion, thus intensifying the process and resulting in economically beneficial application. Besides Pd- or Pd-Ag alloys membranes [208–211] and silica-based membranes [212,213], zeolite membranes are subject of intensive research interest for the present problem.

Considering the application of MFI zeolite membrane, several methods for modification of zeolite membranes prior to application in PBMR for WGS reaction are proposed. For instance, Tang *et al.* modified the pores of MFI zeolite membranes by a so called *in-situ* catalytic cracking deposition (CCD) of silane precursors [124,152,214]. The formed deposits reduce the effective pore size to below 0.36 nm hindering the entry of CO_2 into the pore channels. The idea behind this modification was the need to obtain controlled mass transport of H_2 over CO_2 since thought the unmodified MFI the transport of H_2 and CO_2, respectively, is controlled by gaseous diffusion resulting in separation factors slightly below the Knudsen factor. Accordingly, thanks to the the deposition of molecular silica species in the zeolite channels, the access of the slightly bigger CO_2 molecules (kinetic diameters of H_2 and CO_2 are 0.289 and 0.33 nm, respectively) was restrained giving rise to H_2/CO_2 permselectivity of 68.3 and equimolar mixture separation factors of nearly 38 at 550 °C [152]. The modified zeolite MFI membrane packed with a $Fe_{1.82}Ce_{0.18}O_3$ catalyst was tested in WGS reaction at temperatures between 400 and 550 °C and near atmospheric pressure. At a reaction temperature of 550 °C (WHSV = 60,000 h^{-1}, H_2O/CO = 1) the membrane reactor configuration exhibited CO conversion (81.7%) exceeding the equilibrium limit (65%) and the performance of the traditional packed-bed reactor (62.5%). However, decreasing the reaction temperatures below 500 °C caused a conversion drop due to kinetic resistance. With regard to the WGS reaction, low H_2O/CO ratios are preferred since the hydrogen partial pressure is large and the driving force for permeation as a result, too. The same group evaluated the impact of elevated pressure (2–6 atm) on the efficiency of modified MFI disc membranes in high temperature (400–550 °C) WGS reaction and demonstrated its positive effect on the CO conversion. Furthermore, the authors pointed towards to the larger driving force for H_2 permeation at increased feed pressure and constant permeate pressure [154]. Figure 5 depicts the influence of reaction pressure and temperature on CO conversion (χ_{co}), hydrogen recovery (R_{H_2}), and permeate side H_2 concentration ($\gamma_{H_{2,p}}$). As shown in Figure 5 (left) a higher driving force for hydrogen permeation is achieved by increasing the feed pressure leading to higher R_{H_2}. At the applied temperatures the separation factor $\alpha_{H_2CO_2}$ is not influenced by adsorption and the gases exhibit ideal gas behavior. The authors ascribed the decrease in $\gamma_{H_{2,p}}$ with increasing R_{H_2} and χ_{co} at high pressure to the decreased value of $(\gamma_{H_2}/\gamma_{CO_2})_{permeate}$ $(=\alpha_{H_2CO_2}/(\gamma_{H_2}/\gamma_{CO_2})_{feed})$ when reducing $(\gamma_{H_2}/\gamma_{CO_2})_{feed}$. Furthermore, it is shown (Figure 5 (right)) that an increased feed pressure could be used to overcome the equilibrium CO conversion ($\chi_{CO,e}$) using the PBMR. Generally, the findings indicate that even membranes with moderate selectivity could be powerful tool for conversion enhancement. Particularly, it was emphasized that the prepared MFI membrane showed good resistance against H_2S and was stable at the applied high temperatures and pressures.Recently, the same authors applied the model of one-dimensional plug-flow reactors (PFR) and related the results with data from MFI-type zeolite membranes. Simulations proved the potential of the membrane reactor, combining a modified MFI membrane and cerium-doped ferrite catalysts, to reach CO conversion above 99.5% at 550 °C and ~50 atm at a ratio of H_2O/CO ~5.0 [155]. Similarly, Lin et al. [156] evaluated the performance of ZSM-5/silicalite bilayer membranes packed with Fe-Cr-Cu catalysts combining experimental and theoretical studies and defined the optimal conditions under which CO conversion of over 95% together with H_2 recovery of over 90% could be achieved.

For low-temperature WGS reactions, Zhang et al. [153] illustrated the benefit of PBMRs coupling a H_2-permselective MFI membrane modified by CCD of methyldiethoxysilane and $CuO/ZnO/Al_2O_3$ catalyst over the conventional packed bed reactor. CO conversion exceeding the equilibrium was obtained at 300 °C attributed to the enhanced permeation of H_2 at the applied temperature on the one hand and the catalyst activity on the other hand. Very recently, the authors proposed the idea of using steam as sweep gas instead of inert gas in order to avoid subsequent separation to obtain pure hydrogen [157]. Figure 6 illustrates schematically this membrane reactor configuration with modified hollow fibre MFI zeolite membranes where the steam is applied in a counter-diffusion towards the reactions side in order to remove H_2. The authors stated that sweeping with pure steam contributes to enhanced conversion combined with direct acquisition of the pure H_2. Moreover, despite the fact that the membrane was characterized by high H_2 permeate flow, sweeping by N_2-steam mixed gas resulted in lower conversion if compared to the experiments where pure steam was applied as sweep. The observed phenomenon was attributed to the dilution of the steam by N_2 leading to reduced H_2O counter-diffusion effect due to the lower driving force though the membrane. Operating under low sweep steam flow rate as well as low feed pressure and H_2O/CO ratio contributed to highly pure H_2 permeate streams.

4.1.4. Hydrogen Permeation in Syngas Production

The benefit of using zeolitic PBMRs over the traditional fixed-bed reactor has also been reported by Liu et al. [215,216] for selective product permeation in CO_2 reforming of methane for syngas production. The authors employed a combination of catalytic composite zeolite membranes either La_2NiO_4/NaA or La_2NiO_4/NaY prepared on γ-Al_2O_3/α-Al_2O_3 support packed with NiO-La_2O_3/γ-Al_2O_3 as catalyst. The idea behind this mixed configuration was the significantly low separation efficiency of the as-prepared inert zeolite membranes for the H_2/CH_4 mixture (binary mixture separation factor of 4.2 at room temperature decreasing to below 2 at temperature over 600 °C) and the arising diffusion of methane through the membranes during reforming. Therefore, the authors introduced the use of the catalytic active zeolite membranes packed with catalyst and managed so further methane reforming to syngas during its permeation through the membrane. The permselective permeation of CO and H_2 across the membrane contributed to enhanced CH_4 and CO_2 conversion of 73.6 and 82.4 mol % at 700 °C *vs.* 45 and 52 mol % over the fixed-bed reactor, respectively. Moreover, the coke deposition and thus the catalyst deactivation were remarkably reduced in the membrane reactor.

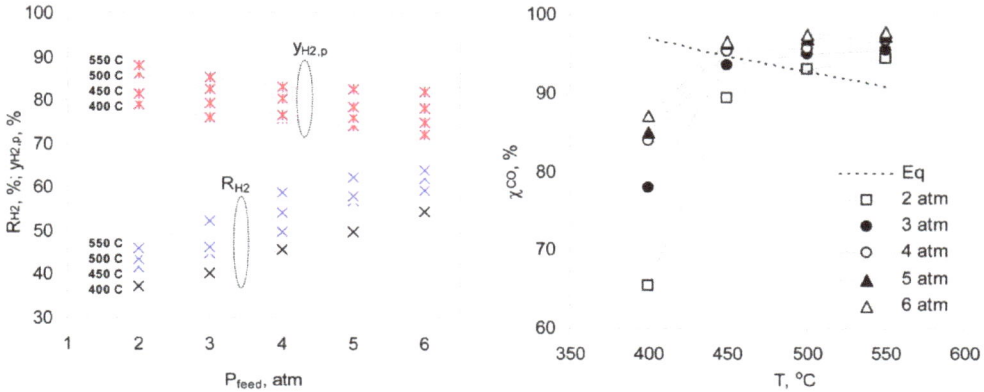

Figure 5. Water gas shift (WGS) reaction in a MFI zeolite membrane reactor (sweep: N_2, at atmospheric pressure, WHSV at 7500 h^{-1} and ratio of $H_2O/CO = 3.5$): Influence of reaction pressure and temperature on CO conversion (χ_{co}), hydrogen recovery (R_{H_2}) and permeate side H_2 concentration ($\gamma_{H_{2,p}}$). Reprinted with permission from [154].

4.1.5. Hydrogenation

One of the most significant current discussions as previously mentioned is the need to reduce the CO_2 concentration in the atmosphere and so mitigate the greenhouse effect. Thereby, the CO_2 utilization as a useful chemical, e.g., through hydrogenation reactions to yield methanol used as fuel or basic chemical, is considered as a promising alternative [190]. Zeolite membrane reactors could be applied in order to assure removal of the condensable products (CH_3OH and H_2O) and to improve the methanol yield in the equilibrium limited reaction. The principle was first theoretically discussed by Barbieri *et al.* [217]. The authors confirmed the benefit of using either hydrophilic or hydrophobic zeolite membranes in terms of improved conversion, methanol selectivity and yield by operating at lower reaction volumes and residence times as well as higher temperatures and lower pressures compared to the conventional tubular reactor. The CO_2 conversion into methanol was later experimentally studied by Gallucci *et al.* in a membrane reactor with a zeolite NaA membrane enclosing a fixed bed of CuO-ZnO/Al_2O_3 catalyst [170]. Generally, the membrane reactor was able to display higher CO_2 conversion and selectivity than the traditional reactor mainly attributed to the selective removal of CH_3OH via the zeolite membrane. Moreover, the authors stressed the positive aspect of the reduced energy demand since the membrane reactor was able to reach the CO_2 conversions of the traditional reactor at milder conditions (e.g., PBMR operating at $H_2/CO_2 = 3$ and temperature of 225 °C could reach the conversion of conventional reactor operating at 265 °C). Furthermore, as stated by the authors the temperature seems to be the important process parameter in the PBMR operation since the methanol separation is mainly due to its capillary condensation inside the pores so that exceeding the critical temperature of methanol will sharply reduce the separation efficiency.

Figure 6. Schematic illustration of a WGS membrane reactor with modified hollow fibre MFI zeolite membrane swept by steam applied in a counter-diffusion towards the reaction side. Reprinted with permission from [157]. Copyright (2015) John Wiley and Sons.

4.1.6. Metathesis of Propene

The selective product removal in the equilibrium limited metathesis of propene to ethene and 2-butene as well as from the simultaneous occurring geometrical isomerization of *cis*-2-butene into *trans*-2-butene was evaluated by the mean of a membrane reactor equipped with a silicalite-1 zeolite membrane supported on stainless steel and 16.4 wt. % Re_2O_7/γ-Al_2O_3 as catalyst [93,94]. The studied reactor benefited from the use of the membrane twice since (i) 13% absolute improvement of the propene conversion compared to the equilibrium value (25.4%) and (ii) 32% increase in the *trans*-2-butene/*cis*-2-butane ratio compared to the equilibrium ratio being 3.2 due to the preferential permeation of *trans*-2-butene were achieved.

However, even though the membrane did not show an absolute separation selectivity for *trans*-2-butene over *cis*-2-butene, it still displayed a balanced performance between sufficient product removal and reactant loss since the preferentially adsorbing *trans*-2-butane was able to block the pores for the permeation of propene. In addition, the use of a supplementary reactor for equilibration of the feed mixture before feeding the zeolite membrane reactor reduced the reactant loss. Furthermore, the authors evaluated the perspectives for industrial applications [93]. It was stated for high permeating silicalite-l membranes this requirement was fulfilled. However, to the best of our knowledge no such industrial plant has ever been built up to now.

4.2. Product Removal: Enhanced Selectivity by Displacing the Chemical Equilibrium

Xylene isomers with a typical composition 18% *p*-xylene, 40% *m*-xylene, 22% *o*-xylene, and 20% ethylbenzene are generally produced from petroleum reformate streams [218]. However, the further use of the obtained isomers requires their separation. Due to the close boiling points of *p*- and *m*-xylene the use of distillation for their separation is not effective so that industrially crystallization and adsorption techniques have been developed, e.g., by using ZSM-5 [219]. The *o*- and *m*-xylenes can be isomerized to obtain more *p*-xylene.

As an energy-efficient alternative to the conventional techniques, membrane reactors coupling xylene isomerization and simultaneous selective recovery represent a research topic gaining in importance in the last years due to the increasing demand of xylenes estimated at approximately 22 Mtones in 2003 [220]. Applying this concept *p*-xylene, the raw material for production of polyester resins could be obtained as product at the permeate side. In a search of an appropriate membrane type, one should consider that *p*-xylene possesses a kinetic diameter of 5.8 Å which is significantly smaller than those of the *m*- and *o*- isomers being about 6.8 Å [221]. Due to the specific pore structure of the MFI-zeolite combining straight, circular pores (0.54 nm × 0.56 nm) and sinusoidal, elliptic pores (0.51 nm × 0.54 nm) [222] it became the membrane material of choice for xylene isomer separation as well as for the application in membrane reactors.

Several groups have studied the separation of xylenes with MFI zeolite membranes demonstrating high permselectivity for *p*-xylene over the other isomers, attributed mainly to the preferential permeation of *p*-xylene, since the zeolite pores expose sterical hindrance for the permeation of the bulkier *m*- and *o*-xylene isomer molecules [104,221,223–225]. In an outstanding study, Lai *et al.* [226] reported on dramatic improvement of the *p*-/*o*-xylene separation by b-oriented silicalite-1 membranes. In Figure 7 a comparison of achievable permeances as well as separation factors (SP) as a function of the operating temperature using (a) *c*-oriented; (b) [h0h]-oriented; (c) *a*- and *b*-oriented; and (d) *b*-oriented MFI films is given. It is clearly evidenced that the oriented membrane possesses the best properties for xylene isomer separation, demonstrating the strong impact of the crystal orientation on the membrane performance for certain separation task and for future applications in membrane reactors as well.

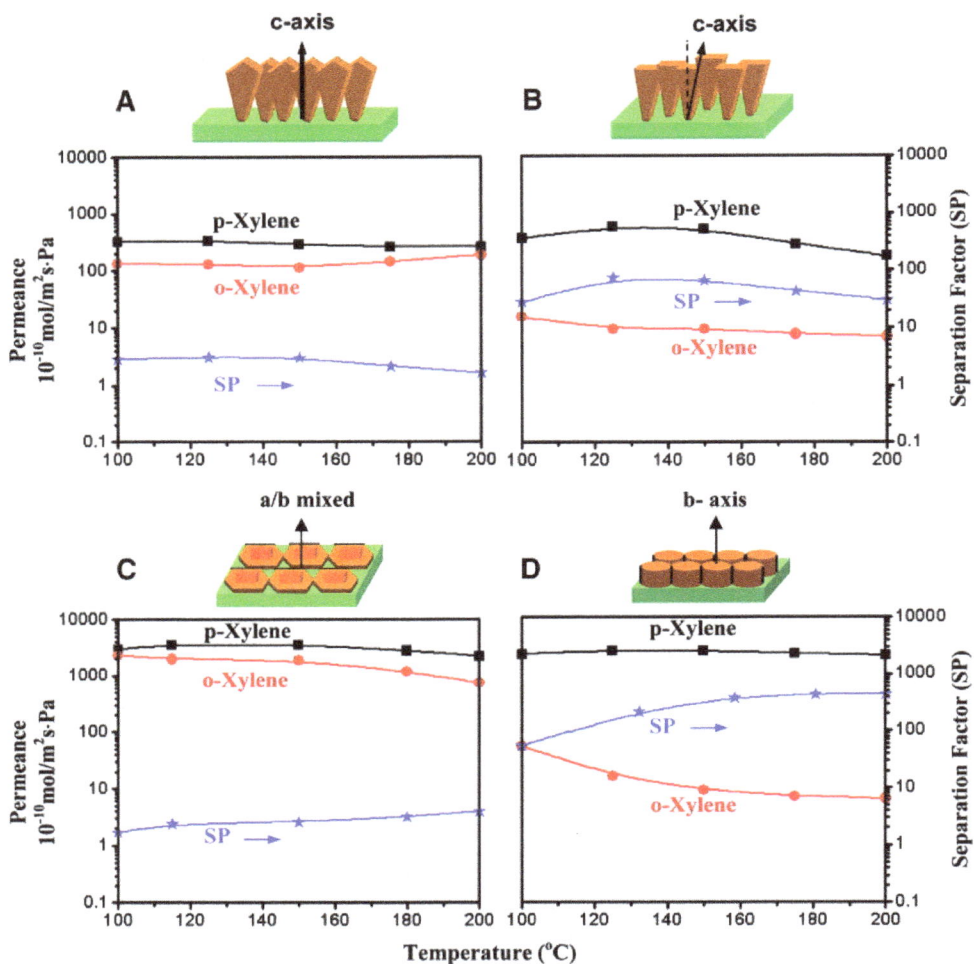

Figure 7. ZSM-5 membrane performance in xylene isomer separation (feed partial pressure of *p*-xylene and *o*-xylene are 0.45 kPa and 0.35 kPa, respectively): *p*-xylene, *o*-xylene permeance, and mixture separation factor (SP) in dependence on temperature for (**A**) *c*-oriented; (**B**) [h0h]-oriented; (**C**) *a*- and *b*-oriented; and (**D**) *b*-oriented film [226]. Reprinted with permission from AAAS.

Considering xylene isomerization in PBMRs, the extraction of the produced *p*-xylene from the reaction zone via zeolite membrane will shift the equilibrium and ensure selectivity enhancement and higher *p*-xylene yields.

In this context, Zhang *et al.* [161] tested silicalite-1/α-Al$_2$O$_3$ zeolite membranes packed together with HZSM-5 catalysts in the isomerization of *m*-xylene to *p*-xylene, applying different packing methods: (i) by depositing the catalyst on the Al$_2$O$_3$ support on the opposite side of the membrane; and (ii) by packing it in the tube in contact with the membrane layer. The latter packing method turned out to be more effective for the studied reaction since it gave rise to higher *p*-xylene yield and selectivity than in conventional reactor ascribed to the immediate removal of *p*-xylene from the reaction zone due to molecular sieving effects. However, since the observed enhancement was found to be strongly dependent on the membrane flux, the authors pointed out the need of membranes exhibiting higher permeation fluxes as a crucial requirement for the desired improvement of the membrane reactor efficiency. Moreover, in order to overcome the resistance diffusion through the catalytic bed, thus enhancing the permeation and intensifying the xylene isomerization, the authors suggested the use of catalytic membranes combining high-catalytic and separation efficiency. The performance of such catalytically active zeolite membranes will be discussed in more details in Section 5.

However, already with the intention to overcome the drawbacks of the supported zeolite films typically suffering from defects occurred during thermal stress causing diminished selectivity, van Dyk *et al.* [158] used a zeolite/alumina nanocomposite membrane of the pore-plugging type (zeolite crystals grown as film inside the pores of the porous tubular support) for the *m*-xylene isomerization in an extractor type zeolite membrane reactor. The tubular membrane used was equipped with the commercial xylene isomerization Pt catalyst, ISOXYL. The selectivity and the *p*-xylene yield of the conventional reactor were 58% and 21% respectively, demonstrating higher selectivity that the equilibrium value of 46% at the expense of the lower yield in comparison to the equilibrium yield of 24.9%. In contrast, by combining the retentate and permeate fractions para-selectivity of 65% and *p*-xylene yield of 23% were reported for the membrane reactor. More interestingly, operating in permeate-only mode, the authors managed to obtain *p*-xylene selectivity of 100%, however at the expense of low productivity. Recently, Daramola *et al.* [162,163] attempted to further reveal the advantages of the nanocomposite architecture over the "film-like" zeolite membranes, applying nanocomposite MFI-alumina membrane tubes prepared via pore-plugging synthesis packed with Pt-HZSM-5 catalyst for *m*-xylene isomerization. The effect of different reactor configurations, namely membrane reactor with catalyst bed packed either in the membrane lumen of the tube or between the membrane tube and the module shell, all operating in the temperature range of 523–673 K was evaluated [162]. Decreasing the operating temperature resulted in linear increase in the *p*-xylene yield for the PBMR configuration due to the effective extraction of *p*-xylene from the reaction zone. On the contrary, packing the catalyst bed outside the membrane layer reduced the effect of the membrane separation resulting in p-xylene yield leveling off with temperature decrease, due to diffusion limitations in the membrane substrate. Moreover, 100% selectivity to p-xylene was reported in the permeate-only mode, whereas the selectivity declined to nearly 48% when considering the retentate and permeate amounts. In a parallel study [163], the authors managed to synthesize higher quality membranes in terms of selectivity (p-xylene/o-xylene > 400). Accordingly, the membrane reactor outperformed the conventional one at the applied reaction temperature of 473 K. Furthermore, the same authors employed nanocomposite MFI-ceramic hollow fibre membranes for the xylene isomer separation, demonstrating nearly 30% p-xylene flux increase and reasonable selectivity [227]. Finally, the utilization of hollow fibre membranes in zeolite membrane reactors for the isomerization of xylene could offer the essential increase in the p-xylene permeation flux in order to overcome one of the considerable limitations of this reactor configuration in the competition with the existing technologies.

4.3. Selectivity Enhancement through Selective Distribution of Reactants or Removal of Intermediate Products

As reviewed in the previous section, the performance of zeolite membranes packed with catalysts is widely evaluated in membrane reactor configurations either for the selective extraction of products in equilibrium limited reactions or for extraction of products inhibiting the catalyst activity. All the efforts were to improve the conversion of the reactions studied. Besides, zeolite membranes could play a further role which is based on upstream separation of a component from the feed mixture or removal of intermediate products.

In this context, the principle of controlling reactants traffic from a feed mixture was demonstrated by Gora *et al.* [171]. In detail, the hydroisomerization of n-hexane was studied in single-pass operation with a membrane reactor combining silicalite-1/TiO$_2$/stainless steel tubular membrane and Pt-chlorinated catalyst. The silicalite-1 membrane was able to selectively permeate n-hexane due to its preferential adsorption from a feed mixture comprising n-hexane and 2-methyl-pentane with a purity of ~99% to the reaction zone, thus revealing to some extent the advantages of this reactor configuration. The same working group proposed a concept for industrial scale heptane hydroisomerization process combining two reactors and a zeolite membrane. According to the simulation data a total feed amount of 907 metric ton per day (existing C$_5$/C$_6$ isomerization technologies operate between 600 and 1200 metric ton per day) is processed forming 220 metric ton per day product with improved research octane number from 57 up to 92 [228]. After economical evaluation, the authors concluded a total investment cost of 40 million euros being significantly higher than the state of the art C$_5$/C$_6$ hydroisomerization process of UOP Penex/Molex plant with an investment of about 23 million euros. Approximately 42% of the total equipment cost belongs to the zeolite membrane.

Further example in this section is the application of membranes with sufficient selectivity for the permeation of valuable intermediate products for yield increase. In this context, the removal of the intermediate products from the reaction before consecutive reactions could be also considered as an example of residence time control. The concept was demonstrated by Piera *et al.* [229] based on zeolitic PBMRs in the oligomerization of *i*-butene, where MFI membranes packed with resin catalysts were used for the selective removal of the formed *i*-octene. In such a way, the formation of the undesired C_{12} and C_{16} hydrocarbons was decreased and at approximately 20 °C significant selectivity increase was obtained resulting in higher *i*-octene yields. Moreover, the authors explained the observed permselectivity to *i*-octene in a mixture of *i*-butene/*i*-octene by the preferential adsorption of *i*-octene and the thereby arising pore-blocking effect for the *i*-butane passage. Increasing the temperature contributed to higher *i*-butane conversion over 90%, however decreased the *i*-octene adsorption and thus the separation selectivity. As a result, the PBMR showed nearly the same conversion as the traditional reactor. Nevertheless, the yield in the PBMR still displayed an absolute enhancement between 20% and 30%.

5. Reaction Processing Using Permselective Catalytic Membrane Reactors (CMR)

So far, in the membrane-assisted reactor applications discussed up to now, the membrane used was inert. To put it differently, the membrane did not display any catalytic function, it showed only permselective properties depending on the dominant transport mechanism (see Section 2.1.) and the catalyst bed was a discrete part of the set-ups tested. The latter was either separated (see Section 2.3.) or packed in a direct contact to the zeolite membrane (PBMR, see Section 4.). In the majority of the applications of PBMRs the zeolite membranes were utilized either for selective product removal or for selective reactant supply. Similarly, in catalytic membrane reactors (CMR), where the zeolite membranes used display both catalytic activity and permselectivity, as shown schematically in Figure 8, e.g., dehydrogenation reactions [144], isomerization [159,160,164,165] and esterification reactions [167,230,231] have been investigated. However, it is the nature of the beast that not all the aforementioned reactions can be carried out in CMRs since zeolites simply are not the candidates of choice for all reactions.

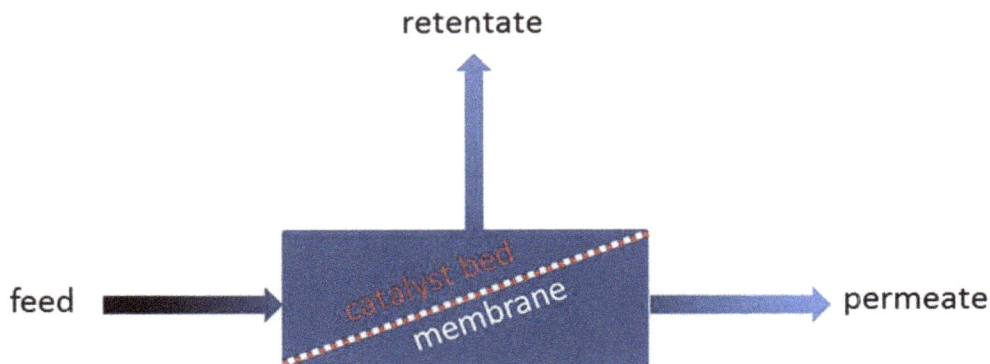

Figure 8. Schematic drawing of a catalytic membrane reactor (CMR).

Xiongfu and coworkers [144] studied the dehydrogenation of ethylbenzene to styrene using either Fe-ZSM-5 or Al-ZSM-5 zeolite membranes synthesized on porous α-Al$_2$O$_3$ tubes in CMRs. Generally, improved conversion compared to the conventional reactor configuration was obtained. However, the MFI membrane with incorporated Fe species showed better conversion compared to the Al-ZSM-5. The authors ascribed the observed effect on the one side to the better adsorption of ethylbenzene on the Fe-ZSM-5 membrane than on its counterpart Al-ZSM-5. On the other hand, the styrene adsorption on the Fe-ZSM-5 membrane was lower that on the Al-ZSM-5 membrane, thus ensuring lower carbon deposition on the membrane and higher ethylbenzene conversion.

The combination of catalytic and separation properties provided by the zeolite membranes has been also evaluated in isomerization reactions. Haag *et al.* [159] proposed the use of H-ZSM-5 zeolite membranes synthesized on top of stainless steel disks. Because of their intrinsic acidic properties, catalytically active sites for the isomerization of xylene were provided.The formed *p*-xylene was selectively separated from the other isomers due to the shape selective properties of the membrane. In contrast to conventional reactors packed with H-ZSM-5 catalyst, slightly enhanced conversions and selectivities were obtained with the membrane reactor configuration. However, the authors pointed out, that difficulties arise, when experimental data obtained in the two reactor configurations is compared since the kinetic properties of the two catalytic materials are dissimilar. On the other hand, Tarditi *et al.* [160] carried isomerization reactions in a membrane reactor with an ion-exchanged Pt/H-ZSM-5 catalytic active membrane and in a membrane reactor equipped with a Ba-ZSM-5 zeolite membrane. Taken as a whole, the two membranes demonstrated reasonably enhanced *p*-xylene selectivity and yield compared to the conventional fixed-bed reactor packed with Pt/silica-alumina commercial catalyst. The *p*-xylene yield increased approximately 28% with the help of the Ba-ZSM-5 membrane, while the

Pt-exchanged membrane achieved a bit lower *p*-xylene relative yield increase of 22% at 370 °C by feeding *m*-xylene. The *p*-xylene flux through the Ba-ZSM-5 membrane was found to be quite dependent on the Ba^{2+} concentration, where an increase in the ion concentration ensured higher *p*-xylene fluxes leading to better extraction from the reaction zone. Moreover, a ternary mixture (65% *m*-xylene, 14.5% *p*-xylene and 20.5% *o*-xylene) isomerization reaction in the fixed bed reactor as well as in the CMR with the fully exchanged Ba-ZSM-5 resulted in 2.86×10^{-8} mol·s^{-1} and 3.74×10^{-8} mol· s^{-1}*p*-xylene production, respectively confirming an increase of 31% in favor of the CMR at 370 °C. Due to experimental limitations, the authors evaluated theoretically using a transport model [232] the effect of the relevant for industrial application high pressure of around 1000 kPa and stated that the *p*-xylene production enhancement could be obtained by operation in CMR despite the selectivity decrease of the membrane at higher pressure. However, experimental evaluation would be still interesting for the verification of the theoretical observations.

Recently, Yeong [164] studied theoretically and experimentally the isomerization of *m*-xylene in acid-functionalized silicalite-1 catalytic membrane reactors. Propylsulfonic acid sites or arenesulfonic acid sites were provided to the inert silicalite-1 membranes via post-synthesis modifications. Higher isomerization activity was obtained with the arenesulfonic acid-functionalized membrane mainly ascribed to its higher acidity and more effective continuous removal of *p*-xylene. According to the results gained in this study, the acid-modified membranes turned out to be more effective in terms of *m*-xylene conversion and *p*-xylene productivity improvement compared with the membranes reported in early studies [158–161]. Moreover, the kinetic parameters reported in this study offer useful platform for further optimization of the catalytic membrane reactor design. Zhang *et al.* [165] prepared H^{+} ion-exchanged MFI zeolite membranes on α-Al$_2$O$_3$ disc support. In earlier works of the same working group [161], the use of that very catalytic active zeolite membrane was proposed as an effective tool for achieving high permeation flux. The catalytic MFI zeolite membrane exhibited notable *p*-xylene selectivity of nearly 92%, however, at a significantly low *m*-xylene conversion of 6.5%, which was mainly ascribed to the limited number of active sites on disk-shaped membrane. Further on, the benefit of membrane reactor combining catalytic active and permselective boron substituted MFI zeolite membrane was demonstrated in the 1-butene double-bond isomerization [166]. Thereby, the incorporated boron in the framework generated Brønsted sites with low acid strength providing the catalytic selectivity. On the other side, the MFI membrane displayed selective permeation for *trans*-2 butene, giving rise to enhanced *trans/cis* ratio. So, in the retentate a *trans/cis* ratio between 1.4 and 1.5 being the same as the equilibrium ratio was found while in the permeate it was increased to a value of 2.2.

Bernal *et al.* [167] was the first who proposed the use of H-ZSM-5 zeolite membrane reactors in the continuous esterification. The idea behind was the lower diffusion resistance offered by a reactor configuration integrating the reaction and separation in once, which could assure the immediate removal of the formed products, thus displacing the equilibrium and giving rise to higher turnover. The catalytic zeolite membrane reactor outperformed both the conventional fixed-bed reactor and the inert zeolite membrane reactor in the conversion of acetic acid with ethanol. The improved performance over the latter was mainly attributed to the absence of a diffusion step from the catalyst bed to the membrane surface. In this way, de la Iglesia *et al.* [231] coupled the catalytic activity of H-ZSM-5 and the selective water separation properties of modernite membranes in two-layered mordenite-ZSM-5 bi-functional membranes thus improving further the performance of the zeolite membrane reactor in the esterification of acetic acid with ethanol. In fact, much more advanced improvement could be reached by simultaneous control of the membrane thickness and the zeolite membrane composition.

Using composite catalytically active H-USY zeolite membranes Peters *et al.* [230] managed to couple reaction and separation for continuous esterification of acetic acid and butanol. Again, the authors pointed that optimization of the catalytic layer will contribute to further enhancement of the reactor performance.

6. Zeolite Membrane Coatings on Catalyst Particles

The potential of combining macroscopic units of zeolite membranes and catalysts in chemical reactors was described for diverse applications in the previous chapters. However, the desired improvements in reaction selectivity and productivity are not all of the important reaction parameters to be optimized. Relatively often zeolite membranes suffer from low permeation flux. To overcome this issue a sufficiently larger membrane area (related to the catalyst volume) is demanded which would lead to space velocities compatible to conventional reactors. The shape selectivity, often being the main property of zeolites membranes, could be used to design novel catalyst materials by adding this feature to conventional catalysts. In this regard, coating permselective zeolite membranes (shell) on particular catalysts (core) may provide selectivity and possible additional catalytic sites. Importantly, such encapsulated catalysts will offer a much larger membrane area per unit reactor volume than conventional membrane reactors. Figure 9 displays a scheme of a catalyst bed filled with such core-shell particles, whereas each of them can be understood as a kind of zeolite membrane microreactor.

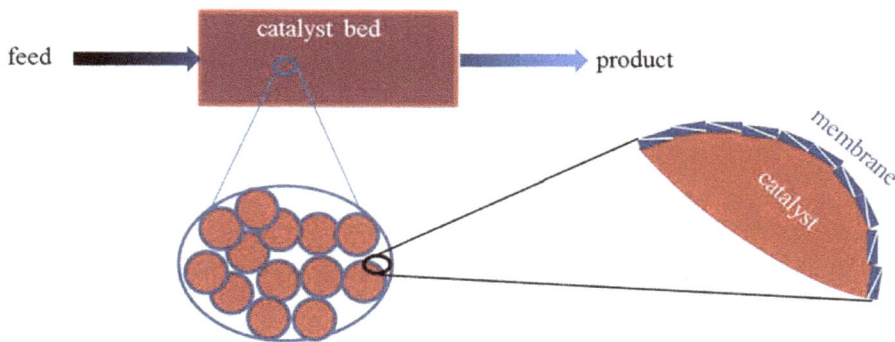

Figure 9. Catalyst bed filled with core-shell particles whereas each of them can be understand as zeolite membrane microreactor.

The application of core-shell catalysts as particle level membrane reactors follows the concept of the traditional membrane reactor so that they could be applied in either reactant-selective or product-selective reactions based on the diffusivities of the reaction components/products within the membrane. In the first case, the controlled supply of reactants could prevent undesirable reaction, due to the selective permeation through the zeolite membrane to the catalyst core. Additionally, the zeolite shell could act as protective barrier against impurities or poisons, e.g., in a direct internal reforming-molten carbonate fuel cell [233]. In the second case, the selective removal of the desired product could shift the equilibrium in thermodynamically limited reactions and enhance the selectivity. Alternatively, the zeolite shell can provide catalytic properties and so encapsulated catalyst can be applied as catalytic membrane e.g., in Fischer-Tropsch synthesis reactions [234]. The reaction selectivity is predetermined to some extent by the membrane synthesis conditions since they affect the membrane thickness and quality. The concept of zeolite membrane coatings on particles was inspired mostly by the pioneering work of Puil *et al.* describing the coating of TiO_2-supported platinum catalyst particles by a silicalite layer for hydrogenation reactions [235] as well as the preparation procedure of MFI and BEA type coatings on preshaped α-alumina supports applied in alkylation reactions [236].

So far, different synthesis techniques have been reported for the preparation of zeolite coated catalyst. Generally, hydrothermal synthesis is applied for the preparation of the zeolite membrane enwrapping the core catalyst. Bouizi *et al.* evaluated the factors governing the formation of core-shell zeolite-zeolite composites by reversing the negative charge of the crystals with 0.5 wt. % aqueous solution of a polycation agent prior to seeding and secondary growth technique [237]. The authors concluded that successful synthesis could be observed for materials displaying compatible framework compositions and close crystallization conditions. Moreover, in order to induce the zeolite matrix seeding turned out to be a crucial step in the preparation technique. On the contrary, a H-β zeolite was coated directly on the surface of Co/Al_2O_3 by one-step hydrothermal synthesis [238,239]. Alternatively, liquid membrane crystallization being a modified vapor transport method was proposed as economic, environmental and highly effective method for synthesis of zeolite-encapsulated catalysts [240]. Furthermore, physical coating was suggested for easy scalable synthesis where the required high temperature and alkaline conditions during hydrothermal synthesis were excluded [241–243]. Recently, steam-assisted crystallization process was recommended as an efficient method for capsuled catalysts preparation [244].

A brief overview of recent publications dealing with zeolite capsuled catalysts is given in Table 2.

Table 2. Overview of the application of packed bed reactor consisting of permselective membrane coated catalyst particles reported in the literature.

Reaction	Reactor Type	Feed	Operating Conditions	Core-Shell Catalyst	X_{mixed} (%) S_{mixed} (%) Y_{mixed} (%)	X_{MR} (%) S_{MR} (%) Y_{MR} (%)	References
Disproportiona-tion of toluene	PLMR product selective	toluene	723–823 K $p = 101.3$ kPa WHSV $= 0.1\ h^{-1}$	silicalite coated on silica-alumina catalyst	$S_{p\text{-xylene}} = 22$	$S_{p\text{-xylene}} > 91$	[245]
Alkylation of toluene	PLMR product selective	toluene, methanol	673 K	silicalite coated on H-ZSM-5 crystals with different Si/Al ratios	$X_{toluene} = 63$ $S_{p\text{-xylene}} = 40$	$X_{toluene} = 42$ $S_{p\text{-xylene}} > 99.9$	[246,247]
Hydro-formylation of 1-hexene	batch type reactor product selective	1-hexene, carbon monoxide, hydrogen	130 °C $H_2/CO = 1$	silicalite-1 coated on Pd–Co/activated carbon	$X = 75.7$ $S_{hexan} = 13.3$ $S_{isomer} = 15.4$ $S_{i\text{-hept.}} = 37.1$ $S_{n\text{-hept.}} = 33.1$	$X = 54$ $S_{hexan} = 28.3$ $S_{isomer} = 21.9$ $S_{i\text{-hept.}} = 13.9$ $S_{n\text{-hept.}} = 35.9$	[248]
Hydrogenation of linear and branched alkenes	PLMR reactant selective	1-hexene, 3,3-dimethyl-but-1-ene	323–373 K $p = 101.3$ kPa	silicalite-1 coated on Pt/TiO$_2$ particles	$X_{1\text{-hex}} > 90$ $X_{3,3\text{-DMB}} > 90$ $S = 1\text{–}1.2$	$X_{1\text{-hex}} > 90$ $X_{3,3\text{-DMB}} < 10$ $S = 12\text{–}20$	[249,250]
Oxidation of CO and n-butane	PLMR reactant selective	air, carbon monoxide and n-butane	483 K $p = 101.3$ kPa	zeolite-4A coated on spherical Pt/γ-Al$_2$O$_3$ particles (two-steps hydrothermal synthesis)	$X_{n\text{-butane}} = 95$ $X_{CO} = 93$	$X_{n\text{-butane}} = 0$ $X_{CO} > 90$	[251]
Shape-selective hydrogenation of xylene isomers	PLMR reactant selective	p-/o-xylene or p-/m- xylene	473 K $p = 1.0$ MPa WHSV $= 1.0\ h^{-1}$	silicalite-1 coated on Pt/Al$_2$O$_3$ pellets	-	$S_{p/o} = 17$ $S_{p/m} = 13.6$	[252]
Steam reforming of methane and toluene	PLMR reactant selective	methane or toluene, steam, helium	780–840 °C $p = 1$ bar $CH_4/H_2O = 1$ $H_2O/C_7H_8 = 7$	Hβ zeolite coated on Ni/Mg/Ce$_{0.6}$Zr$_{0.4}$O$_2$ pellets	X_{CH4} increases with temperature up to ~20X_{C7H8} ~58	X_{CH4} increases with temperature up to ~30 X_{C7H8} ~22	[242]

Table 2. *Cont.*

Reaction	Reactor Type	Feed	Operating Conditions	Core-Shell Catalyst	X_{mixed} (%) S_{mixed} (%) Y_{mixed} (%)	X_{MR} (%) S_{MR} (%) Y_{MR} (%)	References
Direct synthesis of middle *i*-paraffins	PLMR catalytic	hydrogen and carbon monoxide	533 K p = 1.0 MPa H_2/CO = 2	H-ZSM-5 coated on Co/SiO$_2$ pellets with different size	X_{CO} = 93.6 S_{CH4} = 16.9 S_{CO2} = 8 C_i/C_n = 0.49	X_{CO} = 89.1 S_{CH4} = 22.4 S_{CO2} = 6.9 C_i/C_n = 0.74	[234,253,254]
Direct synthesis of *i*-paraffins	PLMR catalytic	hydrogen and carbon monoxide	533 K p = 1 MPa H_2/CO = 2	H-β zeolite coated on Co/Al$_2$O$_3$ catalyst pellets with different size	X_{CO} = 80.8 S_{CH4} = 16.6 S_{CO2} = 3.9 C_i/C_n = 1.4	X_{CO} = 74.3 S_{CH4} = 13.6 S_{CO2} = 2.7 C_i/C_n = 2.3	[239]
Direct synthesis of *i*-paraffins	PLMR catalytic	hydrogen and carbon monoxide	533 K p = 1 MPa H_2/CO = 2	H-ZSM-5 coated on Ru/SiO$_2$ catalyst pellets with different size	X_{CO} = 82.1 S_{CH4} = 17.1 S_{CO2} = 5 C_i/C_n = 0.42	X_{CO} = 81.7 S_{CH4} = 20.5 S_{CO2} = 6.1 C_i/C_n = 1.5	[255]
Direct synthesis of *i*-paraffins	PLMR catalytic	hydrogen and carbon monoxide	533 K p = 1 MPa H_2/CO = 2	H-ZSM-5 coated on Pd/SiO$_2$	-	X_{CO} = 86.1 S_{CH4} = 37.4 S_{CO2} = 7.0 C_i/C_n = 1.88	[256]
Direct synthesis of middle *i*-paraffins	PLMR catalytic	hydrogen and carbon monoxide	573 K p = 1.0 MPa H_2/CO = 1	H-ZSM-5 crystalized on fused-iron catalyst pellet	X_{CO} = 96.7 S_{CH4} = 12.8 S_{CO2} = 44.7 C_i/C_n = 2.31	X_{CO} = 96.9 S_{CH4} = 8.7 S_{CO2} = 33.9 C_i/C_n = 4.17	[257]
Synthesis of gasoline-range *i*-paraffins	PLMR catalytic	hydrogen and carbon monixide	483–533 K p = 2.0 MPa H_2/CO = 2 GHSV = 1000 h^{-1}	H-ZSM-5 coated on CoZr catalyst particles	X_{CO} = 97.4 S_{18+} = 16 S_{18+} = 5.6 $S_{i\text{-}C5\text{-}11}$ = 16.7	X_{CO} = 82.3 S_{CH4} = 14.8 S_{18+} = 0.3 $S_{i\text{-}C5\text{-}11}$ = 24.7	[258]
Direct synthesis of light *i*-paraffins	PLMR catalytic	hydrogen and carbon monoxide	553 K p = 1 Mpa I$_2$/CO – 2	H-ZSM-5 zeolite coated on Co/SiO$_2$	X_{CO} = 98.5 S_{CH4} = 23.7 S_{CO2} = 16 S_n = 53.4 S_i = 36.2	X_{CO} = 99.1 S_{CH4} = 20.1 S_{CO2} = 18.2 S_n = 47.6 S_i = 43.8	[259]

Table 2. *Cont.*

Reaction	Reactor Type	Feed	Operating Conditions	Core-Shell Catalyst	X_{mixed} (%) S_{mixed} (%) Y_{mixed} (%)	X_{MR} (%) S_{MR} (%) Y_{MR} (%)	References
Direct synthesis of middle *i*-paraffins	PLMR catalytic	hydrogen and carbon monoxide	300 °C $p = 1$ MPa $H_2/CO = 1$	H-ZSM-5 coated on Fe/SBA-15	$X_{CO} = 63.9$ $S_{CO2} = 43.8$ $S_{CH4} = 19.2$ $S_n = 56$ $S_i = 33.9$	$X_{CO} = 57.6$ $S_{CO2} = 37.3$ $S_{CH4} = 15.3$ $S_n = 36.7$ $S_i = 46.5$	[244]
Direct synthesis of *i*-paraffins	PMLR catalytic	carbon monoxide, hydrogen	280 °C $p = 1$ MPa $H_2/CO = 1$	Silicalite-1 and H-ZSM-5 coated on Fe/SiO$_2$ dual-membrane coated catalyst	$X_{CO} = 60$ $S_{CO2} = 29.9$ $S_{CH4} = 7$ $S_i = 12.9$	$X_{CO} = 54.8$ $S_{CO2} = 33.8$ $S_{CH4} = 14.9$ $S_i = 29.8$	[260]
Dimethyl ether direct synthesis	PLMR catalytic	hydrogen, carbon monoxide, carbon dioxide and argon	523 K $p = 5.0$ MPa	H-ZSM-5 coated on Cu/ZnO/Al$_2$O$_3$	$X_{CO} = 58.07$ $S_{MeOH} = 57.29$ $S_{DME} = 40.51$	$X_{CO} = 30.4$ $S_{MeOH} = 21.43$ $S_{DME} = 78.57$	[261]
Dimethyl ether direct synthesis	PLMR catalytic	hydrogen, carbon monoxide, carbon dioxide and argon	573–623 K $p = 5.0$ MPa	Double layer H-ZSM-5/Silicalite-1 membrane coated on Cr/ZnO core catalyst	$X_{CO} = 45.16$ $S_{MeOH} = 12.12$ $S_{DME} = 0.47$	$X_{CO} = 9.53$ $S_{MeOH} = 21.23$ $S_{DME} = 50.84$	[262]
Dimethyl ether direct synthesis	PLMR catalytic	hydrogen, carbon monoxide, carbon dioxide, argon	523 K $p = 5.0$ MPa	Double layer H-ZSM-5/Silicalite-1 membrane coated on Pd/SiO$_2$ core catalyst	$X_{CO} = 12.84$ $S_{CH4} = 1.47$ $S_{MeOH} = 16.51$ $S_{DME} = 48.40$	$X_{CO} = 9.48$ $S_{CH4} = 16.8$ $S_{MeOH} = 4.76$ $S_{DME} = 68.70$	[263]
Dimethyl ether direct synthesis	PLMR catalytic	hydrogen, carbon monoxide, carbon dioxide, argon	350 °C $p = 5$ MPa	SAPO-46 zeolite shell encapsulated Cr/ZnO catalyst	$X_{CO} = 4.7$ $S_{CH4} = 3.7$ $S_{MeOH} = 71.7$ $S_{DME} = 16.5$	$X_{CO} = 6.9$ $S_{CH4} = 4.7$ $S_{MeOH} = 52.2$ $S_{DME} = 37.0$	[243]

Table 2. *Cont.*

Reaction	Reactor Type	Feed	Operating Conditions	Core-Shell Catalyst	X_{mixed} (%) S_{mixed} (%) Y_{mixed} (%)	$X_{MR.}$ (%) $S_{MR.}$ (%) $Y_{MR.}$ (%)	References
Dimethyl ether direct synthesis	PLMR catalytic	hydrogen, carbon monoxide, carbon dioxide, argon	250 °C p = 5 MPa	SAPO11 coated on Cu/ZnO/Al$_2$O$_3$	X_{CO} = 64.9 S_{MeOH} = 51.4 S_{DME} = 46.6 Y_{DME} = 30.2	X_{CO} = 92 S_{MeOH} = 9.2 S_{DME} = 90.3 Y_{DME} = 83.1	[241]
Carbon dioxide hydrogenation to dimethyl ether	PLMR	carbon dioxide and hydrogen	270 °C p = 3.0 MPa SV = 1800 mL·g·cat^{-1}·h^{-1} H$_2$/CO = 3	H-ZSM-5 coated on CuO-ZnO-Al$_2$O$_3$ nanoparticles	X_{CO2} = ~24 S_{DME} = ~26 Y_{DME} = ~6	X_{CO2} = 48.3 S_{DME} = 48.5 Y_{DME} = 23.4	[264]

Nishiyama *et al.* [249] coated a silicalite-1 membrane on spherical Pt/TiO$_2$ particles applying a hydrothermal synthesis and obtained core-shell catalysts displaying reactant selectivity due to the adsorption-based permselective properties of the zeolite membrane. The authors impregnated the beforehand prepared Pt/TiO$_2$ with a solution of 0.4 wt. % cationic polyethyleneimine in order to charge its surface positively and thus facilitate the adsorption of the silicalite-1 seeds by subsequent immersing in 1.0 wt. % silicalite-1 seed solution. The final membrane zeolite coating crystallization was performed in a closed vessel at 180 °C for 24 h. Figure 10 depicts the SEM images revealing the dense uniformly formed silicalite-1 shell with a thickness of approximately 40 µm on the surface of the core Pt/TiO$_2$ spheres.

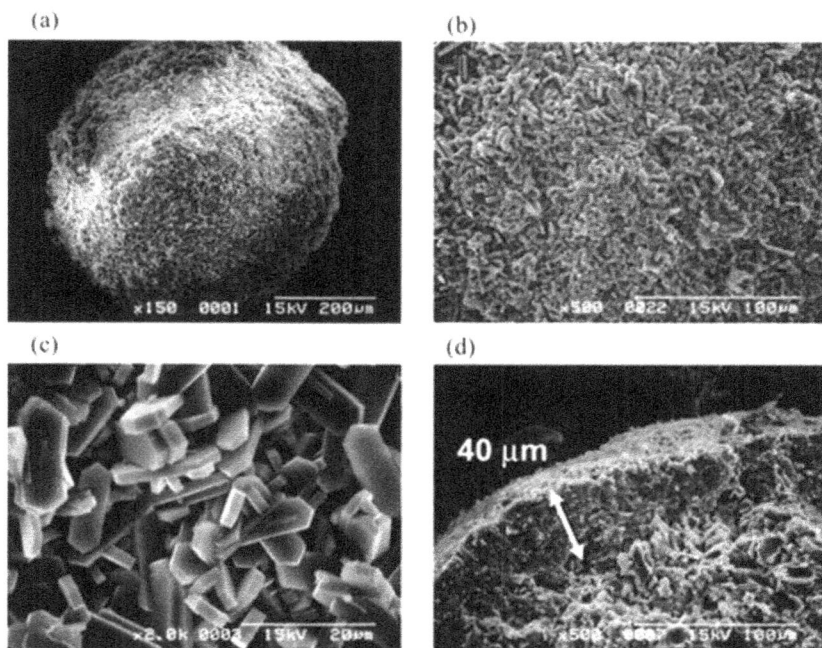

Figure 10. SEM images of Pt/TiO$_2$ particles coated with silicalite-1. Reprinted with permission from [249]. Copyright (2004) American Chemical Society.

The thereby prepared silicate-1 membrane layer permeated preferentially the reactant 1-hexene from a mixture comprising 1-hexene and dibranched 3,3-dimethylbut-1-ene towards the Pt/TiO$_2$ catalyst leading to improved hydrogenation selectivities for 1-hexene. At a reaction temperature of 323 K the ratio of hydrogenation selectivities for the linear/branched alkene mixture was between

12 and 20. At 373 K the selectivity range was increased to 18–30 revealing the positive impact of the selective permeation of 1-hexene across the silicalite-1 membrane towards the Pt/TiO_2 particles in principle and the temperature influence in addition. By applying a hydrothermal synthesis including two crystallization steps, the same working group prepared thinner silicite-1 membranes with improved quality. In that case, the authors adopted first a short crystallization step under rotation in order to increase the number density of crystal nuclei at the particle surface responsible for the small crystal size. Then, athinner membrane was crystallized in a second synthesis at 453 K for 48 h without rotation. The so prepared core-shell catalyst exhibited even higher hydrogenation selectivities (selectivity of 1-hexene/3,3-dimethylbut-1-ene at 323 K of 35 and 80 at 373 K) due to higher permselectivity of the linear/branched alkene mixture through the membrane revealing the enormous impact of the membrane quality in terms of the thickness on the reaction rate [250]. Moreover, it was shown that due to the coating procedure the reaction rate of the process was changed from kinetic-controlled to diffusion-controlled, pointing once again that the permeation of 1-hexene and 3,3-dimethylbut-1-ene is the rate controlling step in the hydrogenation reaction. In addition, reduced catalyst deactivation ascribed to the catalyst protection against poisoning impurities from the feed by the zeolite membrane was reported.

The concept of reactant selectivity was further demonstrated by Zhong *et al.* [251] by applying defect-free zeolite-4A coated on Pt/γ-Al_2O_3 particles prepared via two-step hydrothermal synthesis in a model oxidation reaction of CO and *n*-butane mixture. The permeation of *n*-butane was obviously restrained due to sieving effects, so that only CO and O_2 were able to pass the zeolite membrane and react in the catalyst core. The authors pointed out that the membrane coated catalyst might be attractive for applications where the hydrocarbon feed streams contain a trace amount of CO and it should be removed in order to prevent the catalyst poisoning. Ren *et al.* [265] suggested the encapsulation of noble metal particles with protective, size-selective zeolite shells as an effective strategy to overcome deactivation problems [266,267] occurring during liquid phase reactions, e.g., for targeted production of fine or intermediate chemicals. As an example the selective oxidation of alcohols was chosen. The significantly improved selectivity and the core catalyst protection (Ag and Pt nanoparticles) by a silicalite-1 membrane coating compared to commercial catalysts was attributed to selective permeation through the zeolite membrane. Under those circumstances, the diffusion of large reactants and poison molecules in the reaction environment was restricted due to the shape selectivity of the membrane resulting respectively in retention of the catalytic activity. Moreover, as the ICP-AES analysis revealed, almost no Pd leaching from the core-shell catalyst after 6 cycles of recycling was observed, pointing out the benefit of coating on the reusability of the catalyst.

The application of zeolite membranes being quite lucrative for shape-selective hydrogenation of xylene isomers was also reported. As discussed in Section 4.2., MFI zeolites are considered as appropriate candidates for the preparation of membranes, since their pore diameters approximates the size of *p*-xylene, while the bulkier *m*- and *o*-xylene isomer molecules cannot pass the material due to sterical hindrance resulting in significant permselectivity of *p*-xylene over the other isomers. Coming back, the combination of molecular sieving and hydrogenation of xylene isomers was demonstrated over silicalite-1 coated Pt/Al_2O_3 catalysts [252]. An excellent para-selectivity was achieved, whereas the hydrogenation of the *o*- und *m*-xylene isomers to 1,3-dimethylcyclohexane and 1,2-dimethylcyclohexane was suppressed since almost exclusively *p*-xylene was passing the membrane to reach the catalytic active sites in the core in order to hydrogenate and produce 1,4-dimethylcyclohexane which diffused across the membrane to the product site.

More recently, Zhang *et al.* [268] has chosen a porous metal-organic framework (MOF), namely zeolitic imidazolate framework-8 (ZIF-8), as a shell enwrapping the Pd/ZSM-5 core. A layer by layer self-assembly of polyelectrolyte was used to overcome the incompatibility between the materials prior to the two-step temperature synthesis employing ice bath for the initial nucleation and temperature of 30 °C for further crystallization. The catalytic performance and molecular-size-selectivity of the core-shell structure was evaluated in hydrogenation of 1-hexene and cyclohexene. By applying 1-hexene as a reactant, full conversion was achieved with *n*-hexane being the only product displaying similar performance as the not enshrouded Pd/ZSM-5 catalyst. On the other hand, the diffusion of cyclohexane into the core catalyst was significantly restrained since its molecular size exceeded the aperture size of ZIF-8 resulting in decreased conversion compared to the Pd/ZSM-5 catalyst. However, the observed conversion of 25.1% was obviously the result of cracks in the ZIF-8 shell.

The controlled traffic of reactants providing improved reforming selectivity was demonstrated very recently by Cimenler [242] in the steam reforming of CH_4 and C_7H_8 where the latter represents a model for tar impurity in feed. Applying H-β zeolite membrane shell on the steam reforming $Ni/Mg/Ce_{0.6}Zr_{0.4}O_2$ catalyst led to decreased C_7H_8 conversion compared to the uncoated catalyst exemplifying the molecular-size selective properties of the membrane. The authors suggested that the shape selective effect could be boosted either by preparing thicker shell membranes or by using dealuminated zeolite.

The concept of selective product removal was illustrated for the selective formation of *p*-xylene in the disproportionation of toluene [245]. Thereby, high *p*-xylene selectivity was obtained due to the selective permeation of the produced *p*-xylene through the silicalite membrane coated on the silica-alumina catalyst causing an equilibrium shift. Even so, the activity of the coated catalyst was lower than that of the non-coated catalyst as a direct result of the relatively thick membrane which was limiting the products diffusion.

Another interesting approach described in the literature is the enwrapping of active zeolite crystals with zeolite membranes. The proposed model for the formation of the silicalite-1 layer on ZSM-5 crystals is illustrated in Figure 11 [269]. Silicalite-1 crystals grow on the ZSM-5 crystal surface perpendicular to the a and c axes under hydrothermal conditions at 453 K for 24 h without agitation. In this context, the benefit of silicalite zeolite membrane coated on H-ZSM-5 crystals was evidenced for the selective p-xylene formation in alkylation of toluene with methanol [246,247,269]. The obtained para-selectivity of up to 99.9% was attributed to the suppressed further isomerization of p-xylene since the silicalite coating reduced the acid sites on the external surface of the H-ZSM-5 catalyst.

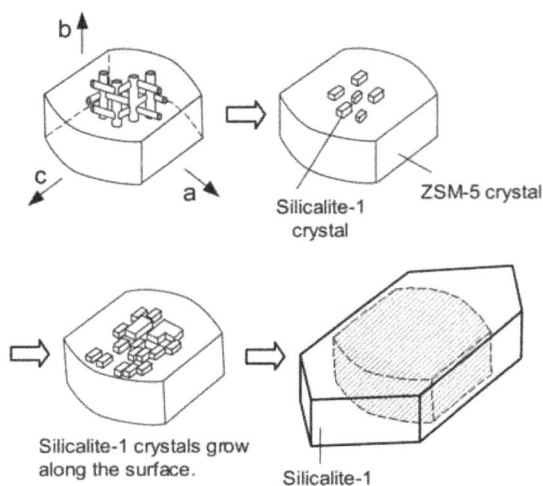

Figure 11. Graphical illustration of the proposed model for the formation of a silicalite-1 layer on a ZSM-5 crystal. Reprinted with permission from [269]. Copyright (2005) John Wiley and Sons.

The toluene conversion remained high even after the coating suggesting catalyst protection against coke formation on the one side and evidencing the positive effect of the thin membrane layer over the crystals on the other side if compared to the relatively thick membrane reported early by Nishiyama et al. [245]. Moreover, FE-SEM and TEM screening indicated that the pores of silicalite-1 were directly connected to the pores of H-ZSM-5 considered as the main reason for the obtained high selectivity and activity [247]. Additionally, the effect of the H-ZSM-5 crystal size was studied revealing a slight decline in the para-selectivity with increasing crystal size, thus pointing out the severity of silicalite layer defect-free growth on larger crystals. Recently, MFI-type zeolite containing aluminum and gallium within its framework (GaAlMFI) was coated with silicalite-1 and applied in propane aromatization reaction [270]. Thereby, the para-selectivity was increased to 80% in comparison to para-selectivity of 57% for the GaAlMFI catalyst without the silicalite-1 coating, remaining however lower than the respective *para*-selectivity in toluene alkylation over silicalite-1/H-ZSM-5 reported above. Going one step further, the same working group evaluated the effect of the synthesis conditions in terms of alkali concentration in the synthesis solution and proton-exchange procedure on the preparation of silicalite-1 coated GaAlMFI [271]. In this context, low TPAOH concentration was reported to improve the *para*-selectivity due to decreased number of acid sites on the external surface of the catalyst. On the other hand, the *para*-selectivity was significantly decreased by repeating the proton-exchanged procedure since Ga and Al species were removed from the framework leading to the formation of non-selective sites on the catalyst surface. Interestingly, during *n*-butane and propane aromatization, the silicalite-1/GaAlMFI performed better in the former case attributed by the authors to the higher formation of naphthalene and its derivatives during the propane aromatization.

Very recently, Zhou *et al.* [272] examined the oriented growth of MFI zeolite shells on ZSM-5 crystals. Ammonia as surface modifier was applied for the pretreatment of the catalyst core in order to facile the formation of *b*-oriented MFI film. The shape-selective core-shell catalyst was evaluated in toluene methylation experiments, demonstrating optimized selectivity and stability against the common ZSM-5 zeolite. However, due to incorporation of small amount of aluminum in the framework near the external surface of the shell, the obtained *para*-selectivity was lower than that in the previous reports [245,246].

Silicalite zeolite membrane was also employed as membrane coating over Pd-Co/activated carbon catalysts for the hydroformylation of 1-hexene with syngas [248]. The silicalite membrane decreased the 1-hexene conversion, while higher *n*- to *i*-heptanal ratios were reported due to the spatial confinement of the zeolite membrane pore channels limiting the diffusion of the *i*-heptanal out of the core. By increasing the membrane thickness via second hydrothermal synthesis, and thus complicating the diffusion of the reactants into the core catalyst, a further decrease in the 1-hexene conversion was observed. However, the ratio of the aldehyde products experienced sharp increase due to the rise in the *n*-heptanal selectivity predominated by the membrane thickness.

6.2. Application as Catalytic Membranes

A further extensively studied area for application of encapsulated catalysts with zeolite membrane as a shell possessing catalytic properties represent the Fischer-Tropsch synthesis reactions. The Fischer-Tropsch synthesis converts syngas (mixture of CO and H_2) to aliphatic hydrocarbons. Typical mixtures of linear, branched and oxygenated hydrocarbons, linear paraffins and α-olefins are the main products [273]. As fuel, they are appropriate only as diesel, while for the use as synthetic gasoline, they must be further hydrocracked and isomerized to branched, light hydrocarbons. In industrial term, the direct production of *i*-paraffins from syngas is of great interest. Physical mixture of Fischer-Tropsch catalyst and zeolite catalyst providing active site for the two reactions was proposed in the literature for the direct synthesis of *i*-paraffins rich hydrocarbons [274–277]. The Fischer-Tropsch catalyst offers thereby the active sites for the conversion of syngas to linear hydrocarbons which then undergo hydrocracking and isomerization to the desired branched hydrocarbons on the acidic sites of zeolites. Even so, a major problem by applying catalyst mixtures is the random distribution of the active sites. The linear hydrocarbons formed on the active sites of the Fischer-Tropsch catalyst can leave the reaction zone without reaching the active sites of the zeolite for further reforming resulting in low selectivity.

In order to improve the migration of the linear hydrocarbons to the active sites of the zeolite, catalysts with core-shell structure have been proposed as a more efficient alternative and extensively studied by the working group of Tsubaki. Accordingly, the selectivity could be significantly improved since the intermediate products are enforced to pass a zeolite membrane in order to desorb from the Fischer-Tropsch catalyst experiencing a higher possibility for further conversion so as to yield the desired product. In detail, the working group of Tsubaki [234,253,254] coated H-ZSM-5 membranes on preshaped Co/SiO_2 pellets and evaluated their performances in *i*-paraffins synthesis from syngas. Thereby, syngas permeated through the membrane to the core catalyst and reacted to straight-chain hydrocarbons which were then hydrocracked and isomerized while passing the zeolite channels as show schematically in Figure 12.

The formation of C_{10+} hydrocarbons was completely suppressed. Moreover, an indication was given that the size of the pellets affects the properties of the capsule catalyst since coating smaller pellets of Co/SiO_2 resulted in a higher *i*-paraffin/*n*-paraffin ratio. However, the prepared core-shell catalysts showed higher methane selectivity compared to the mechanical mixture of Co/SiO_2 and H-ZSM-5 due to the lower diffusion efficiency for CO caused by the different diffusion rates of CO and H_2 in the zeolite pores. The same working group managed to coat H-beta membranes onto the surface of Co/Al_2O_3 pellets and observed excellent performance for the direct synthesis of *i*-paraffins [239]. Not only the formation of C_{12+} hydrocarbons was completely suppressed, but also lower methane selectivity was recorded due to the hydrophilicity provided by the zeolite coating. The presence of water favors higher CO concentrations passing the membrane leading to decreased H_2/CO ratios in the catalyst core and thus to lower CH_4 selectivity.

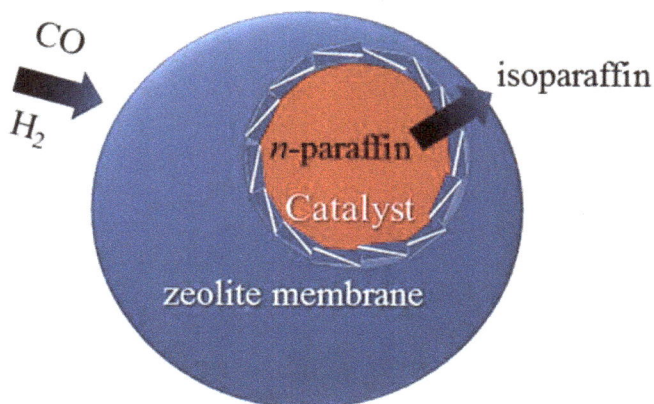

Figure 12. Schematic drawing of a core/shell catalyst in the synthesis of *i*-paraffins from syngas, where the core catalyst is responsible for the Fischer-Tropsch synthesis and the zeolite membrane for the further hydrocracking and isomerization of *n*-paraffins to *i*-paraffins.

Based on early literature reports revealing the improved selectivity of C_{5+} hydrocarbons due to addition of Zr as promotor to Co-based Fischer-Tropsch catalysts [278–280], Huang *et al.* [258] prepared H-ZSM-5 enwrapped CoZr catalyst particles and evaluated their activity for the direct synthesis of gasoline-ranged *i*-paraffins. The authors employed aluminum isopropoxide as Al source during membrane synthesis and confirmed its beneficial impact with respect to the minimized coke deposition. The core-shell catalyst exhibited lower activity compared to the solo CoZr catalyst or the physical mixture of CoZr and zeolite powder. Nevertheless, lower methane selectivity and facilitated selectivity towards *n*- and *i*-C_{5-11} hydrocarbons were reported.

Significantly decreased methane selectivity compared to Co-based core catalysts was observed on Fe-based catalyst pellets coated with H-ZSM-5 zeolite membranes [257]. Fused iron-based capsule catalyst covered with H-modernite zeolite-shell, synthesized without organic template exhibited increased CO conversion and nearly 8 times higher *i*-paraffin/*n*-paraffin ratio compared to the core catalyst [281]. However, since the fused iron (FI) catalyst is lacking in surface hydroxyl groups, its surface was modified by initial threatment with a solution of an organic adhesive, 3-aminopropyltrimethoxysilane (APTES), and ethanol as shown in Figure 13. In the next step, the pretreated catalyst was immersed in a silicalite-1/ethanol solution so that the surface could adsorb silicalite-1 since this layer provides Si-OH groups for sticking the zeolite shell. In the final step, modernite (MOR) shell was crystallized without the use of template at 180 °C for 48 h under rotation. The obtained HMOR/FI core/shell catalyst exhibited increased

51

CO conversion and higher selectivity to middle *i*-paraffins than the other catalyst as compared in Figure 13. Despite its remarkable performance, the necessity of product diffusion improvement was pointed out.

Very recently, Xing *et al.* [244] applied SBA-15 as support for iron-based catalyst cores which were micro-capsuled by H-ZSM-5 with sizes of about 1–2 μm. The mesoporous silica enables high activity and stability of the catalyst cores thanks to confinement effects already discussed in the literature for Fischer-Tropsch synthesis reactions [282–284]. The achieved high *i*-paraffin selectivity of 46.5% was attributed to the improved diffusion rate of reactants and products provided by the mesopore channels of the catalyst core as well as to the micro-pores of the zeolite membrane combined with acidic sites responsible for further hydrocracking and isomerization of the heavy hydrocarbons. However, experience has shown that zeolite membrane synthesis onto Fischer-Tropsch catalyst might be difficult to control since it requires alkaline conditions which could lead to catalyst damage or badly coating [255,257,281]. Another weakness is the already discussed increased methane selectivity. Larger metal particle size or larger amount of metal loading over the used supports in the design of Fischer-Tropsch catalysts is suggested in the literature as a possible route to overcome these drawbacks [284–287].

Very recently, Jin *et al.* [260] developed a novel dual-membrane coating Fe/SiO$_2$ catalyst core with silicalite-1 and H-ZSM-5 zeolite membranes following a previously reported synthesis procedure for double-shell capsule catalyst [262]. The silicalite-1 synthesized under close-to-neutral condition was acting as catalyst protective membrane for the core with high iron loading, whereas the H-ZSM-5 was the active membrane for the synthesis of *i*-paraffins. The authors stressed that attempts to coat H-ZSM-5 directly over high iron loading Fe/SiO$_2$ core failed due to the strong alkaline conditions required for its synthesis. The *i*-paraffin selectivity of 29.8% being higher than that of Fe/SiO$_2$ catalyst (12.9%) and the physical mixture (16.6%) was attributed to the combination of the hydrogenation and isomerization of the formed olefins at the core and the hydrocracking and isomerization of the heavy hydrocarbons in the dual-membrane catalyst. Moreover, relatively low methane selectivity (14.9%) compared to the other reported core-shell catalysts was obtained [234,256,281], indeed being still higher than that of the core catalyst and the physical mixture. Additionally, Yang *et al.* reported the benefit of controlling the pellet size for the preparation of H-ZSM-5 zeolite capsule catalysts on small Ru/SiO$_2$ pellets [255]. It was demonstrated that catalyst with smaller pellet size promoted the growth of the zeolite capsule while increasing the zeolite membrane thickness led to high activity and remarkable *i*-paraffin selectivity.

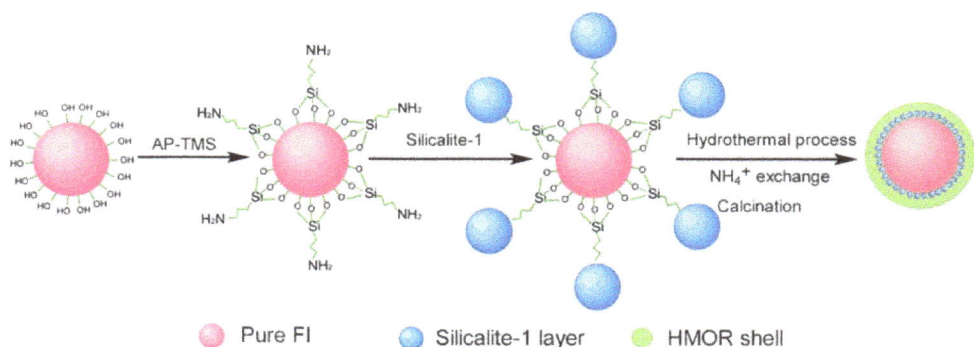

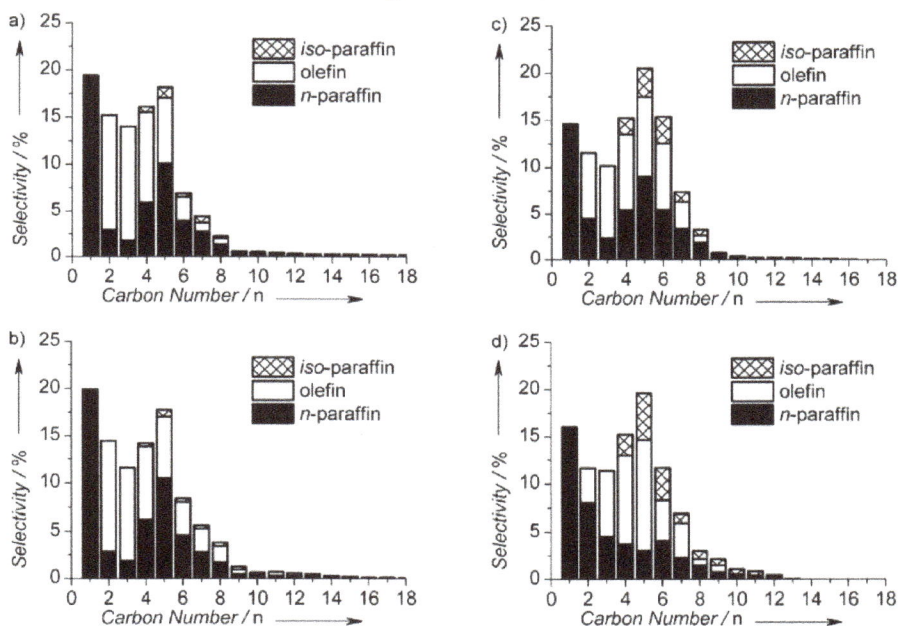

Figure 13. Scheme of the synthesis procedure of HMOR/FI capsule catalysts (FI = Fused Iron) without template and product distributions during Fischer-Tropsch synthesis (operating conditions: $H_2/CO = 1/1$, 1.0 MPa, 300 °C, 10 $g_{FI} \cdot h^{-1} \cdot mol^{-1}$) obtained using (a) pure FI; (b) Silicalite-1/FI; (c) HMOR + FI; and (d) HMOR/FI. Reprinted with permission from [281]. Copyright (2013) John Wiley and Sons.

Another research incentive comes from the request to design appropriate catalysts for the direct production of dimethyl ether (DME) being considered as a basic feedstock and a clean fuel [288]. Beside the dehydration of methanol (produced from syngas) to DME, a more desired and thermodynamically favorable alternative is the direct conversion of syngas to DME over hybrid catalysts [289–293].

Generally, the hybrid catalysts usually utilized in such consecutive reactions comprise two types of active sites, namely for the methanol synthesis and acid sites for the dehydration of methanol to DME. Nevertheless, the core-shell catalysts were proposed as a better alternative to the conventional hybrid catalysts [261]. The main advantage of the capsule catalysts, pointed by the authors, is the feasibility to better control the sequential reactions. In detail, the syngas conversion to methanol takes place on the active sites in the core and the further methanol dehydrogenation to produce DME occurs in the zeolite shell. Therefore, zeolite membranes were coated on $Cu/Zn/Al_2O_3$ applying two different hydrothermal techniques, namely H-ZSM-5 zeolite synthesis containing aluminum sources in the precursor solution and close-to-neutral silicalite-1 zeolite synthesis where the core catalyst was used as aluminum source. The so prepared zeolite capsuled catalysts exhibited extraordinary dimethyl ether selectivity compared to the physical mixture of core catalyst and zeolite powder. Moreover, no further dehydrogenation of the desired product to alkane or alkene occurred on the capsule catalysts. It should be noted that the hybrid catalyst experienced significantly higher conversion than the capsule catalyst. However, the hybrid catalyst provides random contact between methanol and the active sites of zeolite catalyst so that the two consecutive reactions occur independently resulting in moderate selectivity for the desired dimethyl ether. On the other hand, the zeolite shell offers better control of the reaction order since it affords unavoidable contact for methanol formed in the core while passing the membrane and as a result enhances the probability for its further conversion into dimethyl ether leading to the desired high selectivity. Although the H-ZSM-5 coated catalyst exhibited excellent dimethyl ether selectivity, indication was given that the core catalyst structure was damaged during the acidic hydrothermal synthesis resulting in much lower CO conversion. In this context, Yang *et al.* [262,263] discussed the challenges and strategies for preparation of H-ZSM-5 zeolite shells on either bimetallic or silica-based catalyst. It has been demonstrated that the dual-layer method is an effective tool for the preparation of defect-free and compact zeolite shells. A silicalite-1 membrane was first synthesized as intermediate layer acting as core catalyst protection, further facilitating the growth of the H-ZSM-5 under the stronger alkaline conditions. The double-layer capsule catalysts with Cr/ZnO or Pd/SiO_2 core achieved a sharply controlled reaction with excellent DME selectivity and no formation of C_{4+} hydrocarbons.

Next to H-ZSM-5, silico-aluminophosphate molecular sieves (SAPO) are considered as attractive catalysts for the dehydration of methanol to DME [294,295]. Respectively, Pinkaew *et al.* [243] proposed a physical coating procedure using SiO_2 as an adhesive for the preparation of a defect-free SAPO-46 shell over Cr/ZnO catalyst core. According to the authors, the developed method, performed under normal ambient conditions, is an effective way to overcome the disadvantages of the conventional hydrothermal synthesis e.g., high temperature and alkaline conditions. Accordingly, the capsuled catalyst employed exhibited slightly higher activity that the mixed catalyst which can be ascribed to the enhanced syngas diffusion rate through the SAPO zeolite shell.

Very recently, Phienluphon *et al.* [241] applied the physical coating technique and synthesized silicoaluminophosphate-11 (SAPO-11) shell over Cu/ZnO/Al$_2$O$_3$ core catalyst. The so created defect-free, uniform and compact tandem catalyst outperformed the mixture catalyst giving rise to CO conversion of 92.0% and DME selectivity of 90.3%, accompanied by extremely low by-product formation.

The promising results achieved with the core-shell catalysts, where core catalyst and zeolite membrane as shell are combined according to the desired reaction as well as the improvement of the synthesis techniques are good prerequisites for meeting the requirements for wide range of application in particle level membrane reactors.

7. Conclusions

We reviewed literature on the application of zeolite membranes enhancing catalytic reactions in terms of conversion, selectivity or yield. The main advantage of zeolite membranes is based on their property to displace equilibria of thermodynamic difficult reactions. But also shape selectivity or simply separation and purification problems can be solved by using zeolite membranes. Thus, it appears that permselective zeolite membranes can be located (i) spatially decoupled; (ii) in a direct contact to a catalyst packed bed; (iii) as membrane and catalyst all in one, or (iv) as a capsule around a catalyst core. The preferred way how to use a zeolite membrane in catalytic reactions is dependent on numerous factors where membrane and catalyst properties should be carefully evaluated.

From macroscopic zeolite membrane reactors a lot of knowledge has been gained over the last decades. However, may be due to two serious drawbacks such arrays had never got entry into the industrial level in the past. First of all it is the difficulty in producing zeolite membranes on a large scale what makes them still economically unattractive. Secondly, the permeation per membrane area does often not meet the requirements of the volume ratio of the catalytic reactor. Since zeolite capsuled catalysts are able to overcome both the issues it is believed these are the most promising candidates to reach the high criteria of industrial standards. Nevertheless, recent developments on membrane synthesis promise better permeation characteristics as well as easier and cheaper upscaling. Hence, supported zeolite membranes could still also be noticeable candidates for intensified industrial processes in the future.

Author Contributions: R.D. wrote the first draft of the review. R.D. and S.W. iteratively brought it to its final state.

Conflicts of Interest: The authors declare no conflict of interest.

References

1. Corma, A. Inorganic Solid Acids and Their Use in Acid-Catalyzed Hydrocarbon Reactions. *Chem. Rev.* **1995**, *95*, 559–614.
2. Corma, A. From Microporous to Mesoporous Molecular Sieve Materials and Their Use in Catalysis. *Chem. Rev.* **1997**, *97*, 2373–2420.
3. Kulprathipanja, S. *Zeolites in Industrial Separation and Catalysis*; Wiley-VCH: Weinheim, Germany, 2010.
4. Ackley, M.W.; Rege, S.U.; Saxena, H. Application of natural zeolites in the purification and separation of gases. *Microporous Mesoporous Mater.* **2003**, *61*, 25–42.
5. Jasra, R.V.; Choudary, N.V.; Bhat, S.G.T. Separation of Gases by Pressure Swin. *Sep. Sci. Technol.* **1991**, *26*, 885–930.
6. Reiß, G. Status and development of oxygen generation processes on molecular sieve zeolites. *Gas Sep. Purif.* **1994**, *8*, 95–99.
7. Tagliabue, M.; Farrusseng, D.; Valencia, S.; Aguado, S.; Ravon, U.; Rizzo, C.; Corma, A.; Mirodatos, C. Natural gas treating by selective adsorption: Material science and chemical engineering interplay. *Chem. Eng. J.* **2009**, *155*, 553–566.
8. Wang, S.; Peng, Y. Natural zeolites as effective adsorbents in water and wastewater treatment. *Chem. Eng. J.* **2010**, *156*, 11–24.
9. Hedström, A. Ion Exchange of Ammonium in Zeolites: A Literature Review. *J. Environ. Eng.* **2001**, *127*, 673–681.
10. Bein, T. Synthesis and Applications of Molecular Sieve Layers and Membranes. *Chem. Mater.* **1996**, *8*, 1636–1653.
11. Tavolaro, A.; Drioli, E. Zeolite membranes. *Adv. Mater.* **1999**, *11*, 975–996.
12. Caro, J.; Noack, M.; Kölsch, P.; Schäfer, R. Zeolite membranes—State of their development and perspective. *Microporous Mesoporous Mater.* **2000**, *38*, 3–24.
13. Bowen, T.C.; Noble, R.D.; Falconer, J.L. Fundamentals and applications of pervaporation through zeolite membranes. *J. Membr. Sci.* **2004**, *245*, 1–33.
14. Caro, J.; Noack, M.; Kölsch, P. Zeolite Membranes: From the Laboratory Scale to Technical Applications. *Adsorption* **2005**, *11*, 215–227.
15. Snyder, M.A.; Tsapatsis, M. Hierarchical Nanomanufacturing: From Shaped Zeolite Nanoparticles to High-Performance Separation Membranes. *Angew. Chem. Int. Ed.* **2007**, *46*, 7560–7573.
16. Caro, J.; Noack, M. Zeolite membranes—Recent developments and progress. *Microporous Mesoporous Mater.* **2008**, *115*, 215–233.
17. Aroon, M.A.; Ismail, A.F.; Matsuura, T.; Montazer-Rahmati, M.M. Performance studies of mixed matrix membranes for gas separation: A review. *Sep. Purif. Technol.* **2010**, *75*, 229–242.
18. Yu, M.; Noble, R.D.; Falconer, J.L. Zeolite Membranes: Microstructure Characterization and Permeation Mechanisms. *Acc. Chem. Res.* **2011**, *44*, 1196–1206.

19. Gascon, J.; Kapteijn, F.; Zornoza, B.; Sebastian, V.; Casado, C.; Coronas, J. Practical Approach to Zeolitic Membranes and Coatings: State of the Art, Opportunities, Barriers, and Future Perspectives. *Chem. Mater.* **2012**, *24*, 2829–2844.

20. Nasir, R.; Mukhtar, H.; Man, Z.; Mohshim, D.F. Material Advancements in Fabrication of Mixed-Matrix Membranes. *Chem. Eng. Technol.* **2013**, *36*, 717–727.

21. Miachon, S.; Dalmon, J.-A. Catalysis in Membrane Reactors: What About the Catalyst? *Top. Catal.* **2004**, *29*, 59–65.

22. Caro, J. 3.01—Basic Aspects of Membrane Reactors. In *Comprehensive Membrane Science and Engineering*; Drioli, E., Giorno, L., Eds.; Elsevier: Oxford, UK, 2010; pp. 1–24.

23. McLeary, E.E.; Jansen, J.C.; Kapteijn, F. Zeolite based films, membranes and membrane reactors: Progress and prospects. *Microporous Mesoporous Mater.* **2006**, *90*, 198–220.

24. Gascon, J.; van Ommen, J.R.; Moulijn, J.A.; Kapteijn, F. Structuring catalyst and reactor—An inviting avenue to process intensification. *Catal. Sci. Technol.* **2015**, *5*, 807–817.

25. Drioli, E.; Romano, M. Progress and New Perspectives on Integrated Membrane Operations for Sustainable Industrial Growth. *Ind. Eng. Chem. Res.* **2001**, *40*, 1277–1300.

26. Coronas, J.; Santamaría, J. Catalytic reactors based on porous ceramic membranes. *Catal. Today* **1999**, *51*, 377–389.

27. Armor, J.N. Membrane catalysis: Where is it now, what needs to be done? *Catal. Today* **1995**, *25*, 199–207.

28. Daramola, M.O.; Aransiola, E.F.; Ojumu, T.V. Potential applications of zeolite membranes in reaction coupling separation processes. *Materials* **2012**, *5*, 2101–2136.

29. Coronas, J.; Santamaria, J. State-of-the-art in zeolite membrane reactors. *Top. Catal.* **2004**, *29*, 29–44.

30. Saracco, G.; Specchia, V. Catalytic Inorganic-Membrane Reactors: Present Experience and Future Opportunities. *Catal. Rev.* **1994**, *36*, 305–384.

31. Michalkiewicz, B.; Koren, Z. Zeolite membranes for hydrogen production from natural gas: State of the art. *J. Porous Mater.* **2015**, *22*, 635–646.

32. Van den Bergh, J.; Nishiyama, N.; Kapteijn, F. Zeolite Membranes in Catalysis: What Is New and How Bright Is the Future? In *Novel Concepts in Catalysis and Chemical Reactors*; Wiley-VCH Verlag GmbH & Co. KGaA: Weinheim, Germany, 2010; pp. 211–237.

33. Tsapatsis, M. Toward High-Throughput Zeolite Membranes. *Science* **2011**, *334*, 767–768.

34. Agrawal, K.V.; Topuz, B.; Pham, T.C.T.; Nguyen, T.H.; Sauer, N.; Rangnekar, N.; Zhang, H.; Narasimharao, K.; Basahel, S.N.; Francis, L.F.; *et al.* Oriented MFI Membranes by Gel-Less Secondary Growth of Sub-100 nm MFI-Nanosheet Seed Layers. *Adv. Mater.* **2015**, *27*, 3243–3249.

35. Pham, T.C.T.; Nguyen, T.H.; Yoon, K.B. Gel-Free Secondary Growth of Uniformly Oriented Silica MFI Zeolite Films and Application for Xylene Separation. *Angew. Chem. Int. Ed.* **2013**, *52*, 8693–8698.

36. Wang, Z.; Yu, T.; Nian, P.; Zhang, Q.; Yao, J.; Li, S.; Gao, Z.; Yue, X. Fabrication of a Highly *b*-Oriented MFI-Type Zeolite Film by the Langmuir-Blodgett Method. *Langmuir* **2014**, *30*, 4531–4534.

37. Sjöberg, E.; Sandström, L.; Öhrman, O.G.W.; Hedlund, J. Separation of CO_2 from black liquor derived syngas using an MFI membrane. *J. Membr. Sci.* **2013**, *443*, 131–137.

38. Zhou, M.; Korelskiy, D.; Ye, P.; Grahn, M.; Hedlund, J. A Uniformly Oriented MFI Membrane for Improved CO_2 Separation. *Angew. Chem. Int. Ed.* **2014**, *53*, 3492–3495.

39. Davis, M.E. Zeolites from a Materials Chemistry Perspective. *Chem. Mater.* **2014**, *26*, 239–245.

40. Peng, Y.; Lu, H.; Wang, Z.; Yan, Y. Microstructural optimization of MFI-type zeolite membranes for ethanol-water separation. *J. Mater. Chem. A* **2014**, *2*, 16093–16100.

41. Iyoki, K.; Itabashi, K.; Okubo, T. Progress in seed-assisted synthesis of zeolites without using organic structure-directing agents. *Microporous Mesoporous Mater.* **2014**, *189*, 22–30.

42. Kosinov, N.; Gascon, J.; Kapteijn, F.; Hensen, E.J.M. Recent developments in zeolite membranes for gas separation. *J. Membr. Sci.* **2016**, *499*, 65–79.

43. Morigami, Y.; Kondo, M.; Abe, J.; Kita, H.; Okamoto, K. The first large-scale pervaporation plant using tubular-type module with zeolite NaA membrane. *Sep. Purif. Technol.* **2001**, *25*, 251–260.

44. Gallego-Lizon, T.; Edwards, E.; Lobiundo, G.; Freitas dos Santos, L. Dehydration of water/*t*-butanol mixtures by pervaporation: Comparative study of commercially available polymeric, microporous silica and zeolite membranes. *J. Membr. Sci.* **2002**, *197*, 309–319.

45. Urtiaga, A.; Gorri, E.D.; Casado, C.; Ortiz, I. Pervaporative dehydration of industrial solvents using a zeolite NaA commercial membrane. *Sep. Purif. Technol.* **2003**, *32*, 207–213.

46. Richter, H.; Voigt, I.; Kühnert, J.-T. Dewatering of ethanol by pervaporation and vapour permeation with industrial scale NaA-membranes. *Desalination* **2006**, *199*, 92–93.

47. Sato, K.; Aoki, K.; Sugimoto, K.; Izumi, K.; Inoue, S.; Saito, J.; Ikeda, S.; Nakane, T. Dehydrating performance of commercial LTA zeolite membranes and application to fuel grade bio-ethanol production by hybrid distillation/vapor permeation process. *Microporous Mesoporous Mater.* **2008**, *115*, 184–188.

48. Rangnekar, N.; Mittal, N.; Elyassi, B.; Caro, J.; Tsapatsis, M. Zeolite membranes—A review and comparison with MOFs. *Chem. Soc. Rev.* **2015**, *44*, 7128–7154.

49. Bai, C.; Jia, M.-D.; Falconer, J.L.; Noble, R.D. Preparation and separation properties of silicalite composite membranes. *J. Membr. Sci.* **1995**, *105*, 79–87.

50. Dong, J.; Lin, Y.S. In Situ Synthesis of P-Type Zeolite Membranes on Porous α-Alumina Supports. *Ind. Eng. Chem. Res.* **1998**, *37*, 2404–2409.

51. Yan, Y.; Davis, M.E.; Gavalas, G.R. Preparation of Zeolite ZSM-5 Membranes by *In-situ* Crystallization on Porous α-Al_2O_3. *Ind. Eng. Chem. Res.* **1995**, *34*, 1652–1661.

52. Tuan, V.A.; Li, S.; Falconer, J.L.; Noble, R.D. In situ Crystallization of Beta Zeolite Membranes and Their Permeation and Separation Properties. *Chem. Mater.* **2002**, *14*, 489–492.

53. Lin, X.; Chen, X.; Kita, H.; Okamoto, K. Synthesis of silicalite tubular membranes by *in situ* crystallization. *AIChE J.* **2003**, *49*, 237–247.

54. Vilaseca, M.; Mateo, E.; Palacio, L.; Prádanos, P.; Hernández, A.; Paniagua, A.; Coronas, J.N.; Santamaría, J. AFM characterization of the growth of MFI-type zeolite films on alumina substrates. *Microporous Mesoporous Mater.* **2004**, *71*, 33–37.

55. Guillou, F.; Rouleau, L.; Pirngruber, G.; Valtchev, V. Synthesis of FAU-type zeolite membrane: An original *in situ* process focusing on the rheological control of gel-like precursor species. *Microporous Mesoporous Mater.* **2009**, *119*, 1–8.

56. Huang, A.; Liang, F.; Steinbach, F.; Caro, J. Preparation and separation properties of LTA membranes by using 3-aminopropyltriethoxysilane as covalent linker. *J. Membr. Sci.* **2010**, *350*, 5–9.

57. Huang, A.; Wang, N.; Caro, J. Seeding-free synthesis of dense zeolite FAU membranes on 3-aminopropyltriethoxysilane-functionalized alumina supports. *J. Membr. Sci.* **2012**, *389*, 272–279.

58. Wang, Z.; Yan, Y. Controlling crystal orientation in zeolite MFI thin films by direct *in situ* crystallization. *Chem. Mater.* **2001**, *13*, 1101–1107.

59. Zhang, F.-Z.; Fuji, M.; Takahashi, M. *In situ* growth of continuous *b*-oriented MFI zeolite membranes on porous *α*-alumina substrates precoated with a mesoporous silica sublayer. *Chem. Mater.* **2005**, *17*, 1167–1173.

60. Wang, Z.; Yan, Y. Oriented zeolite MFI monolayer films on metal substrates by *in situ* crystallization. *Microporous Mesoporous Mater.* **2001**, *48*, 229–238.

61. Li, S.; Demmelmaier, C.; Itkis, M.; Liu, Z.; Haddon, R.C.; Yan, Y. Micropatterned oriented zeolite monolayer films by direct *in situ* crystallization. *Chem. Mater.* **2003**, *15*, 2687–2689.

62. Xomeritakis, G.; Gouzinis, A.; Nair, S.; Okubo, T.; He, M.; Overney, R.M.; Tsapatsis, M. Growth, microstructure, and permeation properties of supported zeolite (MFI) films and membranes prepared by secondary growth. *Chem. Eng. Sci.* **1999**, *54*, 3521–3531.

63. Bernal, M.A.P.; Xomeritakis, G.; Tsapatsis, M. Tubular MFI zeolite membranes made by secondary (seeded) growth. *Catal. Today* **2001**, *67*, 101–107.

64. Nair, S.; Lai, Z.; Nikolakis, V.; Xomeritakis, G.; Bonilla, G.; Tsapatsis, M. Separation of close-boiling hydrocarbon mixtures by MFI and FAU membranes made by secondary growth. *Microporous Mesoporous Mater.* **2001**, *48*, 219–228.

65. Tomita, T.; Nakayama, K.; Sakai, H. Gas separation characteristics of DDR type zeolite membrane. *Microporous Mesoporous Mater.* **2004**, *68*, 71–75.

66. Liu, Y.; Yang, Z.; Yu, C.; Gu, X.; Xu, N. Effect of seeding methods on growth of NaA zeolite membranes. *Microporous Mesoporous Mater.* **2011**, *143*, 348–356.

67. Wohlrab, S.; Meyer, T.; Stöhr, M.; Hecker, C.; Lubenau, U.; Oßmann, A. On the performance of customized MFI membranes for the separation of *n*-butane from methane. *J. Membr. Sci.* **2011**, *369*, 96–104.

68. Fan, W.; Snyder, M.A.; Kumar, S.; Lee, P.-S.; Yoo, W.C.; McCormick, A.V.; Lee Penn, R.; Stein, A.; Tsapatsis, M. Hierarchical nanofabrication of microporous crystals with ordered mesoporosity. *Nat. Mater.* **2008**, *7*, 984–991.

69. Chen, H.; Wydra, J.; Zhang, X.; Lee, P.-S.; Wang, Z.; Fan, W.; Tsapatsis, M. Hydrothermal Synthesis of Zeolites with Three-Dimensionally Ordered Mesoporous-Imprinted Structure. *J. Am. Chem. Soc.* **2011**, *133*, 12390–12393.

70. Roth, W.J.; Nachtigall, P.; Morris, R.E.; Čejka, J. Two-Dimensional Zeolites: Current Status and Perspectives. *Chem. Rev.* **2014**, *114*, 4807–4837.

71. Diaz, U.; Corma, A. Layered zeolitic materials: An approach to designing versatile functional solids. *Dalton Trans.* **2014**, *43*, 10292–10316.

72. Morawetz, K.; Reiche, J.; Kamusewitz, H.; Kosmella, H.; Ries, R.; Noack, M.; Brehmer, L. Zeolite films prepared via the Langmuir-Blodgett technique. *Colloids Surf. A* **2002**, *198–200*, 409–414.

73. Pan, M.; Lin, Y.S. Template-free secondary growth synthesis of MFI type zeolite membranes. *Microporous Mesoporous Mater.* **2001**, *43*, 319–327.

74. Huang, A.; Lin, Y.S.; Yang, W. Synthesis and properties of A-type zeolite membranes by secondary growth method with vacuum seeding. *J. Membr. Sci.* **2004**, *245*, 41–51.

75. Li, G.; Kikuchi, E.; Matsukata, M. The control of phase and orientation in zeolite membranes by the secondary growth method. *Microporous Mesoporous Mater.* **2003**, *62*, 211–220.

76. Lovallo, M.C.; Gouzinis, A.; Tsapatsis, M. Synthesis and characterization of oriented MFI membranes prepared by secondary growth. *AIChE J.* **1998**, *44*, 1903–1913.

77. Choi, J.; Ghosh, S.; Lai, Z.; Tsapatsis, M. Uniformly a-oriented MFI zeolite films by secondary growth. *Angew. Chem.* **2006**, *118*, 1172–1176.

78. Gouzinis, A.; Tsapatsis, M. On the preferred orientation and microstructural manipulation of molecular sieve films prepared by secondary growth. *Chem. Mater.* **1998**, *10*, 2497–2504.

79. Lai, Z.; Tsapatsis, M.; Nicolich, J.P. Siliceous ZSM-5 Membranes by Secondary Growth of b-Oriented Seed Layers. *Adv. Funct. Mater.* **2004**, *14*, 716–729.

80. Liu, Y.; Li, Y.; Yang, W. Fabrication of Highly *b*-Oriented MFI Film with Molecular Sieving Properties by Controlled in-Plane Secondary Growth. *J. Am. Chem. Soc.* **2010**, *132*, 1768–1769.

81. Hedlund, J.; Schoeman, B.; Sterte, J. Ultrathin oriented zeolite LTA films. *Chem. Commun.* **1997**.

82. Pham, T.C.T.; Kim, H.S.; Yoon, K.B. Growth of Uniformly Oriented Silica MFI and BEA Zeolite Films on Substrates. *Science* **2011**, *334*, 1533–1538.

83. Xu, W.; Dong, J.; Li, J.; Li, J.; Wu, F. A novel method for the preparation of zeolite ZSM-5. *Chem. Commun.* **1990**, 755–756.

84. Nishiyama, N.; Ueyama, K.; Matsukata, M. Synthesis of defect-free zeolite-alumina composite membranes by a vapor-phase transport method. *Microporous Mater.* **1996**, *7*, 299–308.

85. Matsufuji, T.; Nishiyama, N.; Matsukata, M.; Ueyama, K. Separation of butane and xylene isomers with MFI-type zeolitic membrane synthesized by a vapor-phase transport method. *J. Membr. Sci.* **2000**, *178*, 25–34.

86. Kikuchi, E.; Yamashita, K.; Hiromoto, S.; Ueyama, K.; Matsukata, M. Synthesis of a zeolitic thin layer by a vapor-phase transport method: Appearance of a preferential orientation of MFI zeolite. *Microporous Mater.* **1997**, *11*, 107–116.

87. Matsufuji, T.; Nishiyama, N.; Ueyama, K.; Matsukata, M. Crystallization of ferrierite (FER) on a porous alumina support by a vapor-phase transport method. *Microporous Mesoporous Mater.* **1999**, *32*, 159–168.

88. Nishiyama, N.; Matsufuji, T.; Ueyama, K.; Matsukata, M. FER membrane synthesized by a vapor-phase transport method: Its structure and separation characteristics. *Microporous Mater.* **1997**, *12*, 293–303.

89. Li, Y.; Yang, W. Microwave synthesis of zeolite membranes: A review. *J. Membr. Sci.* **2008**, *316*, 3–17.

90. Weisz, P.B. The Science of the Possible. *Chemtech* **1982**, *12*, 689–690.

91. Boudart, M. Surface Time Yields of Membrane Reactors. In *CaTTech*; Springer Science & Business: Berlin, Germany, 1997.

92. Yan, S.; Maeda, H.; Kusakabe, K.; Morooka, S. Thin Palladium Membrane Formed in Support Pores by Metal-Organic Chemical Vapor Deposition Method and Application to Hydrogen Separation. *Ind. Eng. Chem. Res.* **1994**, *33*, 616–622.

93. Graaf, J.M.V.; Zwiep, M.; Kapteijn, F.; Moulijn, J.A. Application of a silicalite-1 membrane reactor in metathesis reactions. *Appl. Catal. A* **1999**, *178*, 225–241.

94. Graaf, J.M.V.D.; Zwiep, M.; Kapteijn, F.; Moulijn, J.A. Application of a zeolite membrane reactor in the metathesis of propene. *Chem. Eng. Sci.* **1999**, *54*, 1441–1445.

95. Moon, W.S.; Park, S.B. Design guide of a membrane for a membrane reactor in terms of permeability and selectivity. *J. Membr. Sci.* **2000**, *170*, 43–51.

96. Battersby, S.; Teixeira, P.W.; Beltramini, J.; Duke, M.C.; Rudolph, V.; Diniz da Costa, J.C. An analysis of the Peclet and Damkohler numbers for dehydrogenation reactions using molecular sieve silica (MSS) membrane reactors. *Catal. Today* **2006**, *116*, 12–17.

97. Choi, S.-W.; Jones, C.W.; Nair, S.; Sholl, D.S.; Moore, J.S.; Liu, Y.; Dixit, R.S.; Pendergast, J.G. Material properties and operating configurations of membrane reactors for propane dehydrogenation. *AIChE J.* **2015**, *61*, 922–935.

98. Baker, R.W. *Membrane Technologies and Applications*; John Wiley & Sons Ltd.: Chichester, UK, 2004.

99. Krishna, R.; van den Broeke, L.J.P. The Maxwell-Stefan description of mass transport across zeolite membranes. *Chem. Eng. J. Biochem. Eng. J.* **1995**, *57*, 155–162.

100. Krishna, R.; Wesselingh, J.A. The Maxwell-Stefan approach to mass transfer. *Chem. Eng. Sci.* **1997**, *52*, 861–911.

101. Kapteijn, F.; Moulijn, J.A.; Krishna, R. The generalized Maxwell-Stefan model for diffusion in zeolites: Sorbate molecules with different saturation loadings. *Chem. Eng. Sci.* **2000**, *55*, 2923–2930.

102. Krishna, R.; Baur, R. Modelling issues in zeolite based separation processes. *Sep. Purif. Technol.* **2003**, *33*, 213–254.

103. Krishna, R. The Maxwell-Stefan description of mixture diffusion in nanoporous crystalline materials. *Microporous Mesoporous Mater.* **2014**, *185*, 30–50.

104. Yuan, W.; Lin, Y.S.; Yang, W. Molecular Sieving MFI-Type Zeolite Membranes for Pervaporation Separation of Xylene Isomers. *J. Am. Chem. Soc.* **2004**, *126*, 4776–4777.

105. Aoki, K.; Kusakabe, K.; Morooka, S. Gas permeation properties of A-type zeolite membrane formed on porous substrate by hydrothermal synthesis. *J. Membr. Sci.* **1998**, *141*, 197–205.

106. Xu, X.; Bao, Y.; Song, C.; Yang, W.; Liu, J.; Lin, L. Synthesis, characterization and single gas permeation properties of NaA zeolite membrane. *J. Membr. Sci.* **2005**, *249*, 51–64.

107. Van de Graaf, J.M.; van der Bijl, E.; Stol, A.; Kapteijn, F.; Moulijn, J.A. Effect of Operating Conditions and Membrane Quality on the Separation Performance of Composite Silicalite-1 Membranes. *Ind. Eng. Chem. Res.* **1998**, *37*, 4071–4083.

108. Vroon, Z.A.E.P.; Keizer, K.; Burggraaf, A.J.; Verweij, H. Preparation and characterization of thin zeolite MFI membranes on porous supports. *J. Membr. Sci.* **1998**, *144*, 65–76.

109. Keizer, K.; Burggraaf, A.J.; Vroon, Z.A.E.P.; Verweij, H. Two component permeation through thin zeolite MFI membranes. *J. Membr. Sci.* **1998**, *147*, 159–172.

110. Coronas, J.; Falconer, J.L.; Noble, R.D. Characterization and permeation properties of ZSM-5 tubular membranes. *AIChE J.* **1997**, *43*, 1797–1812.

111. Jiang, M.; Eic, M.; Miachon, S.; Dalmon, J.-A.; Kocirik, M. Diffusion of *n*-butane, isobutane and ethane in a MFI-zeolite membrane investigated by gas permeation and ZLC measurements. *Sep. Purif. Technol.* **2001**, *25*, 287–295.

112. Barrer, R.M. Porous crystal membranes. *J. Chem. Soc. Faraday Trans.* **1990**, *86*, 1123–1130.

113. Bakker, W.J.W.; Kapteijn, F.; Poppe, J.; Moulijn, J.A. Permeation characteristics of a metal-supported silicalite-1 zeolite membrane. *J. Membr. Sci.* **1996**, *117*, 57–78.

114. Neubauer, K.; Lubenau, U.; Hecker, C.; Lücke, B.; Paschek, D.; Wohlrab, S. Abreicherung von Flüssiggas aus Erdgas mittels Zeolithmembranen; Depletion of Liquefied Petroleum Gas from Natural Gas by Zeolite Membranes. *Chem. Ing. Tech.* **2013**, *85*, 713–722.

115. Dragomirova, R.; Stohr, M.; Hecker, C.; Lubenau, U.; Paschek, D.; Wohlrab, S. Desorption-controlled separation of natural gas alkanes by zeolite membranes. *RSC Adv.* **2014**, *4*, 59831–59834.

116. Neubauer, K.; Dragomirova, R.; Stöhr, M.; Mothes, R.; Lubenau, U.; Paschek, D.; Wohlrab, S. Combination of membrane separation and gas condensation for advanced natural gas conditioning. *J. Membr. Sci.* **2014**, *453*, 100–107.

117. Baker, R.W.; Lokhandwala, K. Natural Gas Processing with Membranes: An Overview. *Ind. Eng. Chem. Res.* **2008**, *47*, 2109–2121.

118. Arruebo, M.; Coronas, J.; Menéndez, M.; Santamaría, J. Separation of hydrocarbons from natural gas using silicalite membranes. *Sep. Purif. Technol.* **2001**, *25*, 275–286.

119. Dragomirova, R.; Kreft, S.; Georgi, G.; Seeburg, D.; Wohlrab, S. Liquefied Petroleum Gas Enrichment by Zeolite Membranes for Low-Temperature Steam Reforming of Natural Gas. In Proceedings of Synthesis Gas Chemistry, 2015-2, DGMK International Conference, Dresden, Germany, 7–9 October 2015; pp. 273–280.

120. Ockwig, N.W.; Nenoff, T.M. Membranes for Hydrogen Separation. *Chem. Rev.* **2007**, *107*, 4078–4110.

121. Varela-Gandía, F.J.; Berenguer-Murcia, Á.; Lozano-Castelló, D.; Cazorla-Amorós, D. Zeolite A/carbon membranes for H_2 purification from a simulated gas reformer mixture. *J. Membr. Sci.* **2011**, *378*, 407–414.

122. Gallo, M.; Nenoff, T.M.; Mitchell, M.C. Selectivities for binary mixtures of hydrogen/methane and hydrogen/carbon dioxide in silicalite and ETS-10 by Grand Canonical Monte Carlo techniques. *Fluid Phase Equilibria* **2006**, *247*, 135–142.

123. Mitchell, M.C.; Autry, J.D.; Nenoff, T.M. Molecular dynamics simulations of binary mixtures of methane and hydrogen in zeolite A and a novel zinc phosphate. *Mol. Phys.* **2001**, *99*, 1831–1837.

124. Tang, Z.; Dong, J.; Nenoff, T.M. Internal Surface Modification of MFI-Type Zeolite Membranes for High Selectivity and High Flux for Hydrogen. *Langmuir* **2009**, *25*, 4848–4852.

125. Wee, S.-L.; Tye, C.-T.; Bhatia, S. Membrane separation process—Pervaporation through zeolite membrane. *Sep. Purif. Technol.* **2008**, *63*, 500–516.

126. Lima, S.Y.; Parkb, B.; Hunga, F.; Sahimia, M.; Tsotsisa, T.T. Design issues of pervaration membrane reactors for esterification. *Chem. Eng. Sci.* **2002**, *57*, 4933–4946.

127. Jafar, J.J.; Budd, P.M.; Hughes, R. Enhancement of esterification reaction yield using zeolite—A vapour permeation membrane. *J. Membr. Sci.* **2002**, *199*, 117–123.

128. Khajavi, S.; Jansen, J.C.; Kapteijn, F. Application of a sodalite membrane reactor in esterification—Coupling reaction and separation. *Catal. Today* **2010**, *156*, 132–139.

129. Hasegawa, Y.; Abe, C.; Mizukami, F.; Kowata, Y.; Hanaoka, T. Application of a CHA-type zeolite membrane to the esterification of adipic acid with isopropyl alcohol using sulfuric acid catalyst. *J. Membr. Sci.* **2012**, *415*, 368–374.

130. Okamoto, K.-I.; Kita, H.; Horii, K. Zeolite NaA Membrane: Preparation, Single-Gas Permeation, and Pervaporation and Vapor Permeation of Water/Organic Liquid Mixtures. *Ind. Eng. Chem. Res.* **2001**, *40*, 163–175.

131. Kita, H.; Horii, K.; Ohtoshi, Y.; Tanaka, K.; Okamoto, K.-I. Synthesis of a zeolite NaA membrane for pervaporation of water/organic liquid mixtures. *J. Mater. Sci. Lett.* **1995**, *14*, 206–208.

132. Hasegawa, Y.; Nagase, T.; Kiyozumi, Y.; Hanaoka, T.; Mizukami, F. Influence of acid on the permeation properties of NaA-type zeolite membranes. *J. Membr. Sci.* **2010**, *349*, 189–194.

133. Inoue, T.; Nagase, T.; Hasegawa, Y.; Kiyozumi, Y.; Sato, K.; Nishioka, M.; Hamakawa, S.; Mizukami, F. Stoichiometric ester condensation reaction processes by pervaporative water removal via acid-tolerant zeolite membranes. *Ind. Eng. Chem. Res.* **2007**, *46*, 3743–3750.

134. Khajavi, S.; Jansen, J.C.; Kapteijn, F. Application of hydroxy sodalite films as novel water selective membranes. *J. Membr. Sci.* **2009**, *326*, 153–160.

135. Zhang, W.; Na, S.; Li, W.; Xing, W. Kinetic Modeling of Pervaporation Aided Esterification of Propionic Acid and Ethanol Using T-Type Zeolite Membrane. *Ind. Eng. Chem. Res.* **2015**, *54*, 4940–4946.

136. Aiouache, F.; Goto, S. Reactive distillation-pervaporation hybrid column for tert-amyl alcohol etherification with ethanol. *Chem. Eng. Sci.* **2003**, *58*, 2465–2477.

137. Thomas, S.; Hamel, C.; Seidel-Morgenstern, A. Basic Problems of Chemical Reaction Engineering and Potential of Membrane Reactors. In *Membrane Reactors*; Wiley-VCH Verlag GmbH & Co. KGaA: Weinheim, Germany, 2010; pp. 1–27.

138. Gallucci, F.; Fernandez, E.; Corengia, P.; van Sint Annaland, M. Recent advances on membranes and membrane reactors for hydrogen production. *Chem. Eng. Sci.* **2013**, *92*, 40–66.

139. Gallucci, F.; Basile, A.; Hai, F.I. Introduction—A Review of Membrane Reactors. In *Membranes for Membrane Reactors*; John Wiley & Sons, Ltd.: Chichester, UK, 2011; pp. 1–61.

140. Téllez, C.; Menéndez, M. Zeolite Membrane Reactors. In *Membranes for Membrane Reactors*; John Wiley & Sons, Ltd.: Chichester, UK, 2011; pp. 243–273.

141. Koros, W.J.; Ma, Y.H.; Shimidzu, T. Terminology for membranes and membrane processes. *Pure Appl. Chem.* **1999**, *68*, 1479–1489.

142. Dittmeyer, R.; Hollein, V.; Daub, K. Membrane reactors for hydrogenation and dehydrogenation processes based on supported palladium. *J. Mol. Catal. A* **2001**, *173*, 135–184.

143. Dittmeyer, R.; Caro, J. Catalytic Membrane Reactors. In *Handbook of Heterogeneous Catalysis*; Wiley-VCH Verlag GmbH & Co. KGaA: Weinhein, Germany, 2008.

144. Xiongfu, Z.; Yongsheng, L.; Jinqu, W.; Huairong, T.; Changhou, L. Synthesis and characterization of Fe-MFI zeolite membrane on a porous g-Al$_2$O$_3$ tube. *Sep. Purif. Technol.* **2001**, *25*, 269–274.

145. Ciavarella, P.; Casanave, D.; Moueddeb, H.; Miachon, S.; Fiaty, K.; Dalmon, J.A. Isobutane dehydrogenation in a membrane reactor—Influence of the operating conditions on the performance. *Catal. Today* **2001**, *67*, 177–184.

146. Illgen, U.; Schäfer, R.; Noack, M.; Kölsch, P.; Kühnle, A.; Caro, J. Membrane supported catalytic dehydrogenation of iso-butane using an MFI zeolite membrane reactor. *Catal. Commun.* **2001**, *2*, 339–345.

147. Bergh, J.V.D.; Gücüyener, C.; Gascon, J.; Kapteijn, F. Isobutane dehydrogenation in a DD$_3$R zeolite membrane reactor. *Chem. Eng. J.* **2011**, *166*, 368–377.

148. Jeong, B.H.; Sotowa, K.I.; Kusakabe, K. Catalytic dehydrogenation of cyclohexane in an FAU-type zeolite membrane reactor. *J. Membr. Sci.* **2003**, *224*, 151–158.

149. Jeong, B.H.; Sotowa, K.I.; Kusakabe, K. Modeling of an FAU-type zeolite membrane reactor for the catalytic dehydrogenation of cyclohexane. *Chem. Eng. J.* **2004**, *103*, 69–75.

150. Kong, C.L.; Lu, J.M.; Yang, H.H.; Wang, J.Q. Catalytic dehydrogenation of ethylbenzene to styrene in a zeolite silicalite-1 membrane reactor. *J. Membr. Sci.* **2007**, *306*, 29–35.

151. Avila, A.M.; Yu, Z.; Fazli, S.; Sawada, J.A.; Kuznicki, S.M. Hydrogen-selective natural mordenite in a membrane reactor for ethane dehydrogenation. *Microporous Mesoporous Mater.* **2014**, *190*, 301–308.

152. Tang, Z.; Kim, S.J.; Reddy, G.K.; Dong, J.H.; Smirniotis, P. Modified zeolite membrane reactor for high temperature water gas shift reaction. *J. Membr. Sci.* **2010**, *354*, 114–122.

153. Zhang, Y.T.; Wu, Z.J.; Hong, Z.; Gu, X.H.; Xu, N.P. Hydrogen-selective zeolite membrane reactor for low temperature water gas shift reaction. *Chem. Eng. J.* **2012**, *197*, 314–321.

154. Kim, S.J.; Xu, Z.; Reddy, G.K.; Smirniotis, P.; Dong, J.H. Effect of Pressure on High-Temperature Water Gas Shift Reaction in Microporous Zeolite Membrane Reactor. *Ind. Eng. Chem. Res.* **2012**, *51*, 1364–1375.

155. Kim, S.J.; Yang, S.W.; Reddy, G.K.; Smirniotis, P.; Dong, J.H. Zeolite Membrane Reactor for High-Temperature Water-Gas Shift Reaction: Effects of Membrane Properties and Operating Conditions. *Energy Fuels* **2013**, *27*, 4471–4480.

156. Dong, X.L.; Wang, H.B.; Rui, Z.B.; Lin, Y.S. Tubular dual-layer MFI zeolite membrane reactor for hydrogen production via the WGS reaction: Experimental and modeling studies. *Chem. Eng. J.* **2015**, *268*, 219–229.

157. Zhang, Y.; Sun, Q.; Gu, X. Pure H_2 production through hollow fiber hydrogen-selective MFI zeolite membranes using steam as sweep gas. *AIChE J.* **2015**.

158. Dyk, L.V.; Lorenzen, L.; Miachon, S.; Dalmon, J.-A. Xylene isomerization in an extractor type catalytic membrane reactor. *Catal. Today* **2005**, *104*, 274–280.

159. Haag, S.; Hanebuth, M.; Mabande, G.T.P.; Avhale, A.; Schwieger, W.; Dittmeyer, R. On the use of a catalytic H-ZSM-5 membrane for xylene isomerization. *Microporous Mesoporous Mater.* **2006**, *96*, 168–176.

160. Tarditi, A.M.; Horowitz, G.I.; Lombardo, E.A. Xylene isomerization in a ZSM-5/SS membrane reactor. *Catal. Lett.* **2008**, *123*, 7–15.

161. Zhang, C.; Hong, Z.; Gu, X.H.; Zhong, Z.X.; Jin, W.Q.; Xu, N.P. Silicalite-1 Zeolite Membrane Reactor Packed with HZSM-5 Catalyst for *meta*-Xylene Isomerization. *Ind. Eng. Chem. Res.* **2009**, *48*, 4293–4299.

162. Daramola, M.O.; Deng, Z.; Pera-Titus, M.; Giroir-Fendler, A.; Miachon, S.; Burger, A.J.; Lorenzen, L.; Guo, Y. Nanocomposite MFI-alumina membranes prepared via pore-pugging synthesis: Application as packed-bed membrane reactors for *m*-xylene isomerization over a Pt-HZSM-5 catalyst. *Catal. Today* **2010**, 261–267.

163. Daramola, M.O.; Burger, A.J.; Giroir-Fendler, A.; Miachon, S.; Lorenzen, L. Extractor-type catalytic membrane reactor with nanocomposite MFI-alumina membrane tube as separation unit: Prospect for ultra-pure *para*-Xylene production from *m*-Xylene isomerization over Pt-HZSM-5 catalyst. *Appl. Catal. A* **2010**, *386*, 109–115.

164. Yeong, Y.F.; Abdullah, A.Z.; Ahmad, A.L.; Bhatia, S. Xylene isomerization kinetic over acid-functionalized silicalite-1 catalytic membranes: Experimental and modeling studies. *Chem. Eng. J.* **2010**, *157*, 579–589.

165. Zhang, C.; Hong, Z.; Chen, J.; Gu, X.H.; Jin, W.; Xu, N. Catalytic MFI zeolite membranes supported on Al_2O_3 substrates for *m*-xylene isomerization. *J. Membr. Sci.* **2012**, *389*, 451–458.

166. Mihályi, R.M.; Patis, A.; Nikolakis, V.; Kollár, M.; Valyon, J. [B]MFI membrane: Synthesis, physico-chemical properties and catalytic behavior in the double-bond isomerization of 1-butene. *Sep. Purif. Technol.* **2013**, *118*, 135–143.

167. Bernal, M.P.; Coronas, J.; Menendez, M.; Santamaria, J. Coupling ofreaction and separation at the microscopic level: Esterification processes in a H-ZSM-5 membrane reactor. *Chem. Eng. Sci.* **2002**, *57*, 1557–1562.

168. De la Iglesia, Ó.; Mallada, R.; Menéndez, M.; Coronas, J. Continuous zeolite membrane reactor for esterification of ethanol and acetic acid. *Chem. Eng. J.* **2007**, *131*, 35–39.

169. Fedosov, D.A.; Smirnov, A.V.; Shkirskiy, V.V.; Voskoboynikov, T.; Ivanova, I.I. Methanol dehydration in NaA zeolite membrane reactor. *J. Membr. Sci.* **2015**, *486*, 189–194.

170. Gallucci, F.; Paturzo, L.; Basile, A. An experimental study of CO_2 hydrogenation into methanol involving a zeolite membrane reactor. *Chem. Eng. Process.* **2004**, *43*, 1029–1036.

171. Gora, L.; Jansen, J.C. Hydroisomerization of C6 with a zeolite membrane reactor. *J. Catal.* **2005**, *230*, 269–281.

172. Mota, S.; Miachon, S.; Volta, J.C.; Dalmon, J.A. Membrane reactor for selective oxidation of butane to maleic anhydride. *Catal. Today* **2001**, *67*, 169–176.

173. Abon, M.; Volta, J.-C. Vanadium phosphorus oxides for *n*-butane oxidation to maleic anhydride. *Appl. Catal. A* **1997**, *157*, 173–193.

174. Mallada, R.; Menéndez, M.; Santamaría, J. Use of membrane reactors for the oxidation of butane to maleic anhydride under high butane concentrations. *Catal. Today* **2000**, *56*, 191–197.

175. Xue, E.; Ross, J. The use of membrane reactors for catalytic *n*-butane oxidation to maleic anhydride with a butane-rich feed. *Catal. Today* **2000**, *61*, 3–8.

176. Cruz-López, A.M.; Guilhaume, N.; Miachon, S.; Dalmon, J.-A. Selective oxidation of butane to maleic anhydride in a catalytic membrane reactor adapted to rich butane feed. *Catal. Today* **2005**, *2005*, 949–956.

177. Pantazidis, A.; Dalmon, J.A.; Mirodatos, C. Oxidative Dehydrogenation of Propane on Catalytic Membrane Reactors. *Catal. Today* **1995**, *25*, 403–408.

178. Julbe, A.; Farrusseng, D.; Jalibert, J.C.; Mirodatos, C.; Guizard, C. Characteristics and performance in the oxidative dehydrogenation of propane of MFI and V-MFI zeolite membranes. *Catal. Today* **2000**, *56*, 199–209.

179. Diakov, V.; Blackwell, B.; Varma, A. Methanol oxidative dehydrogenation in a catalytic packed-bed membrane reactor: Experiments and model. *Chem. Eng. Sci.* **2002**, *57*, 1563–1569.

180. Diakov, V.; Varma, A. Optimal Feed Distribution in a Packed-Bed Membrane Reactor: The Case of Methanol Oxidative Dehydrogenation. *Ind. Eng. Chem. Res.* **2004**, *43*, 309–314.

181. Chommeloux, B.; Cimaomo, S.; Jolimaitre, E.; Uzio, D.; Magnoux, P.; Sanchez, J. New membrane for use as hydrogen distributor for hydrocarbon selective hydrogenation. *Microporous Mesoporous Mater.* **2008**, *109*, 28–37.

182. Torres, M.; López, L.; Domínguez, J.M.; Mantilla, A.; Ferrat, G.; Gutierrez, M.; Maubert, M. Olefins catalytic oligomerization on new composites of beta-zeolite films supported on α-Al_2O_3 membranes. *Chem. Eng. J.* **2003**, *92*, 1–6.

183. Torres, M.; Gutiérrez, M.; Mugica, V.; Romero, M.; López, L. Oligomerization of isobutene with a,β-zeolite membrane: Effect of the acid properties of the catalytic membrane. *Catal. Today* **2011**, *166*, 205–208.

184. Aguado, S.; Coronas, J.; Santamaría, J. Use of Zeolite Membrane Reactors for the Combustion of VOCs Present in Air at Low Concentrations. *Chem. Eng. Res. Des.* **2005**, *83*, 295–301.

185. Tanaka, K.; Yoshikawa, R.; Ying, C.; Kita, H.; Okamoto, K. Application of zeolite membranes to esterification reactions. *Catal. Today* **2001**, *67*, 121–125.

186. Tanaka, K.; Yoshikawa, R.; Ying, C.; Kita, H.; Okamoto, K. Application of zeolite T membrane to vapor-permeation-aided esterification of lactic acid with ethanol. *Chem. Eng. Sci.* **2002**, *57*, 1577–1584.

187. Salomón, M.A.; Coronas, J.; Menéndez, M.; Santamaría, J. Synthesis of MTBE in zeolite membrane reactors. *Appl. Catal. A* **2000**, *200*, 201–210.

188. Pera-Titus, M.; Llorens, J.; Cunill, F. Technical and economical feasibility of zeolite NaA membrane-based reactors in liquid-phase etherification reactions. *Chem. Eng. Process. Process Intensif.* **2009**, *48*, 1072–1079.

189. Aresta, M.; Galatola, M. Life cycle analysis applied to the assessment of the environmental impact of alternative synthetic processes. The dimethylcarbonate case: Part 1. *J. Clean Prod.* **1999**, *7*, 181–193.

190. Aresta, M.; Dibenedetto, A.; Tommasi, I. Developing Innovative Synthetic Technologies of Industrial Relevance Based on Carbon Dioxide as Raw Material. *Energy Fuels* **2001**, *15*, 269–273.

191. Diban, N.; Urtiaga, A.M.; Ortiz, I.; Ereña, J.; Bilbao, J.; Aguayo, A.T. Influence of the membrane properties on the catalytic production of dimethyl ether with *in situ* water removal for the successful capture of CO_2. *Chem. Eng. J.* **2013**, *234*, 140–148.

192. Olah, G.A.; Goeppert, A.; Prakash, G.K.S. Chemical Recycling of Carbon Dioxide to Methanol and Dimethyl Ether: From Greenhouse Gas to Renewable, Environmentally Carbon Neutral Fuels and Synthetic Hydrocarbons. *J. Org. Chem.* **2009**, *74*, 487–498.

193. Aguayo, A.T.; Ereña, J.; Sierra, I.; Olazar, M.; Bilbao, J. Deactivation and regeneration of hybrid catalysts in the single-step synthesis of dimethyl ether from syngas and CO_2. *Catal. Today* **2005**, *106*, 265–270.

194. Diban, N.; Urtiaga, A.M.; Ortiz, I.; Ereña, J.; Bilbao, J.; Aguayo, A.T. Improved Performance of a PBM Reactor for Simultaneous CO_2 Capture and DME Synthesis. *Ind. Eng. Chem. Res.* **2014**, *53*, 19479–19487.

195. Rohde, M.P.; Unruh, D.; Schaub, G. Membrane application in Fischer-Tropsch synthesis reactors—Overview of concepts. *Catal. Today* **2005**, *106*, 143–148.

196. Rohde, M.P.; Schaub, G.; Khajavi, S.; Jansen, J.C.; Kapteijn, F. Fischer-Tropsch synthesis with *in situ* H_2O removal—Directions of membrane development. *Microporous Mesoporous Mater.* **2008**, *115*, 123–136.

197. Espinoza, R.L.; du Toit, E.; Santamaria, J.; Menendez, M.; Coronas, J.; Irusta, S. Use of membranes in Fischer-Tropsch reactors. In *Studies in Surface Science and Catalysis*; Avelino Corma, F.V.M.S.M., José Luis, G.F., Eds.; Elsevier: Amsterdam, The Netherlands, 2000; Volume 130, pp. 389–394.

198. Khajavi, S.; Kapteijn, F.; Jansen, J.C. Synthesis of thin defect-free hydroxy sodalite membranes: New candidate for activated water permeation. *J. Membr. Sci.* **2007**, *299*, 63–72.

199. Rahimpour, M.R.; Mirvakili, A.; Paymooni, K. A novel water perm-selective membrane dual-type reactor concept for Fischer-Tropsch synthesis of GTL (gas to liquid) technology. *Energy* **2011**, *36*, 1223–1235.

200. Rahimpour, M.R.; Mirvakili, A.; Paymooni, K.; Moghtaderi, B. A comparative study between a fluidized-bed and a fixed-bed water perm-selective membrane reactor with *in situ* H₂O removal for Fischer-Tropsch synthesis of GTL technology. *J. Nat. Gas Sci. Eng.* **2011**, *3*, 484–495.

201. Cavani, F.; Trifiro, F. Classification of industrial catalysts and catalysis for the petrochemical industry. *Catal. Today* **1997**, *34*, 269–279.

202. Casanave, D.; Giroirfendler, A.; Sanchez, J.; Loutaty, R.; Dalmon, J.A. Control of Transport-Properties with a Microporous Membrane Reactor to Enhance Yields in Dehydrogenation Reactions. *Catal. Today* **1995**, *25*, 309–314.

203. Casanave, D.; Ciavarella, P.; Fiaty, K.; Dalmon, J.A. Zeolite membrane reactor for isobutane dehydrogenation: Experimental results and theoretical modelling. *Chem. Eng. Sci.* **1999**, *54*, 2807–2815.

204. Dyk, L.V.; Miachon, S.; Lorenzen, L.; Torres, M.; Fiaty, K.; Dalmon, J.-A. Comparison of microporous MFI and dense Pd membrane performances in an extractor-type CMR. *Catal. Today* **2003**, *82*, 167–177.

205. Gokhale, Y.V.; Noble, R.D.; Falconer, J.L. Effects of reactant loss and membrane selectivity on a dehydrogenation reaction in a membrane-enclosed catalytic reactor. *J. Membr. Sci.* **1995**, *103*, 235–242.

206. Vaezi, M.J.; Babaluo, A.A.; Shafiei, S. Modeling of ethylbenzene dehydrogenation membrane reactor to investigate the potential application of a microporous dydroxy sodalite membrane. *J. Chem. Pet. Eng.* **2015**, *49*, 51–62.

207. Rhodes, C.; Hutchings, G.J.; Ward, A.M. Water-gas shift reaction: Finding the mechanistic boundary. *Catal. Today* **1995**, *23*, 43–58.

208. Tosti, S.; Basile, A.; Chiappetta, G.; Rizzello, C.; Violante, V. Pd-Ag membrane reactors for water gas shift reaction. *Chem. Eng. J.* **2003**, *93*, 23–30.

209. Basile, A.; Chiappetta, G.; Tosti, S.; Violante, V. Experimental and simulation of both Pd and Pd/Ag for a water gas shift membrane reactor. *Sep. Purif. Technol.* **2001**, *25*, 549–571.

210. Peters, T.A.; Stange, M.; Klette, H.; Bredesen, R. High pressure performance of thin Pd-23%Ag/stainless steel composite membranes in water gas shift gas mixtures; influence of dilution, mass transfer and surface effects on the hydrogen flux. *J. Membr. Sci.* **2008**, *316*, 119–127.

211. Augustine, A.S.; Ma, Y.H.; Kazantzis, N.K. High pressure palladium membrane reactor for the high temperature water-gas shift reaction. *Int. J. Hydrogen Energy* **2011**, *36*, 5350–5360.

212. Giessler, S.; Jordan, L.; Diniz da Costa, J.C.; Lu, G.Q. Performance of hydrophobic and hydrophilic silica membrane reactors for the water gas shift reaction. *Sep. Purif. Technol.* **2003**, *32*, 255–264.

213. Battersby, S.; Duke, M.C.; Liu, S.; Rudolph, V.; Costa, J.C.D.D. Metal doped silica membrane reactor: Operational effects of reaction and permeation for the water gas shift reaction. *J. Membr. Sci.* **2008**, *316*, 46–52.

214. Gu, X.; Tang, Z.; Dong, J. On-stream modification of MFI zeolite membranes for enhancing hydrogen separation at high temperature. *Microporous Mesoporous Mater.* **2008**, *111*, 441–448.

215. Liu, B.S.; Au, C.T. A La$_2$NiO$_4$-Zeolite Membrane Reactor for the CO$_2$ Reforming of Methane to Syngas. *Catal. Lett.* **2001**, *77*, 67–74.

216. Liu, B.S.; Gao, L.Z.; Au, C.T. Preparation, characterization and application of a catalytic NaA membrane for CH$_4$/CO$_2$ reforming to syngas. *Appl. Catal. A* **2002**, *235*, 193–206.

217. Barbieri, G.; Marigliano, G.; Golemme, G.; Drioli, E. Simulation of CO$_2$ hydrogenation with CH$_3$OH removal in a zeolite membrane reactor. *Chem. Eng. J.* **2002**, *85*, 53–59.

218. Lee, G.-S.; McCain, J.; Bhasin, M. Synthetic Organic Chemicals. In *Kent and Riegel's Handbook of Industrial Chemistry and Biotechnology*; Kent, J., Ed.; Springer US: New York, NY, USA, 2007; pp. 345–403.

219. Yan, T.Y. Separation of *p*-xylene and ethylbenzene from C8 aromatics using medium-pore zeolites. *Ind. Eng. Chem. Res.* **1989**, *28*, 572–576.

220. Daramola, M.O.; Burger, A.J.; Pera-Titus, M.; Giroir-Fendler, A.; Miachon, S.; Dalmon, J.A.; Lorenzen, L. Separation and isomerization of xylenes using zeolite membranes: A short overview. *Asia-Pac. J. Chem. Eng.* **2010**, *5*, 815–837.

221. Xomeritakis, G.; Lai, Z.; Tsapatsis, M. Separation of Xylene Isomer Vapors with Oriented MFI Membranes Made by Seeded Growth. *Ind. Eng. Chem. Res.* **2001**, *40*, 544–552.

222. Baerlocher, C.; McCusker, L.B.; Olson, D.H. MFI-Pnma. In *Atlas of Zeolite Framework Types*, 6th ed.; Baerlocher, C., Olson, L.B.M.H., Eds.; Elsevier Science B.V.: Amsterdam, The Netherlands, 2007; pp. 212–213.

223. Sakai, H.; Tomita, T.; Takahashi, T. *p*-Xylene separation with MFI-type zeolite membrane. *Sep. Purif. Technol.* **2001**, *25*, 297–306.

224. Gu, X.; Dong, J.; Nenoff, T.M.; Ozokwelu, D.E. Separation of *p*-xylene from multicomponent vapor mixtures using tubular MFI zeolite mmbranes. *J. Membr. Sci.* **2006**, *280*, 624–633.

225. Gump, C.J.; Tuan, V.A.; Noble, R.D.; Falconer, J.L. Aromatic Permeation through Crystalline Molecular Sieve Membranes. *Ind. Eng. Chem. Res.* **2001**, *40*, 565–577.

226. Lai, Z.; Bonilla, G.; Diaz, I.; Nery, J.G.; Sujaoti, K.; Amat, M.A.; Kokkoli, E.; Terasaki, O.; Thompson, R.W.; Tsapatsis, M.; *et al.* Microstructural optimization of a zeolite membrane for organic vapor separation. *Science* **2003**, *300*, 456–460.

227. Daramola, M.O.; Burger, A.J.; Pera-Titus, M.; Giroir-Fendler, A.; Miachon, S.; Lorenzen, L.; Dalmon, J.A. Nanocomposite MFI-ceramic hollow fibre membranes via pore-plugging synthesis: Prospects for xylene isomer separation. *J. Membr. Sci.* **2009**, *337*, 106–112.

228. Maloncy, M.L.; Maschmeyer, T.; Jansen, J.C. Technical and economical evaluation of a zeolite membrane based heptane hydroisomerization process. *Chem. Eng. J.* **2005**, *106*, 187–195.

229. Piera, E.; Tellez, C.; Coronas, J.; Menendez, M.; SaIntamaria, J. Use of zeolite membrane reactors for selectivity enhancement: Application to the liquid-phase oligomerization of *i*-butene. *Catal. Today* **2001**, *67*, 127–138.

230. Peters, T.A.; Benes, N.E.; Keurentjes, J.T.F. Zeolite-Coated Ceramic Pervaporation Membranes; Pervaporation-Esterification Coupling and Reactor Evaluation. *Ind. Eng. Chem. Res.* **2005**, *44*, 9490–9496.

231. De la Iglesia, Ó.; Irusta, S.; Mallada, R.; Menéndez, M.; Coronas, J.; Santamaría, J. Preparation and characterization of two-layered mordenite-ZSM-5 bi-functional membranes. *Microporous Mesoporous Mater.* **2006**, *93*, 318–324.

232. Tarditi, A.M.; Lombardo, E.A.; Avila, A.M. Xylene Permeation Transport through Composite Ba-ZSM-5/SS Tubular Membranes: Modeling the Steady-State Permeation. *Ind. Eng. Chem. Res.* **2008**, *47*, 2377–2385.

233. Zhou, J.L.; Zhang, X.F.; Zhang, J.; Liu, H.O.; Zhou, L.; Yeung, K.L. Preparation of alkali-resistant, Sil-1 encapsulated nickel catalysts for direct internal reforming-molten carbonate fuel cell. *Catal. Commun.* **2009**, *10*, 1804–1807.

234. He, J.; Yoneyama, Y.; Xu, B.; Nishiyama, N.; Tsubaki, N. Designing a Capsule Catalyst and Its Application for Direct Synthesis of Middle Isoparaffins. *Langmuir* **2005**, *21*, 1699–1702.

235. Puil, N.V.D.; Creyghton, E.J.; Rodenburg, E.C.; Sie, T.S.; Bekkum, H.v.; Jansen, J.C. Catalytic testing of TiO_2/platinum/silicalite-1 composites. *Faraday Trans.* **1996**, *96*, 4609–4615.

236. Van der Puil, N.; Dautzenberg, F.M.; van Bekkum, H.; Jansen, J.C. Preparation and catalytic testing of zeolite coatings on preshaped alumina supports. *Microporous Mesoporous Mater.* **1999**, *27*, 95–106.

237. Bouizi, Y.; Rouleau, L.; Valtchev, V.P. Factors Controlling the Formation of Core-Shell Zeolite-Zeolite Composites. *Chem. Mater.* **2006**, *18*, 4959–4966.

238. Li, X.; He, J.; Meng, M.; Yoneyama, Y.; Tsubaki, N. One-step synthesis of H-β zeolite-enwrapped Co/Al_2O_3 Fischer-Tropsch catalyst with high spatial selectivity. *J. Catal.* **2009**, *265*, 26–34.

239. Bao, J.; He, J.; Zhang, Y.; Yoneyama, Y.; Tsubaki, N. A Core/Shell Catalyst Produces a Spatially Confined Effect and Shape Selectivity in a Consecutive Reaction. *Angew. Chem. Int. Ed.* **2008**, *47*, 353–356.

240. Li, C.; Xu, H.; Kido, Y.; Yoneyama, Y.; Suehiro, Y.; Tsubaki, N. A capsule catalyst with a zeolite membrane prepared by direct liquid membrane crystallization. *ChemSusChem* **2012**, *5*, 862–866.

241. Phienluphon, R.; Pinkaew, K.; Yang, G.H.; Li, J.; Wei, Q.H.; Yoneyama, Y.; Vitidsant, T.; Tsubaki, N. Designing core ($Cu/ZnO/Al_2O_3$)-shell (SAPO-11) zeolite capsule catalyst with a facile physical way for dimethyl ether direct synthesis from syngas. *Chem. Eng. J.* **2015**, *270*, 605–611.

242. Cimenler, U.; Joseph, B.; Kuhn, J.N. Molecular-size selective H-β zeolite-encapsulated Ce-Zr/Ni-Mg catalysts for steam reforming. *Appl. Catal. A* **2015**, *505*, 494–500.

243. Pinkaew, K.; Yang, G.; Vitidsant, T.; Jin, Y.; Zeng, C.; Yoneyama, Y.; Tsubaki, N. A new core-Shell-like capsule catalyst with SAPO-46 zeolite shell encapsulated Cr/ZnO for the controlled tandem synthesis of dimethyl ether from syngas. *Fuel* **2013**, *111*, 727–732.

244. Xing, C.; Sun, J.; Chen, Q.; Yang, G.; Muranaka, N.; Lu, P.; Shen, W.; Zhu, P.; Wei, Q.; Li, J.; *et al.* Tunable isoparaffin and olefin yields in Fischer-Tropsch synthesis achieved by a novel iron-based micro-capsule catalyst. *Catal. Today* **2015**, *251*, 41–46.

245. Nishiyama, N.; Miyamoto, M.; Egashira, Y.; Ueyama, K. Zeolite membrane on catalyst particles for selective formation of *p*-xylene in the disproportionation of toluene. *Chem. Commun.* **2001**, 1746–1747.

246. Vu, D.V.; Miyamoto, M.; Nishiyama, N.; Egashira, Y.; Ueyama, K. Selective formation of para-xylene over H-ZSM-5 coated with polycrystalline silicalite crystals. *J. Catal.* **2006**, *243*, 389–394.

247. Vu, D.V.; Miyamoto, M.; Nishiyama, N.; Ichikawa, S.; Egashira, Y.; Ueyama, K. Catalytic activities and structures of silicalite-1/H-ZSM-5 zeolite composites. *Microporous Mesoporous Mater.* **2008**, *115*, 106–112.

248. Li, X.; Zhang, Y.; Meng, F.; San, X.; Yang, G.; Meng, M.; Takahashi, M.; Tsubaki, N. Hydroformylation of 1-Hexene on Silicalite-1 Zeolite Membrane Coated Pd-Co/AC Catalyst. *Top Catal.* **2010**, *53*, 608–614.

249. Nishiyama, N.; Ichioka, K.; Park, D.-H.; Egashira, Y.; Ueyama, K.; Gora, L.; Zhu, W.; Kapteijn, F.; Moulijn, J.A. Reactant-Selective Hydrogenation over Composite Silicalite-1-Coated Pt/TiO$_2$ Particles. *Ind. Eng. Chem. Res.* **2004**, *43*, 1211–1215.

250. Nishiyama, N.; Ichioka, K.; Miyamoto, M.; Egashira, Y.; Ueyama, K.; Gora, L.; Zhu, W.D.; Kapteijn, F.; Moulijn, J.A. Silicalite-1 coating on Pt/TiO$_2$ particles by a two-step hydrothermal synthesis. *Microporous Mesoporous Mater.* **2005**, *83*, 244–250.

251. Zhong, Y.; Chen, L.; Luo, M.; Xie, Y.; Zhu, W. Fabrication of zeolite-4A membranes on a catalyst particle level. *Chem. Commun.* **2006**, 2911–2912.

252. Wu, Y.; Chai, Y.; Li, J.; Guo, H.; Wena, L.; Liu, C. Preparation of silicalite-1@Pt/alumina core-shell catalyst for shape-selective hydrogenation of xylene isomers. *Catal. Commun.* **2015**, *64*, 110–113.

253. He, J.; Liu, Z.; Yoneyama, Y.; Nishiyama, N.; Tsubaki, N. Multiple-Functional Capsule Catalysts: A Tailor-Made Confined Reaction Environment for the Direct Synthesis of Middle Isoparaffins from Syngas. *Chemistry* **2006**, *12*, 8296–8304.

254. Yang, G.H.; He, J.J.; Yoneyama, Y.; Tan, Y.S.; Han, Y.Z.; Tsubaki, N. Preparation, characterization and reaction performance of H-ZSM-5/cobalt/silica capsule catalysts with different sizes for direct synthesis of isoparaffins. *Appl. Catal. A* **2007**, *329*, 99–105.

255. Yang, G.H.; Tan, Y.S.; Han, Y.Z.; Qiu, J.S.; Tsubaki, N. Increasing the shell thickness by controlling the core size of zeolite capsule catalyst: Application in iso-paraffin direct synthesis. *Catal. Commun.* **2008**, *9*, 2520–2524.

256. Yang, G.H.; He, J.J.; Zhang, Y.; Yoneyama, Y.; Tan, Y.S.; Han, Y.Z.; Vitidsant, T.; Tsubaki, N. Design and modification of zeolite capsule catalyst, a confined reaction field, and its application in one-step isoparaffin synthesis from syngas. *Energy Fuels* **2008**, *22*, 1463–1468.

257. Bao, J.; Yang, G.H.; Okada, C.; Yoneyama, Y.; Tsubaki, N. H-type zeolite coated iron-based multiple-functional catalyst for direct synthesis of middle isoparaffins from syngas. *Appl. Catal. A* **2011**, *394*, 195–200.

258. Huang, X.; Hou, B.; Wang, J.G.; Li, D.B.; Jia, L.T.; Chen, J.G.; Sun, Y.H. CoZr/H-ZSM-5 hybrid catalysts for synthesis of gasoline-range isoparaffins from syngas. *Appl. Catal. A* **2011**, *408*, 38–46.

259. Yang, G.; Xing, C.; Hirohama, W.; Jin, Y.; Zeng, C.; Suehiro, Y.; Wang, T.; Yoneyama, Y.; Tsubaki, N. Tandem catalytic synthesis of light isoparaffin from syngas via Fischer-Tropsch synthesis by newly developed core-shell-like zeolite capsule catalysts. *Catal. Today* **2013**, *215*, 29–35.

260. Jin, Y.; Yang, G.; Chen, Q.; Niu, W.; Lu, P.; Yoneyama, Y.; Tsubaki, N. Development of dual-membrane coated Fe/SiO$_2$ catalyst for efficient synthesis of isoparaffins directly from syngas. *J. Membr. Sci.* **2015**, *475*, 22–29.

261. Yang, G.; Tsubaki, N.; Shamoto, J.; Yoneyama, Y.; Zhang, Y. Confinement effect and synergistic function of H-ZSM-5/Cu-ZnO-Al$_2$O$_3$ capsule catalyst for one-step controlled synthesis. *J. Am. Chem. Soc.* **2010**, *132*, 8129–8136.

262. Yang, G.; Thongkam, M.; Vitidsant, T.; Yoneyama, Y.; Tan, Y.; Tsubaki, N. A double-shell capsule catalyst with core-shell-like structure for one-step exactly controlled synthesis of dimethyl ether from CO$_2$ containing syngas. *Catal. Today* **2011**, *171*, 229–235.

263. Yang, G.; Wang, D.; Yoneyama, Y.; Tan, Y.; Tsubaki, N. Facile synthesis of H-type zeolite shell on a silica substrate for tandem catalysis. *Chem. Commun.* **2012**, *48*, 1263–1265.

264. Liu, R.; Tian, H.; Yang, A.; Zha, F.; Ding, J.; Chang, Y. Preparation of HZSM-5 membrane packed CuO-ZnO-Al$_2$O$_3$ nanoparticles for catalysing carbon dioxide hydrogenation to dimethyl ether. *Appl. Surf. Sci.* **2015**, *345*, 1–9.

265. Ren, N.; Yang, Y.H.; Shen, J.; Zhang, Y.H.; Xu, H.L.; Gao, Z.; Tang, Y. Novel, efficient hollow zeolitically microcapsulized noble metal catalysts. *J. Catal.* **2007**, *251*, 182–188.

266. Besson, M.; Gallezot, P. Deactivation of metal catalysts in liquid phase organic reactions. *Catal. Today* **2003**, *81*, 547–559.

267. Papp, A.; Miklos, K.; Forgo, M.; Molnar, A. Heck coupling by Pd deposited onto organic-inorganic hybrid supports. *J. Mol.Catal. A* **2005**, *229*, 107–116.

268. Zhang, T.; Zhang, X.; Yan, X.; Lin, L.; Liu, H.; Qiu, J.; Yeung, K.L. Core-shell Pd/ZSM-5@ZIF-8 membrane micro-reactors with sizeselectivity properties for alkene hydrogenation. *Catal. Today* **2014**, *236*, 41–48.

269. Miyamoto, M.; Kamei, T.; Nishiyama, N.; Egashira, Y.; Ueyama, K. Single crystals of ZSM-5/silicalite composites. *Adv. Mater.* **2005**, *17*, 1985–1988.

270. Mabuchi, K.; Miyamoto, M.; Oumi, Y.; Uemiya, S. Selective formation of *p*-xylene in aromatization of propane over silicalite-1-coated GaAlMFI. *J. Jpn. Pet. Inst.* **2011**, *54*, 275–276.

271. Miyamoto, M.; Mabuchi, K.; Kamada, J.; Hirota, Y.; Oumi, Y.; Nishiyama, N.; Uemiya, S. para-Selectivity of silicalite-1 coated MFI type galloaluminosilicate in aromatization of light alkanes. *J. Porous Mater.* **2015**, *22*, 769–778.

272. Zhou, Y.X.; Tong, W.Y.; Zou, W.; Qi, X.L.; Kong, D.J. Manufacture of *b*-Oriented ZSM-5/Silicalite-1 Core/Shell Structured Zeolite Catalyst. *Synth. React. Inorg. Met.-Org. Nano-Met. Chem.* **2015**, *45*, 1356–1362.

273. Van der Laan, G.P.; Beenackers, A.A.C.M. Kinetics and Selectivity of the Fischer-Tropsch Synthesis: A Literature Review. *Catal. Rev.* **1999**, *41*, 255–318.

274. Li, X.H.; Asami, K.; Luo, M.F.; Michiki, K.; Tsubaki, N.; Fujimoto, K. Direct synthesis of middle iso-paraffins from synthesis gas. *Catal. Today* **2003**, *84*, 59–65.

275. Li, X.H.; Luo, M.F.; Asami, K. Direct synthesis of middle iso-paraffins from synthesis gas on hybrid catalysts. *Catal. Today* **2004**, *89*, 439–446.

276. Botes, F.G.; Böhringer, W. The addition of HZSM-5 to the Fischer-Tropsch process for improved gasoline production. *Appl. Catal. A* **2004**, *267*, 217–225.

277. Yoneyama, Y.; He, J.; Morii, Y.; Azuma, S.; Tsubaki, N. Direct synthesis of isoparaffin by modified Fischer-Tropsch synthesis using hybrid catalyst of iron catalyst and zeolite. *Catal. Today* **2005**, *104*, 37–40.

278. Moradi, G.R.; Basir, M.M.; Taeb, A.; Kiennemann, A. Promotion of Co/SiO$_2$ Fischer-Tropsch catalysts with zirconium. *Catal. Commun.* **2003**, *4*, 37–32.

279. Xiong, H.; Zhang, Y.; Liew, K.; Li, J. Catalytic performance of zirconium-modified Co/Al$_2$O$_3$ for Fischer-Tropsch synthesis. *J. Mol. Catal. A* **2005**, *231*, 145–151.

280. Koizumi, N.; Seki, H.; Hayasaka, Y.; Oda, Y.; Shindo, T.; Yamada, M. Application of liquid phase deposition method for preparation of Co/ZrO$_x$/SiO$_2$ catalyst with enhanced Fischer-Tropsch synthesis activity: Importance of Co-Zr interaction. *Appl. Catal. A* **2011**, *398*, 168–178.

281. Lin, Q.; Yang, G.; Li, X.; Yoneyama, Y.; Wan, H.; Tsubaki, N. A Catalyst for One-step Isoparaffin Production via Fischer-Tropsch Synthesis: Growth of a H-Mordenite Shell Encapsulating a Fused Iron Core. *ChemCatChem* **2013**, *5*, 3101–3106.

282. Panpranot, J.; Goodwin, J.G.; Sayari, A. CO hydrogenation on Ru-promoted Co/MCM-41 catalysts. *J. Catal.* **2002**, *211*, 530–539.

283. Panpranot, J.; Goodwin, J.G.; Sayari, A. Effect of H$_2$ partial pressure on surface reaction parameters during CO hydrogenation on Ru-promoted silica-supported Co catalysts. *J. Catal.* **2003**, *213*, 78–85.

284. Cano, L.A.; Cagnoli, M.V.; Fellenz, N.A.; Bengoa, J.F.; Gallegos, N.G.; Alvarez, A.M.; Marchetti, S.G. Fischer-Tropsch synthesis. Influence of the crystal size of iron active species on the activity and selectivity. *Appl. Catal. A* **2010**, *379*, 105–110.

285. Bezemer, G.L.; Bitter, J.H.; Kuipers, H.P.; Oosterbeek, H.; Holewijn, J.E.; Xu, X.; Kapteijn, F.; van Dillen, A.J.; de Jong, K.P. Cobalt particle size effects in the Fischer-Tropsch reaction studied with carbon nanofiber supported catalysts. *J. Am. Chem. Soc.* **2006**, *128*, 3956–3964.

286. Khodakov, A.Y. Fischer-Tropsch synthesis: Relations between structure of cobalt catalysts and their catalytic performance. *Catal. Today* **2009**, *144*, 251–257.

287. Zhang, Q.H.; Deng, W.P.; Wang, Y. Recent advances in understanding the key catalyst factors for Fischer-Tropsch synthesis. *J. Energy Chem.* **2013**, *22*, 27–38.

288. Arcoumanis, C.; Bae, C.; Crookes, R.; Kinoshita, E. The potential of di-methyl ether (DME) as an alternative fuel for compression-ignition engines: A review. *Fuel* **2008**, *87*, 1014–1030.

289. Mao, D.S.; Yang, W.M.; Xia, J.C.; Zhang, B.; Song, Q.Y.; Chen, Q.L. Highly effective hybrid catalyst for the direct synthesis of dimethyl ether from syngas with magnesium oxide-modified HZSM-5 as a dehydration component. *J. Catal.* **2005**, *230*, 140–149.

290. Ramos, F.S.; de Farias, A.M.D.; Borges, L.E.P.; Monteiro, J.L.; Fraga, M.A.; Sousa-Aguiar, E.F.; Appel, L.G. Role of dehydration catalyst acid properties on one-step DME synthesis over physical mixtures. *Catal. Today* **2005**, *101*, 39–44.

291. Moradi, G.R.; Nosrati, S.; Yaripor, F. Effect of the hybrid catalysts preparation method upon direct synthesis of dimethyl ether from synthesis gas. *Catal. Commun.* **2007**, *8*, 598–606.

292. Jin, D.; Zhu, B.; Hou, Z.; Fei, J.; Lou, H.; Zheng, X. Dimethyl ether synthesis via methanol and syngas over rare earth metals modified zeolite Y and dual Cu-Mn-Zn catalysts. *Fuel* **2007**, *86*, 2707–2713.

293. Mao, D.; Xia, J.; Zhang, B.; Lu, G. Highly efficient synthesis of dimethyl ether from syngas over the admixed catalyst of CuO-ZnO-Al$_2$O$_3$ and antimony oxide modified HZSM-5 zeolite. *Energy Convers. Manag.* **2010**, *51*, 1134–1139.

294. Pop, G.; Bozga, G.; Ganea, R.; Natu, N. Methanol Conversion to Dimethyl Ether over H-SAPO-34 Catalyst. *Ind. Eng. Chem. Res.* **2009**, *48*, 7065–7071.

295. Dai, W.L.; Kong, W.B.; Wu, G.J.; Li, N.; Li, L.D.; Guan, N.J. Catalytic dehydration of methanol to dimethyl ether over aluminophosphate and silico-aluminophosphate molecular sieves. *Catal. Commun.* **2011**, *12*, 535–538.

Effects of Dealumination and Desilication of Beta Zeolite on Catalytic Performance in *n*-Hexane Cracking

Yong Wang, Toshiyuki Yokoi, Seitaro Namba and Takashi Tatsumi

Abstract: Catalytic cracking of *n*-hexane to selectively produce propylene on Beta zeolite was carried out. The H-Beta (HB) (Si/Al = 77) zeolite showed higher catalytic stability and propylene selectivity than the Al-rich HB (Si/Al = 12), due to its smaller number of acid sites, especially Lewis acid sites (LAS). However, catalytic stability and propylene selectivity in high *n*-hexane conversions were still not satisfactory. After dealumination with HNO_3 treatment, catalytic stability was improved and propylene selectivity during high *n*-hexane conversions was increased. On the other hand, catalytic stability was not improved after desilication with NaOH treatment, although mesopores were formed. This may be related to the partially destroyed structure. However, propylene selectivity in high *n*-hexane conversions was increased after alkali treatment. We successfully found that the catalytic stability was improved and the propylene selectivity in high *n*-hexane conversions was further increased after the NaOH treatment followed by HNO_3 treatment. This is due to the decrease in the number of acid sites and the increase in mesopores which are beneficial to the diffusion of coke precursor.

Reprinted from *Catalysts*. Cite as: Wang, Y.; Yokoi, T.; Namba, S.; Tatsumi, T. Effects of Dealumination and Desilication of Beta Zeolite on Catalytic Performance in *n*-Hexane Cracking. *Catalysts* **2016**, *6*, 8.

1. Introduction

Light alkenes, especially propylene which is mainly supplied as a by-product of ethylene production through thermal cracking of naphtha, are gaining more and more significance in the chemical industry. However, thermal cracking is an energy intensive process, where the product distribution, especially propylene/ethylene ratio, is hard to control. Therefore, catalytic cracking of naphtha has been drawing more attention [1]. Compared to the thermal steam cracking, the catalytic cracking of naphtha over acidic zeolite catalysts gives a high propylene/ethylene ratio, since the transformation of long-chain alkanes to short-chain alkenes occurs at least partly *via* the carbeniumion/β-scission mechanism, *i.e.*, classical bimolecular cracking [2,3].

Among the zeolite catalysts examined for catalytic cracking, ZSM-5 (MFI-type) zeolite with three dimensional 10-membered ring (10-MR) channels has been recognized as a prime candidate because of its high thermal and hydrothermal stabilities and its considerable resistance to deactivation caused by coking as well as its strong acidity [4]. Yoshimura *et al.* reported that $La_2O_3/P/ZSM-5$ zeolite (Si/Al = 100) exhibited a high naphtha catalytic cracking activity and a high yield of *ca.* 60% to light olefins (ethylene and propylene) at 923 K, and the propylene/ethylene ratio was approximately 0.7 [1]. Recently, we carried out catalytic cracking of *n*-hexane as a model reaction of naphtha cracking over H-ZSM-5 zeolites (Si/Al = 45) with different crystal sizes, and found that the catalytic stability could be improved by decreasing the crystal size [5]. However, the propylene selectivity at *ca.* 100% *n*-hexane conversion slightly decreased from *ca.* 32 to 26 C-% at 923 K.

To improve the propylene selectivity in catalytic cracking, zeolite catalysts with other structures have been considered. Recently, dealuminated MCM-68 (MSE-type) zeolite (Si/Al = 51) with multidimensional 10-MR or 12-MR micropores has been reported to exhibit a high propylene selectivity of *ca.* 45–50 C-% and a high durability to coke formation during *n*-hexane cracking, regardless of the reaction temperature [6]. However, the *n*-hexane conversion was only changed from *ca.* 7%–75% by changing the reaction temperature from 723 to 873 K. Thus, the selectivity to propylene was uncertain when the *n*-hexane conversion approached 100%. More recently, we reported that dealuminated MCM-22 (MWW-type) zeolite (Si/Al = 62) with two dimensional sinusoidal 10-MR channels and 12-MR supercages exhibited a high propylene selectivity of *ca.* 40 C-% with a high *n*-hexane conversion of 95% and a high catalytic stability at 923 K [7].

Beta (*BEA-type) zeolite with three dimensional 12-MR channels has been an important catalyst in the petroleum industry, especially in the alkylation process of benzene for the production of ethylbenzene and cumene [8,9]. Additionally, Beta zeolite also shows an excellent catalytic performance in FCC processes [10]. It is well-known tha t the Brønsted acid sites (BAS) in the zeolite catalyst act as active sites in alkane cracking, and that Lewis acid sites (LAS) can enhance the cracking reaction [11,12]. However, a large amount of acid sites, especially the LAS, would accelerate the secondary reactions of propylene and butenes and the hydride transfer, leading to coke formation followed by deactivation [7]. Thus, dealumination of zeolite catalyst leads to an increase in propylene selectivity and the improvement of catalytic stability. It was reported that dealumination of Beta zeolite occurred very easily compared to other ones such as ZSM-5 and MCM-22, due to the presence of stacking defects [13]. The dealumination of Beta zeolite by different methods such as acid treatment, steaming treatment, and chemical treatments had already been studied in the literature [14–16]. However, the effect of dealumination of Beta zeolite on the catalytic performance of alkane cracking was seldom reported [6]. More recently, we have reported that the dealuminated organic structure-directing agent (OSDA) free Beta zeolite showed a better catalytic performance in *n*-hexane cracking [17].

On the other hand, desilication proceeded by a controlled extraction of Si from the zeolite framework in alkali medium has been proven an efficient method to induce significant mesoporosity [18,19]. Recently, we found that the catalytic stability in n-hexane cracking can be improved by alkali treatment of ZSM-5 due to the formation of mesopores, which facilitates the diffusion of the coke precursor [20]. In addition, LAS were generated by higher concentrated NaOH treatment, leading to the increase in the selectivities to benzene, toluene and xylene (BTX) at high reaction temperatures ($\geqslant$873 K). However, selectivity to propylene did not increase. Similarly, Bjorgen *et al.* reported that product distribution and catalytic stability in the methanol to gasoline (MTG) reaction were also altered dramatically with desilication of ZSM-5, as a result of formation of LAS [21]. Up to now, there are many reports on the effect of desilication of ZSM-5 zeolite on the catalytic performance in various reactions [20–23]. However, only a few papers on the desilication of Beta zeolite were reported [24,25]. Moreover, the effect of desilication of Beta zeolite on the catalytic performance of alkane cracking has never been reported.

In this study, the dealuminated and desilicated H-Beta zeolites (hereinafter, designated as "HB") were prepared via acid and alkali treatment, respectively. The effects of dealumination and desilication on the physicochemical properties (including porosity and acidity) and catalytic properties (including activity, selectivity and catalytic life) of HB zeolites in the cracking of n-hexane as a model compound of naphtha were investigated. In addition, as a control, the physicochemical properties and catalytic properties of Al-rich HB were investigated.

2. Results and Discussion

2.1. A Summary of Catalysts' Characterization

The acid-treated HB samples were denoted as HB(77)-NT(a M, b h), where a and b were the HNO_3 concentration and treatment time, respectively. The alkali-treated HB samples were denoted as HB(77)-AT(c M), where c was the NaOH concentration. The alkali-acid-treated samples were denoted as HB(77)-AT(c M)-NT (d M), where d was the HNO_3 concentration.

2.1.1. XRD

Figure 1 shows the XRD patterns of the Al-rich HB(12), HB(77) and post-treated HB zeolites. Both of the HB(12) and HB(77) had a pure Beta phase and relatively high crystallinities (Figure 1A). Little change was observed in the diffractions after HNO_3 treatment, regardless of the HNO_3 concentration and treatment time (Figure 1A), meaning that the structure remained even after the acid treatment.

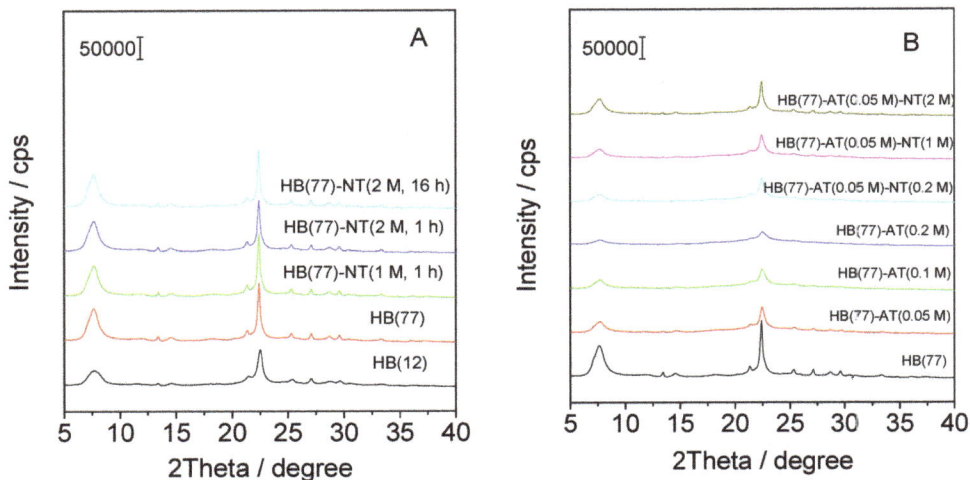

Figure 1. XRD patterns of HB(12), HB(77), acid-treated HB (**A**); alkali-treated and alkali-acid-treated HB (**B**) catalysts.

On the other hand, it was apparent that the crystallinities of the alkali-treated HB were dependent on the alkali treatment conditions, *i.e.*, NaOH concentration. The crystallinity was gradually decreased with an increase in the NaOH concentration (Figure 1B), meaning that the structure was partially destroyed due to desilication (and/or dealumination), especially after 0.2 M NaOH treatment. In addition, the crystallinity was slightly decreased after 0.05 M NaOH treatment followed by 0.2 or 1 M HNO_3 treatment. However, it was slightly recovered after 0.05 M NaOH treatment followed by 2 M HNO_3 treatment (Figure 1B).

2.1.2. SEM

The SEM images revealed that HB(12) was composed of dispersed nanoparticles with *ca.* 20 nm (Figure 2a,b). The HB(77) particles were uniform and exhibited an average particle size of *ca.* 200–300 nm (Figure 2c). At high magnification, these particles appeared to be assemblies of very small Beta zeolite crystals (*ca.* 20 nm) (Figure 2d). The morphology and size of HB particles were almost unchanged after acid treatment (Figure 2e,f). On the other hand, the edge of the aggregated HB particles became blurry and some of the large particles were broken into small ones after alkali treatment (Figure 3a–d). This may be related to the formation of amorphous species (Figure 1B). Moreover, it can be observed that some small particles formed after alkali treatment followed by acid treatment (Figure 3e,f). These facts suggested that alkali treatment (and subsequent acid treatment) affected the particle morphology.

78

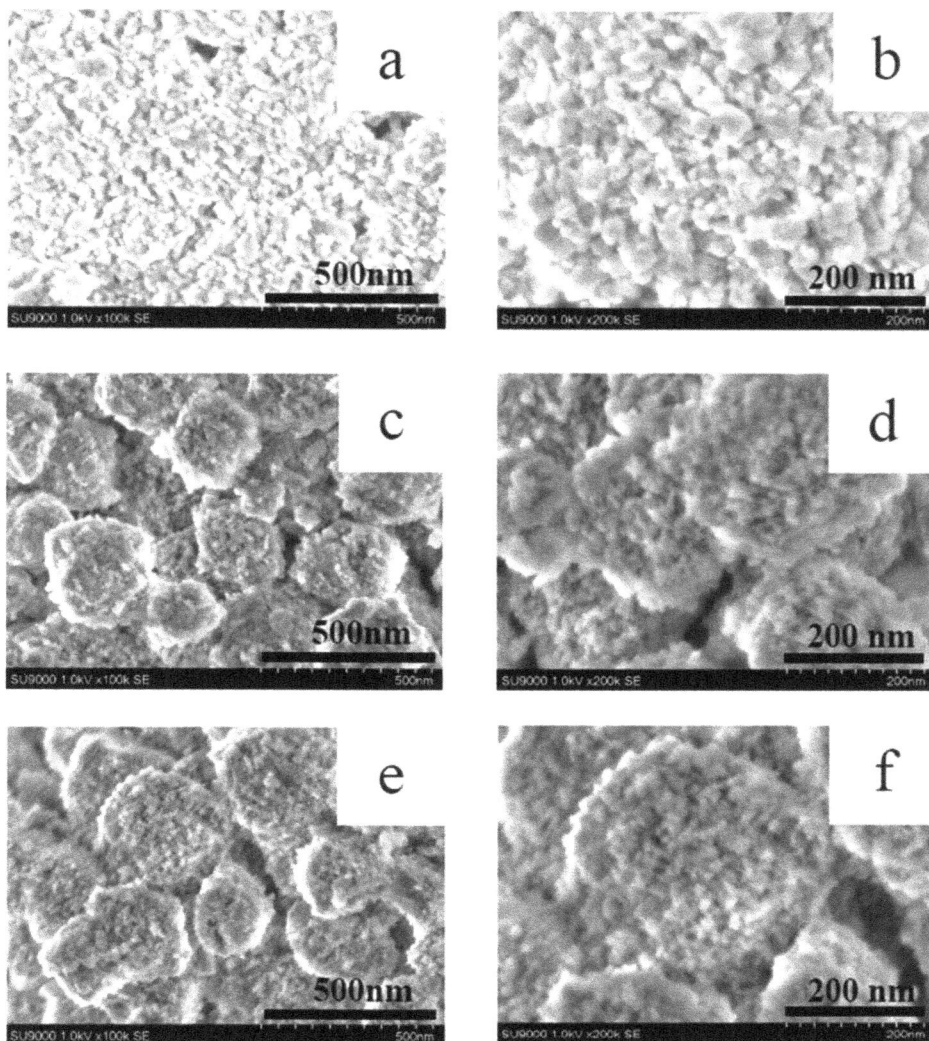

Figure 2. SEM images of HB(12) (**a,b**); HB(77) (**c,d**); and HB(77)-NT(2 M, 16 h) (**e,f**) catalysts.

2.1.3. N_2 Adsorption-Desorption

Figures 4 and 5 show the N_2 adsorption-desorption isotherms and the corresponding pore size distributions of the HB(12), HB(77) and post-treated HB zeolites. Both of the HB(12) and HB(77) displayed the type I + IV isotherm, suggesting they contained a hierarchical porous system consisting of micropores and mesopores (Figure 4A). Moreover, the high N_2 adsorption at high P/P_0 (>0.9) for these two samples was attributed to the filling of the mesopores, originating from the presence of interparticle voids in the materials [26]. Correspondingly, broad peaks centered at *ca.* 4 or 3 nm in the pore size distribution patterns were observed for the HB(12) and HB(77), respectively (Figure 4B).

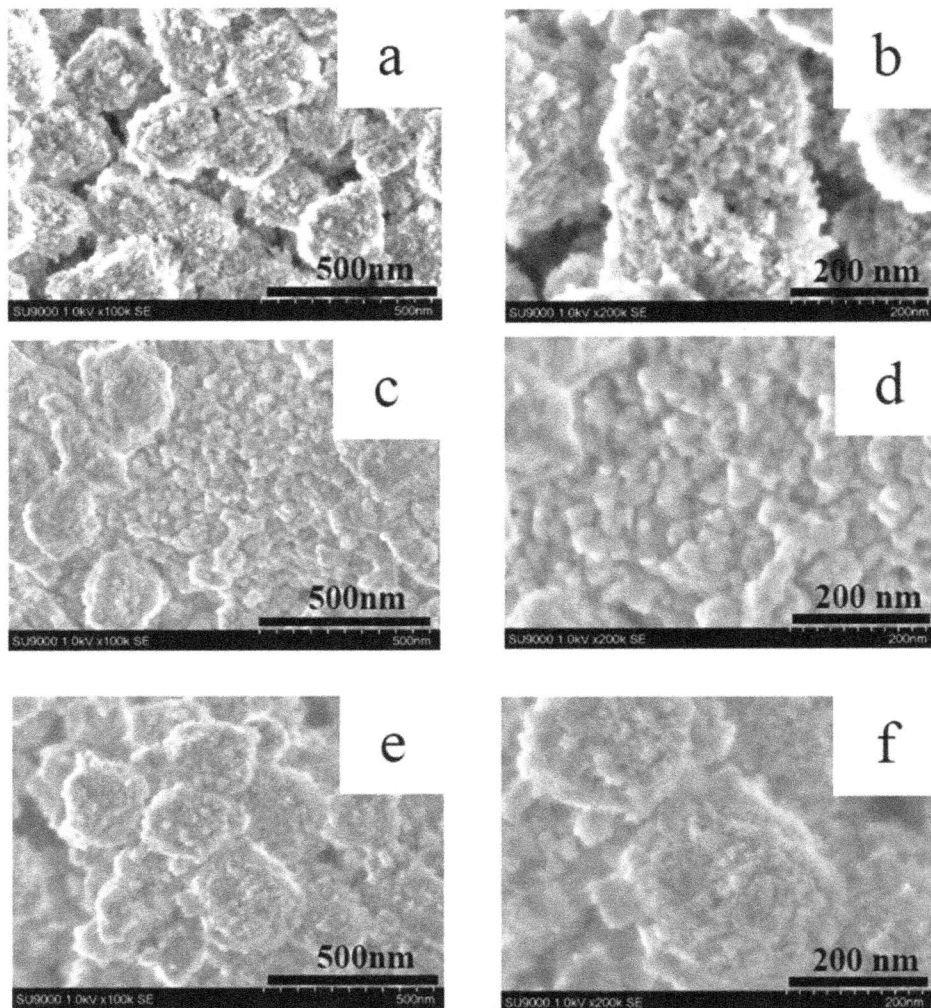

Figure 3. SEM images of HB(77)-AT(0.05 M) (**a,b**); HB(77)-AT(0.2 M) (**c,d**); and HB(77)-AT(0.05 M)-NT(1 M) (**e,f**) catalysts.

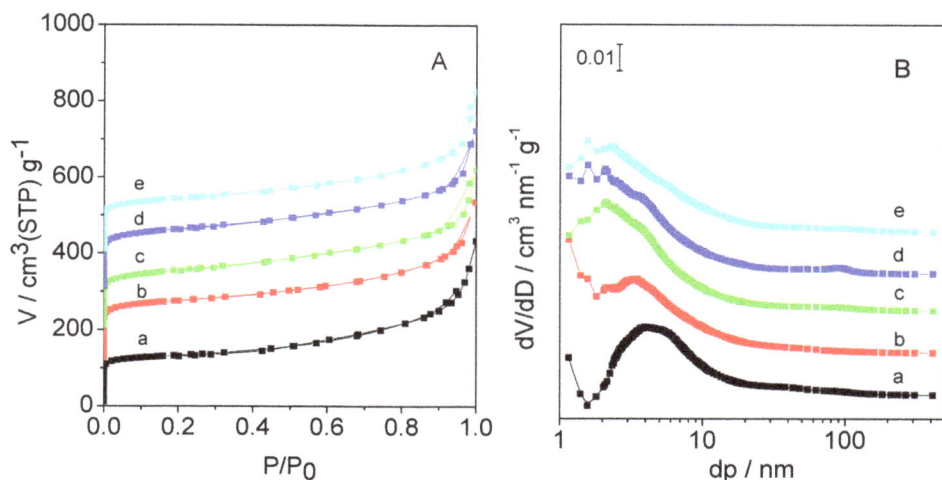

Figure 4. N_2 adsorption and desorption isotherms (**A**) and pore size distribution (**B**) of HB(12) (**a**); HB(77) (**b**); HB(77)-NT(1 M, 1 h) (**c**); HB(77)-NT(2 M, 1 h) (**d**); and HB(77)-NT(2 M, 16 h) (**e**) catalysts.

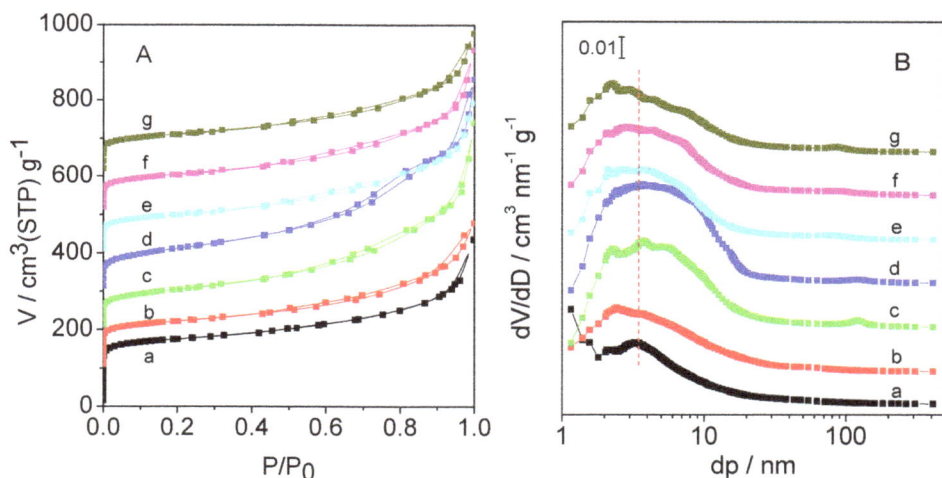

Figure 5. N_2 adsorption-desorption isotherms (**A**) and pore size distribution (**B**) of HB(77) (**a**); HB(77)-AT(0.05 M) (**b**); HB(77)-AT(0.1 M) (**c**); HB(77)-AT(0.2M) (**d**); HB(77)-AT(0.05 M)-NT(0.2 M) (**e**); HB(77)-AT(0.05 M)-NT(1 M) (**f**); and HB(77)-AT(0.05 M)-NT(2 M) (**g**) catalysts.

Compared to the HB(77), the acid-treated HB(77) only showed a small difference at the higher P/P_0 (>0.9), meaning that the porosity properties almost remained after HNO_3 treatment (Figure 4A). Note that the peak in the pore size distribution pattern was slightly broadened and the center of the peak was slightly shifted to the low value after the acid treatment (Figure 4B). On the other hand, the alkali-treated HB(77), especially HB(77)-AT(0.2 M), displayed a new weak hysteresis loop at P/P_0 of 0.5–0.9, indicating that new mesopores were formed by NaOH treatment (Figure 5A). Correspondingly, the peaks in the pore size distribution pattern were broadened, suggesting the presence of inhomogeneous mesopores (Figure 5B).

The isotherms of the alkali-acid-treated HB, especially HB(77)-AT(0.05 M)-NT(1 M), showed increasing step due to multilayer adsorption at P/P_0 (>0.5) and slightly enlarged hysteresis loops at P/P_0 of 0.5–0.9, indicating that the mesopores were increased by the subsequent HNO_3 treatment. A reason may be that amorphous Si and/or Al species formed by alkali treatment were partially removed by the subsequent acid treatment. Correspondingly, broad peaks centered at $ca.$ 3–4 nm in the pore size distribution pattern were also observed for the alkali-acid-treated HB(77) samples (Figure 5B).

The textural properties of all HB samples obtained from N_2 adsorption isotherms are listed in Tables 1 and 2. Both of the HB(12) and HB(77) had high surface area, micropore volume and external surface area. After HNO_3 treatment, the surface area and the micropore volume of acid-treated HB(77) were slightly decreased, while the external surface area was slightly increased (Table 1).

Table 1. Physicochemical properties of HB(12), HB(77) and acid-treated HB catalysts.

Catalysts	Si/Al [a]	Acid Amount [b]/ $mmol \cdot g^{-1}$	Textual Properties				
			S_{BET} / $m^2 \cdot g^{-1}$	S_{ext} [d] / $m^2 \cdot g^{-1}$	V_{total} [c] / $cm^3 \cdot g^{-1}$	V_{micro} [d] / $cm^3 \cdot g^{-1}$	V_{meso} [e] / $cm^3 \cdot g^{-1}$
HB(12)	12	0.76	510	157	0.61	0.18	0.43
HB(77)	77	0.16	670	170	0.64	0.22	0.42
HB(77)-NT(1 M, 1 h)	80	0.16	650	176	0.63	0.20	0.43
HB(77)-NT(2 M, 1 h)	104	0.12	611	200	0.62	0.18	0.44
HB(77)-NT(2 M, 16 h)	160	0.07	590	210	0.61	0.17	0.44

[a] Molar ratio determined by ICP; [b] Determined by NH_3-TPD; [c] Total volume for pores below $p/p_0 = 0.99$; [d] Calculated by t-plot method; [e] $V_{meso} = V_{total} - V_{micro}$.

On the other hand, the surface area and micropore volume were decreased after the 0.05 M NaOH treatment, while the external surface area was slightly increased. Note that the total pore volume was slightly decreased, however, the mesopore volume was almost unchanged (Table 2). By increasing the NaOH concentration to 0.1 or 0.2 M, the total volume and mesopore volume increased, meaning that a large amount of mesopores were formed by these treatments. However, the surface area and micropore volume were further decreased, suggesting the severe destruction of structure (Table 2).

Table 2. Physicochemical properties of HB(77), alkali-treated and alkali-acid-treated catalysts.

Catalysts	Si/Al [a]	Acid Amount [b] / $mmol \cdot g^{-1}$	Textual Properties				
			S_{BET} / $m^2 \cdot g^{-1}$	S_{ext} [d] / $m^2 \cdot g^{-1}$	V_{total} [c] / $cm^3 \cdot g^{-1}$	V_{micro} [d] / $cm^3 \cdot g^{-1}$	V_{meso} [e] / $cm^3 \cdot g^{-1}$
HB(77)	77	0.16	670	170	0.64	0.22	0.42
HB(77)-AT(0.05 M)	71	0.14	446	212	0.55	0.12	0.43
HB(77)-AT(0.1 M)	51	0.13	360	259	0.74	0.05	0.69
HB(77)-AT(0.2 M)	52	0.14	376	300	0.77	0.04	0.73
HB(77)-AT(0.05 M)-NT(0.2 M)	137	0.08	354	214	0.57	0.06	0.51
HB(77)-AT(0.05 M)-NT(1 M)	116	0.07	374	220	0.61	0.07	0.54
HB(77)-AT(0.05 M)-NT(2 M)	105	0.06	376	210	0.56	0.08	0.48

[a] Molar ratio determined by ICP; [b] Determined by NH_3-TPD; [c] Total volume for pores below $p/p_0 = 0.99$; [d] Calculated by t-plot method; [e] $V_{meso} = V_{total} - V_{micro}$.

The surface area and micropore volume were further decreased after 0.05 M NaOH treatment followed by 0.2 M HNO_3 treatment. The possible reason for the decreases is that the framework of zeolite became unstable by alkali treatment, and it would be easily destroyed by the subsequent acid treatment. However, the total volume remained and the mesopore volume slightly increased.

2.1.4. Elemental Analysis

As shown in Table 1, the Si/Al ratio of the acid-treated HB(77) samples was gradually increased with an increase in the HNO_3 concentration or treatment time. The bulk Si/Al ratio of 160 was achieved by 2 M HNO_3 treatment for 16 h. These facts suggest that the direct acid treatment is an efficient dealumination method for Beta zeolite. This may be due to the presence of stacking defects and three dimensional 12-MR channels, which are beneficial to the removal of Al atoms from the framework and/or extra-framework [13].

On the other hand, as shown in Table 2, the Si/Al ratio was slightly decreased to 71 after 0.05 M NaOH treatment. By increasing the NaOH concentration to 0.1 or 0.2 M, the Si/Al ratio was decreased to *ca.* 50, meaning that the extraction of Si atom was more easily than Al atoms, especially during the higher concentrated NaOH treatment. Note that the Si/Al ratio was drastically increased to 137 after 0.05 M NaOH treatment followed by 0.2 M HNO_3 treatment. This result was different from the facts that the Si/Al ratio was slightly increased after the direct HNO_3 treatment (Si/Al = 80 for HB(77)-NT(1 M)). Thus, it can be concluded that the alkali treatment even under mild conditions (*i.e.*, 0.05 M NaOH treatment at room temperature for 0.5 h) would lead to the formation of many stacking defects or some mesopores which can facilitate the removal of Al atoms from the framework or extra-framework.

2.1.5. NH$_3$-TPD

Figures 6 and 7 show the NH$_3$-TPD profiles of the HB(12), HB(77) and post-treated HB samples. All these profiles were composed of two desorption peaks, the so-called "*l*-peak" and "*h*-peak". The *l*-peak corresponds to NH$_3$ adsorbed on non-acidic -OH groups and on NH$_4^+$, which forms by the reaction of NH$_3$ and Brønsted acid sites. On the other hand, the *h*-peak corresponds to NH$_3$ absorbed on true acid sites [27,28]. Thus, the acid amounts were estimated by the *h*-peak areas and are listed in Tables 1 and 2.

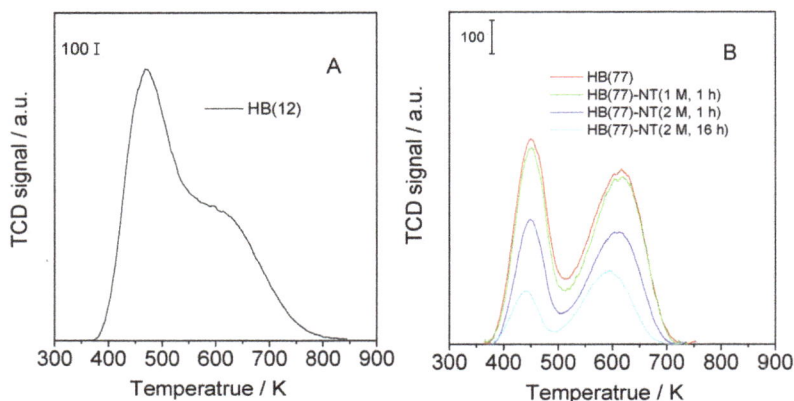

Figure 6. NH$_3$-TPD profiles of HB(12) (**A**); HB(77) and acid-treated HB (**B**) catalysts.

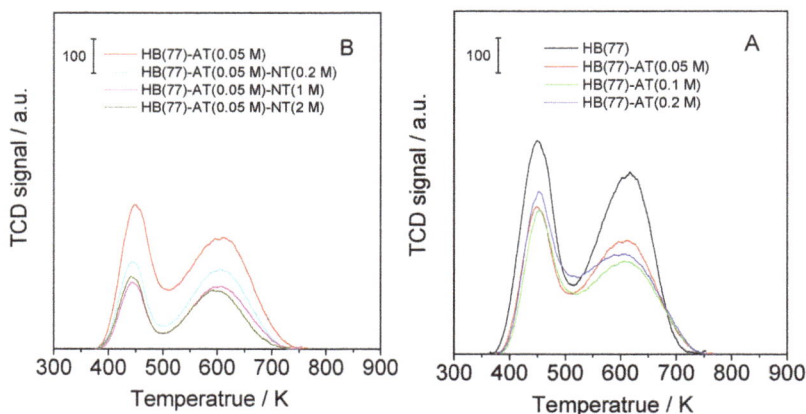

Figure 7. NH$_3$-TPD profiles of HB(77), alkali-treated HB (**A**); and alkali-acid-treated HB (**B**) catalysts.

Obviously, the HB(12) had the highest acid amount among all the HB samples. In addition, the *l*-peak and *h*-peak were seriously overlapped for the HB(12), however, they can be easily separated for the HB(77) (Figure 6). This fact suggests that the HB(12) had a wider distribution of acid strength than the HB(77). As shown in Figure 6B and Table 1, the acid amount was decreased after the HNO_3 treatment, indicating dealumination occurred during the acid treatment. Moreover, the acid amount was gradually decreased along with the HNO_3 concentration or treatment time. On the other hand, as shown in Figure 7A and Table 2, the acid amount was also decreased after NaOH treatment, indicating dealumination occurred accompanied with desilication during the alkali treatment procedure. Note that the amount of acid of HB(77)-AT(0.2 M) was slightly higher than that of HB(77)-AT(0.1 M) (Figure 7A, Table 2). This may be due to the increase in the LAS generated by the higher concentrated NaOH treatment (see in Section 2.1.6). In addition, the acid amount was further decreased after 0.05 M NaOH treatment followed by HNO_3 treatment (Figure 7B, Table 2), suggesting further dealumination.

2.1.6. Pyridine Adsorption FT-IR

To clarify the acidic properties in detail, the pyridine adsorption FT-IR spectra of the HB(12), HB(77) and post-treated HB samples were measured. In the region of hydroxyl stretching vibration (Figure 8A), the HB(12) exhibited three main bands at 3781, 3741 and 3603 cm^{-1}, which are attributed to AlOH hydroxyls, isolated external silanols and structural Si(Al)OH hydroxyls, respectively [29]. The HB(77) exhibited two main bands at 3730 and 3603 cm^{-1}, which are attributed to isolated internal silanols and structural Si(Al)OH hydroxyls, respectively [29]. After HNO_3 treatment, the band at 3603 cm^{-1} decreased in intensity, however, the band at 3730 cm^{-1} increased. Additionally, HB(77)-NT(2 M, 16 h) showed a broad band at 3700–3500 cm^{-1} ascribed to hydrogen-bonded hydroxyl groups [29]. These facts suggest that the silanols were increased by the dealumination with HNO_3 treatment. On the other hand, the band at 3603 cm^{-1} was slightly decreased in intensity after NaOH treatment (Figure 9A), indicating that the dealumination occurred accompanied by desilication during the alkali treatment. However, the band at 3741 cm^{-1} increased in intensity after NaOH treatment. The increase in the isolated external silanols was consistent with the increase in the external surface area after alkali treatment (Table 2). In addition, the band at 3603 cm^{-1} further decreased in intensity after 0.05 M NaOH treatment followed by 0.2 M HNO_3 treatment, meaning that further dealumination occurred. Note that the band at 3730 cm^{-1} decreased in intensity for HB(77)-AT(0.05 M)-NT(2 M). This may be related to the repair of the defect sites by the amorphous Si and/Al species generated by the alkali treatment in the acidic medium [30]. It can also be explained for the increases in its crystallinity,

Si/Al ratio and microporous volume (Figures 1B and 5, Table 2). However, the repair of the defect sites did not increase its acid amount (Figure 7B, Table 2).

In the region of pyridine ring stretching vibration (Figures 8B and 9B), all the HB samples exhibited the bands at 1542 and 1453 cm^{-1}, which are attributed to BAS and LAS, respectively [29]. Compared to the HB(12), the HB(77) had a smaller amount of acid sites especially the LAS. After acid treatment, both of the amounts of BAS and LAS were gradually decreased with an increase in the HNO$_3$ concentration or treatment time. Moreover, the LAS were removed more readily than the BAS during the acid treatment [7]. Note that a new band around 1446 cm^{-1} was observed for HB(77)-NT(2 M, 16 h). This band could be the pyridine molecules coordinated with the LAS and also interaction through the H bond with the BAS, this interaction causing a shift of the corresponding bands to lower frequencies [29]. Thus, it was concluded that both acid amount and acid type of HB(77) were changed after the acid treatment.

After the NaOH treatment, the amount of BAS gradually decreased while that of the LAS gradually increased along with the NaOH concentration, indicating that dealumination occurred accompanied by desilication, and that the extra-framework Al atoms generated by the dealumination acted as the LAS. The similar results were also reported in our previous work on the desilication of ZSM-5 zeolite by alkali treatment [20]. However, the amount of LAS was drastically decreased after the subsequent HNO$_3$ treatment for HB(77)-AT(0.05 M), suggesting that the LAS generated by alkali treatment were easily removed [20]. Moreover, both the amount of BAS and LAS gradually decreased with an increase in HNO$_3$ concentration.

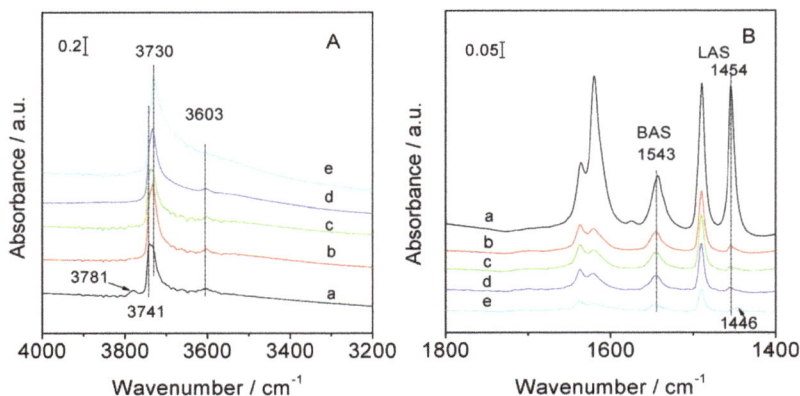

Figure 8. IR spectra of HB(12) (**a**); HB(77) (**b**); HB(77)-NT(1 M, 1 h) (**c**); HB(77)-NT(2 M, 1 h) (**d**); and HB(77)-NT(2 M, 16 h) (**e**) catalysts before (**A**) and after (**B**) pyridine was absorbed and then desorbed at 423 K for 1 h.

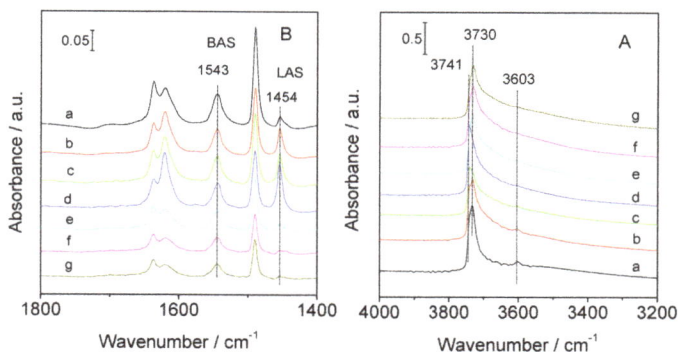

Figure 9. IR spectra of HB(77) (**a**); HB(77)-AT(0.05 M) (**b**); HB(77)-AT(0.1 M) (**c**); HB(77)-AT(0.2M) (**d**); HB(77)-AT(0.05 M)-NT(0.2 M) (**e**); HB(77)-AT(0.05 M)-NT(1 M) (**f**); and HB(77)-AT(0.05 M)-NT(2 M) (**g**) catalysts before (**A**) and after (**B**) pyridine was absorbed and then desorbed at 423 K for 1 h.

2.2. Catalytic Cracking of n-Hexane over Different HB Zeolite Catalysts

2.2.1. Al-Rich HB(12) Catalysts

Figure 10 shows the change in n-hexane conversion with time on stream (TOS) at different reaction temperatures for the Al-rich HB(12) catalyst. The initial n-hexane conversion at TOS of 15 min was increased along with the reaction temperature. At the high temperature of 923 K, the initial n-hexane conversion was nearly 100%; however, it sharply decreased to *ca.* 40% at TOS of 90 min. Moreover, the n-hexane conversion further decreased to *ca.* 23% at TOS of 210 min. This value was a little higher than that of thermal cracking under these reaction conditions [7], meaning that the catalyst was almost deactivated.

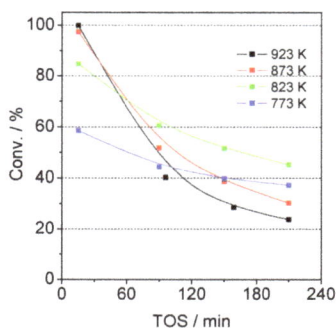

Figure 10. Change in n-hexane conversion with time on stream (TOS) at different reaction temperatures for HB(12) catalysts. Reaction conditions: Cat., 0.1 g; $P_{n\text{-hexane}} = 6$ kPa; $W/F_{n\text{-hexane}} = 64$ g h/mol; temp., 773–923 K.

87

Figure 11 shows the changes in the products selectivities at the initial stage with reaction temperature for the Al-rich HB(12) catalyst. The selectivity to propylene was almost constant at *ca.* 24 C-% in the temperature ranging between 773 and 873 K, and it was decreased to *ca.* 17 C-% at 923 K. Similarly, the selectivities to butenes were also gradually decreased along with the temperature, accompanied by a dramatic increase in the selectivities to BTX. These facts suggest that propylene and butenes are easily transformed into aromatics at high reaction temperature. The selectivities to propane and butanes were drastically decreased along with the reaction temperature, suggesting that the subsequent reactions of propane and butanes took place at high temperatures, resulting in the formation of alkenes, BTX and so on [31]. On the other hand, the selectivity to ethylene was steeply increased along with the reaction temperature. Because ethylene is formed via primary carbenium ions [12], the apparent activation energy for the ethylene formation should be high. Thus, the higher reaction temperature must be of benefit to the ethylene formation. In addition, the selectivities to methane and ethane, which are formed solely via the monomolecular mechanism [12], were also increased along with the reaction temperature. These findings clearly indicate that the cracking via the monomolecular mechanism is more predominant at the high reaction temperature.

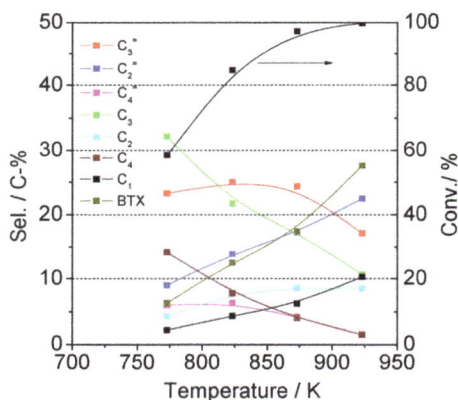

Figure 11. Effect of reaction temperature on the catalytic cracking of *n*-hexane for HB(12) catalyst. Reaction conditions: Cat., 0.1 g; $P_{n\text{-hexane}}$ = 6 kPa; $W/F_{n\text{-hexane}}$ = 64 g h/mol; temperature 773–923 K, TOS = 15 min. Products abbreviations: propylene ($C_3^=$); ethylene ($C_2^=$); butenes ($C_4^=$); propane (C_3); ethane (C_2); butanes (C_4); methane (C_1); benzene, toluene and xylene (BTX).

Figure 12 shows the changes in the products' selectivities with *n*-hexane conversion for the Al-rich HB(12) catalysts at 923 K. The selectivities to propylene and butenes were drastically decreased while the selectivities to BTX were drastically increased with an increase in the *n*-hexane conversion, suggesting that BTX are formed mainly from propylene and butenes. The propylene selectivity was decreased from *ca.* 32 *to* 17 C-% when the *n*-hexane conversion was increased from *ca.* 88% to 100%. On the other hand, the selectivity to ethylene was increased while the selectivities to butanes were decreased with an increase in the *n*-hexane conversion. These facts suggest that ethylene would be formed not only by *n*-hexane cracking but also by butane cracking. Meanwhile, the selectivities to methane and ethane were slightly increased with an increase in the *n*-hexane conversion.

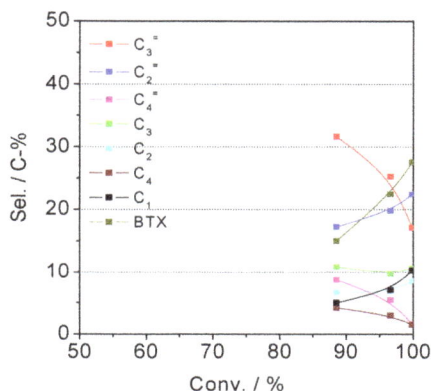

Figure 12. Change in products' selectivities with *n*-hexane conversion on HB(12) catalysts. Reaction conditions: $P_{n\text{-hexane}} = 6$ kPa; $W/F_{n\text{-hexane}} = 12.8\text{–}64$ g h/mol; temperature 923 K, TOS = 15 min. Products abbreviations: see Figure 11.

2.2.2. Comparison of HB(77) and Acid-Treated HB Catalysts

Figure 13 shows the change in *n*-hexane conversion with TOS for the HB(77) and acid-treated HB(77) catalysts at 923 K. A slight decrease in the *n*-hexane conversion from 100% at TOS of 15 min to *ca.* 95% at TOS of 90 min was observed for the HB(77). Obviously, the catalytic stability of the HB(77) was higher than that of the Al-rich HB(12) (Figure 10). However, the *n*-hexane conversion was decreased to *ca.* 73% at TOS of 210 min, and the catalytic stability was still not satisfactory. Thus, the catalytic performances of dealuminated HB zeolites with a much lower amount of acid were considered.

All the acid-treated HB(77) catalysts showed high initial *n*-hexane conversions (>97%). Compared to the HB(77), HB(77)-NT(1 M, 1 h) showed slightly higher catalytic stability. By increasing the HNO$_3$ concentration and treatment time, the catalytic stability was further improved. A slight decrease in *n*-hexane conversion from *ca.* 97% at TOS of 15 min to *ca.* 93% at TOS of 210 min was observed for HB(77)-NT(2 M, 16 h).

89

Figure 14 shows the changes in the products' selectivities with n-hexane conversion for the HB(77) and acid-treated HB(77) catalysts. Compared to the Al-rich HB(12) (Figure 12), the HB(77) showed a high selectivity to propylene of *ca.* 33 C-% with a n-hexane conversion of nearly 100% (Figure 14A), and it also showed high selectivities to butenes, butanes and propane and lower selectivities to BTX and ethylene. When the HB(77) was dealuminated by acid treatment, the selectivities to propylene and butenes at high n-hexane conversions were further increased, however, those to ethylene and BTX further were decreased. For example, the constant selectivity to propylene of *ca.* 46 C-% even at a high n-hexane conversion of *ca.* 97% was achieved for the HB(77)-NT(2 M, 16 h) (Figure 14D).

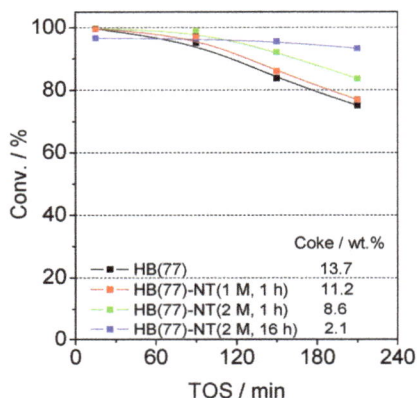

Figure 13. Change in n-hexane conversion with time on stream (TOS) for HB(77) and acid-treated HB catalysts. Reaction conditions: Cat., 0.1 g; $P_{n\text{-hexane}}$ = 6 kPa; $W/F_{n\text{-hexane}}$ = 64 g h/mol; temperature 923 K.

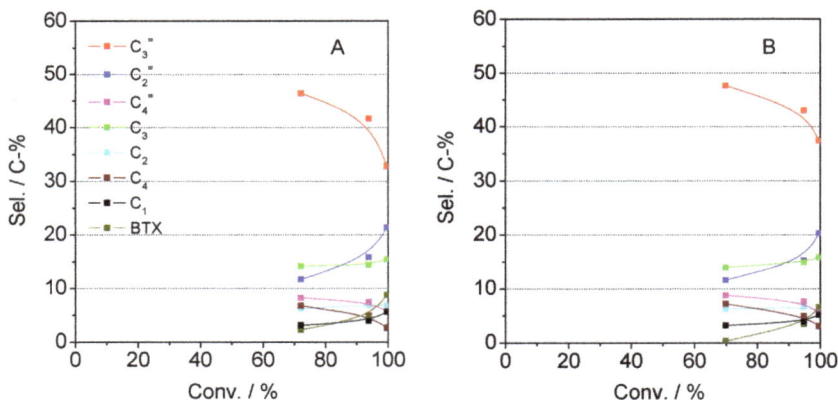

Figure 14. *Cont.*

90

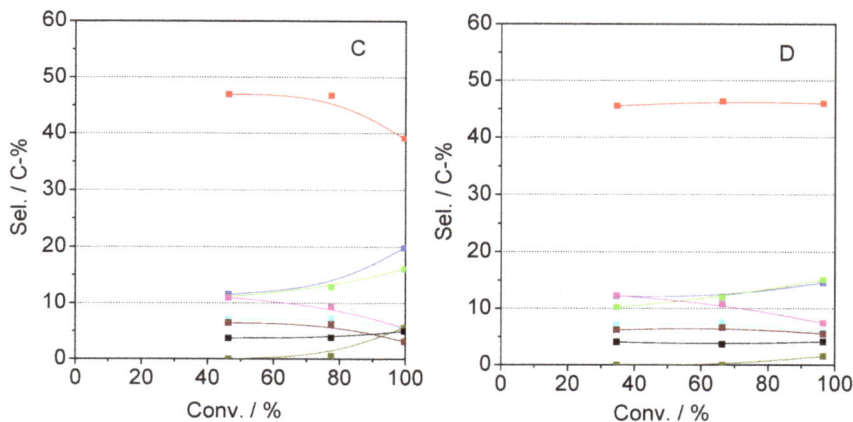

Figure 14. Changes in products' selectivities with *n*-hexane conversion on HB(77) (**A**); HB(77)-NT(1 M, 1 h) (**B**); HB(77)-NT(2 M, 1 h) (**C**); and HB(77)-NT(2 M, 16 h) (**D**) catalysts. Reaction conditions: see Figure 12. Products abbreviations: see Figure 11.

2.2.3. Comparison of HB(77), Alkali-Treated HB, and Alkali-Acid-Treated HB Catalysts

Figure 15 shows the change in *n*-hexane conversion with TOS for the HB(77), alkali-treated, and alkali-acid-treated HB(77) catalysts at 923 K. Compared to the HB(77), HB-AT(0.05 M) and HB-AT(0.1 M) still showed a nearly 100% initial *n*-hexane conversion. However, the catalytic stability was not improved after these treatments. Note that HB(77)-AT(0.2 M) showed a slightly low initial *n*-hexane conversion of *ca.* 87% and a relatively slow deactivation rate (Figure 15A). On the other hand, the catalytic stability of alkali-treated HB(77) can be improved by subsequent HNO_3 treatment, although the initial *n*-hexane conversion slightly decreased. For example, a slight decrease in *n*-hexane conversion from *ca.* 88% at TOS of 15 min to *ca.* 85% at TOS of 210 min was observed for HB-AT(0.05 M)-NT(1 M) (Figure 15B).

Figure 16 shows the changes in products' selectivities with *n*-hexane conversion for the HB(77), alkali-treated and alkali-acid-treated HB(77) catalysts. Compared to the HB(77) (Figure 16A), HB(77)-AT(0.05 M) showed a slightly higher propylene selectivity of *ca.* 36 C-% at the *n*-hexane conversion of nearly 100% (Figure 16B). Meanwhile, the HB-AT(0.05 M) also showed higher selectivities to BTX and lower selectivity to propane than the HB(77) in high *n*-hexane conversions. By increasing the NaOH concentration to 0.2 M, the selectivities to propylene and butenes in high *n*-hexane conversions were increased while the selectivities to ethylene and propane were decreased. The HB(77)-AT(0.2 M) showed a higher selectivity to propylene of *ca.* 40 C-% even with a high *n*-hexane conversion of *ca.* 97% (Figure 16C). On the other hand, when the HB(77) was treated by 0.05 M NaOH followed by HNO_3

treatment, the selectivities to propylene and butenes in high n-hexane conversions drastically increased, however, those to ethylene and BTX drastically decreased. For example, the HB(77)-AT(0.05 M)-NT(1 M) showed a higher selectivity of *ca.* 48 C-% to propylene in a high n-hexane conversion of *ca.* 90% (Figure 16D).

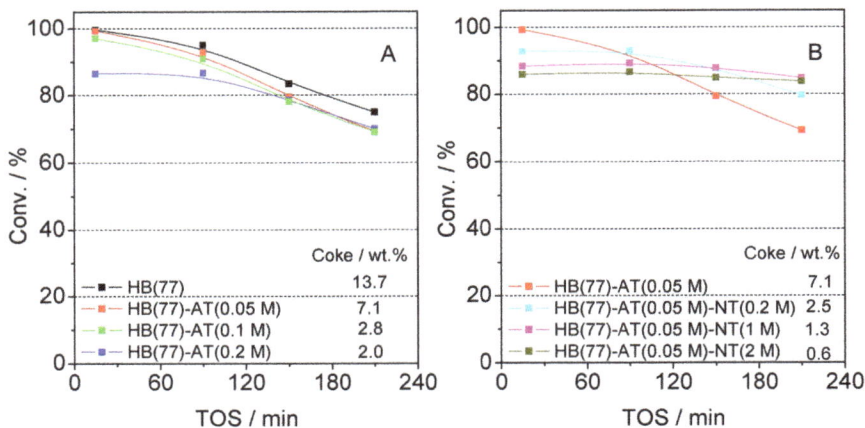

Figure 15. Change in n-hexane conversion with time on stream (TOS) for HB(77), alkali-treated HB (**A**) and alkali-acid-treated HB (**B**) catalysts. Reaction conditions: see Figure 13.

In summary, the HB(77) with a lower amount of acid showed higher selectivity to propylene with high n-hexane conversions and a higher catalytic stability at a high reaction temperature than the Al-rich HB(12). The catalytic performance of HB(77) including propylene selectivity and catalytic stability can be further improved by acid treatment. On the other hand, selectivity to propylene in high n-hexane conversions can also be improved by alkali treatment; however, catalytic stability cannot be improved. Furthermore, both selectivity to propylene in high n-hexane conversions and catalytic stability can be improved after alkali treatment followed by acid treatment, although the initial n-hexane conversion was slightly decreased.

2.3. Effect of Dealumination and Desilication on the Catalytic Performance

In this section, catalytic properties, especially catalytic stability and propylene selectivity in high n-hexane conversions for different HB catalysts, were related to their physicochemical properties (including porosity and acidity), and the effects of dealumination and desilication on their catalytic performances in n-hexane cracking were discussed.

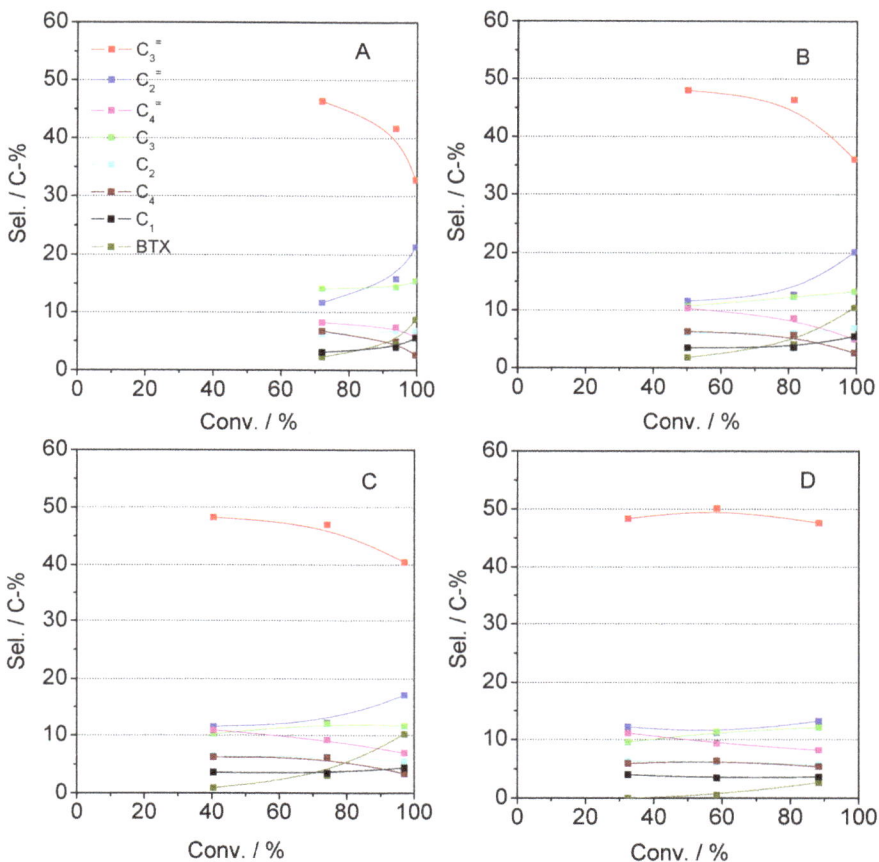

Figure 16. Changes in products' selectivities with *n*-hexane conversion on HB(77) (**A**); HB(77)-AT(0.05 M) (**B**); HB(77)-AT(0.2 M) (**C**); and HB(77)-AT(0.05 M)-NT(1 M) (**D**) catalysts. Reaction conditions: see Figure 12. Products abbreviations: see Figure 11.

2.3.1. Effect of Dealumination

As shown in Figures 10–12 Al-rich HB(12) showed a poor catalytic stability, a low propylene selectivity and a high BTX selectivity at the high temperature of 923 K. The inferior catalytic performance can be explained by two possible reasons. One reason may be related to the large amount of acid sites (Table 1), which would accelerate the secondary reactions of propylene and butenes and the hydride transfer, leading to a large amount of coke being formed (163 mg/g-catalyst at TOS of 210 min). The other reason may be related to the special structure of Beta zeolite. The polymerization of olefins and the successive formation of coke components may predominantly occur in the three dimensional 12-MR micropores and their wide intersections [6]. Note

93

that the HB(12) composed of nanoparticles had a higher external surface area and mesopore volume (Figure 2, Table 1) which are beneficial to the products' diffusion. Therefore, we concluded that the poor catalytic performance of HB(12) was mainly related to its acidic properties.

Compared to Al-rich HB(12), HB(77) showed high selectivities to propylene and butenes, however, low selectivities to BTX, ethylene, propane and butanes (Figures 13 and 14A) were demonstrated. It could be due to the smaller amount of acid sites, especially the LAS (Table 1, Figures 6 and 8B). Note that, although the HB(77) had superior catalytic stability, the amount of coke formed on the HB(77) (137 mg/g-catalyst at TOS of 210 min) was comparable to that of Al-rich HB(12).

As shown in Figures 13 and 14 catalytic stability and propylene selectivity with high n-hexane conversions can be greatly improved by dealumination with acid treatment. Note that crystal morphology and size were almost unchanged and the textural properties remained after the acid treatment (Figure 2, Table 1). The improvement in catalytic performance would be due to a further decrease in the amount of acid sites, especially the LAS (Table 1, Figures 6B and 8B), resulting in suppression of the hydride transfer and coke formation (e.g., 21 mg/g-catalyst at TOS of 210 min for HB(77)-NT(2 M, 16 h)). This conclusion was consistent with our previous work on n-hexane cracking over dealuminated H-MCM-22 zeolites [7].

2.3.2. Effect of Desilication

As shown in Figure 15A, catalytic stability was not improved after 0.05 or 0.1 M NaOH treatment; however, it was slightly improved after 0.2 M NaOH treatment, although the initial n-hexane conversion was slightly decreased. This may be related to the partially destroyed structure (Figures 1B and 5, Table 2). On the other hand, the catalytic stability was remarkably improved after 0.05 M NaOH treatment followed by HNO_3 treatment, although the initial n-hexane conversion was slightly decreased (Figure 15B). This may be due to the decrease in the number of acid sites (Table 2, Figure 9), leading to a smaller amount of coke being formed (e.g., 13 mg/g-catalyst at TOS of 210 min for HB(77)-AT(0.05 M)-NT(1 M)). In addition, the increase in the mesopores formed by the subsequent acid treatment would be beneficial to the diffusion of coke precursor (Table 2).

As to the products' distribution, the propylene selectivity was slightly increased after 0.05 M NaOH treatment. Meanwhile, the BTX selectivities were increased while the propane selectivity was decreased (Figure 16B). This fact suggested that the dehydrogenation of propane and subsequent aromatization of olefins were improved by 0.05 M NaOH treatment, due to the increase in the amount of LAS (Figure 9B). Similar results were observed in our work on n-hexane cracking over alkali-treated H-ZSM-5 zeolites [20]. By increasing the NaOH concentration to 0.2 M, the propane selectivity was further decreased. This is due to the further increase in the amount of LAS which accelerated the dehydrogenation of propane. However, the BTX selectivities remained, and the selectivities to propylene and butenes increased while the selectivity to ethylene decreased. It can be explained by two possible reasons. One possible reason would be related to the decrease in the amount of BAS (Figure 9B), which suppressed the secondary reactions of propylene and butenes. Another possible reason may be due to the strong increase in the mesopores after 0.2 M NaOH treatment (Table 2, Figure 5), which was beneficial to the diffusion of the formed olefins and the resultant lower amount of coke being formed (20 mg/g-catalyst at TOS of 210 min). This finding differed from our previous results on n-hexane cracking over alkali-treated H-ZSM-5 zeolites, in which the amount of coke used for alkali-treated catalyst was larger than that for the untreated one [20].

On the other hand, the selectivities to propylene and butenes in high n-hexane conversions were remarkably increased; however, those to ethylene and BTX were drastically decreased after 0.05 M NaOH treatment followed by acid treatment. These findings indicated that the secondary reactions to form BTX from propylene and butenes and the subsequent cracking reaction of butanes and/or propane to form ethylene could be further suppressed, due to the dramatic decrease in the number of acid sites, especially the LAS (Table 2, Figure 9). In addition, another possible reason would be related to the increase in the mesopores during the subsequent acid treatment (Table 2, Figure 5), which was beneficial to the diffusion of the formed olefins.

3. Experimental Section

3.1. Catalyst Preparation

3.1.1. Dealumination of H-Beta

The H-Beta (Si/Al = 12) and H-Beta (Si/Al = 77) zeolites, denoted as HB(12) and HB(77), were supplied from the Catalysis Society of Japan and Sud-Chemie AG. Munchen, Germany, respectively. The HB(77) was used as the parent material and dealuminated by acid treatment. Specifically, 1 g HB(77) was suspended and stirred in a 1–2 M HNO_3 (60%, Wako, Osaka, Japan) solution at 393 K for 1 or 16 h. The weight ratio of solid to liquid ratio was 1:30. The acid-treated HB samples were denoted as HB(77)-NT(a M, b h), where a and b were the HNO_3 concentration and treatment time, respectively.

3.1.2. Desilication of H-Beta

The HB(77) zeolite was used as the parent material and desilicated by alkali treatment. Specifically, 1 g HB(77) was suspended and stirred in a 0.05–0.2 M NaOH (97%, Wako, Osaka, Japan) solution at room temperature for 0.5 h. The weight ratio of solid to liquid ratio was 1:30. The samples were filtered, washed, and dried at 373 K and then converted to the H-type ones with 1 M NH_4NO_3 (99%, Wako, Osaka, Japan) twice at 353 K for 2 h, followed by calcination in air at 773 K for 2 h. The alkali-treated HB samples were denoted as HB(77)-AT(c M), where c was the NaOH concentration. On the other hand, the alkali-treated HB sample was further dealuminated by acid treatment. Specifically, 1 g HB(77)-AT(0.05 M) was suspended and stirred in a 0.2–2 M HNO_3 solution at 393 K for 1 h. The weight ratio of solid to liquid ratio was 1:30. Thus obtained samples were filtered, washed, and dried at 373 K. The samples were denoted as HB(77)-AT(c M)-NT (d M), where d was the HNO_3 concentration.

3.2. Catalyst Characterization

The catalysts were characterized by various techniques. The Si/Al ratio was determined by inductively coupled plasma-atomic emission spectrometer (ICP-AES) on a Shimadzu ICPE-9000 (Shimadzu, Kyoto, Japan). The X-ray diffraction (XRD) patterns were recorded on a Rint-Ultima III (Rigaku, Tokyo, Japan) using a Cu Kα X-ray source. The N_2 adsorption was carried out at 77 K on a Belsorp-mini II (BEL, Osaka, Japan). The crystal morphology and size were measured by field emission scanning electron microscopy (FE-SEM) on a Hitachi SU-9000 microscope (Hitachi, Tokyo, Japan). Thermogravimetric analyses (TGA) were conducted on a Rigaku Thermo plus EVO II equipment (Rigaku, Tokyo, Japan) in air atmosphere.

Temperature-programmed desorption of ammonia (NH_3-TPD) profiles were recorded on a BEL Multitrack TPD equipment (BEL, Osaka, Japan). IR spectra were collected on a Jasco FT/IR-6100 spectrometer (Jasco, Tokyo, Japan) equipped with a TGS detector. The IR spectra of adsorbed pyridine were recorded as follows: a self-supported wafer (9.6· mg· cm^{-2} thickness and 2 cm diameter) was set in a quartz IR cell sealed with CaF_2 windows, where it was evacuated at 723 K for 2 h before pyridine adsorption. The adsorption was carried out by exposing the wafer to pyridine vapor (1.3 kPa) at 423 K for 0.5 h. Physisorbed pyridine was removed by evacuation at 423 K for 1 h. IR spectra were collected at 423 K.

3.3. Catalytic Cracking of n-Hexane

The catalytic cracking of n-hexane was carried out with a fixed-bed flow reactor under atmospheric pressure. Typically, 0.1 g of catalyst (18–30 mesh size) was put into a quartz tubular reactor (6 mm inner diameter) and activated in a dry air flow of 20 mL· min^{-1} at 923 K for 1 h. The height of catalyst bed was $ca.$ 1 cm. The temperature was measured by a thermocouple in the top of the catalyst bed. After the reactor temperature was adjusted to the desired one, n-hexane vapor diluted in the helium was fed into the reactor. The initial partial pressure of n-hexane was set at 6 kPa. To investigate the effect of contact time on n-hexane catalytic cracking, the catalyst weight to the n-hexane flow rate ratio ($W/F_{n\text{-hexane}}$) was varied from 12.8 to 64 g h/mol.

The reaction products were analyzed with an on-line gas chromatography (GC-14B, Shimadzu, Kyoto, Japan) with an FID detector and a HP-AL/S column (30 m × 0.535 mm × 15 μm). The n-hexane conversion was calculated based on the peak area in the GC spectra before and after reaction. The selectivities to hydrocarbon products represent the products' distribution and were calculated using area normalization method. Because hydrocarbons generally have molar response factors that are equal to number of carbon atoms in their molecule when using FID, the relative correction factors for different hydrocarbons are approximately 1.0 based on the carbon numbers. Coke amount was calculated by thermogravimetry analysis (TGA). The weight loss from 673 to 1073 K in each TG profile was defined as the contents of coke in the used catalyst.

4. Conclusions

HB(77) exhibited higher catalytic stability and propylene selectivity than Al-rich HB(12), due to its smaller number of acid sites, especially the LAS. Dealumination of HB(77) by acid treatment improved catalytic stability and increased propylene selectivity in high n-hexane conversions. This may be attributed to the decrease in the number of acid sites, especially the LAS which accelerate the secondary reactions of propylene and butenes, the hydride transfer, and the resultant coke formation.

On the other hand, the desilication of HB(77) by alkali treatment did not improve catalytic stability, although mesopores were formed. This may be related to the partially destroyed structure. However, alkali treatment increased propylene selectivity in high n-hexane conversions. Furthermore, alkali treatment followed by acid treatment improved catalytic stability and increased propylene selectivity in high n-hexane conversions. This may be due to the decrease in the number of acid sites and the increase in mesopores which are beneficial to the diffusion of coke precursor.

Acknowledgments: This work was supported by the green sustainable chemistry project of New Energy and Industrial Technology Development Organization (NEDO).

Author Contributions: The present work was conducted under the supervision of Seitaro Namba and Takashi Tatsumi. Toshiyuki Yokoi did the main research consulting and paper editing, and is named the correspondent author of the submitted work. Yong Wang did the experimental work and paper writing.

Conflicts of Interest: The authors declare no conflict of interest.

References

1. Yoshimura, Y.; Kijima, N.; Hayakawa, T.; Murata, K.; Suzuki, K.; Mizukami, F.; Matano, K.; Konishi, T.; Oikawa, T.; Saito, M.; *et al.* Catalytic cracking of naphtha to light olefins. *Catal. Surv. Jpn.* **2000**, *4*, 157–167.

2. Wojciechowski, B.W. The reaction mechanism of catalytic cracking: quantifying activity, selectivity, and catalyst decay. *Catal. Rev. Sci. Eng.* **1998**, *40*, 209–328.

3. Kissin, Y.V. Chemical mechanism of catalytic cracking over solid acidic catalysts: Alkanes and alkenes. *Catal. Rev. Sci. Eng.* **2001**, *43*, 85–146.

4. Rahimi, N.; Karimzadeh, R. Catalytic cracking of hydrocarbons over modified ZSM-5 zeolites to produce light olefins: A review. *Appl. Catal. A* **2011**, *398*, 1–17.

5. Mochizuki, H.; Yokoi, T.; Imai, H.; Watanabe, R.; Namba, S.; Kondo, J.N.; Tatsumi, T. Facile control of crystallite size of ZSM-5 catalyst for cracking of hexane. *Microporous Mesoporous Mater.* **2011**, *145*, 165–171.

6. Inagaki, S.; Takechi, K.; Kubota, Y. Selective formation of propylene by hexane cracking over MCM-68 zeolite catalyst. *Chem. Commun.* **2010**, *46*, 2662–2664.

7. Wang, Y.; Yokoi, T.; Namba, S.; Kondo, J.N.; Tatsumi, T. Catalytic cracking of *n*-hexane for producing propylene on MCM-22 zeolites. *Appl. Catal. A* **2015**, *504*, 192–202.

8. Bellussi, G.; Pazzuconi, G.; Perego, C.; Girotti, G.; Terzoni, G. Liquid-phase alkylation of benzene with light olefins catalyzed by β-zeolites. *J. Catal.* **1995**, *157*, 227–234.

9. Wichterlová, B.; Čejka, J.; Žilková, N. Selective synthesis of cumene and *p*-cymene over Al and Fe silicates with large and medium pore structures. *Microporous Mater.* **1996**, *6*, 405–414.

10. Corma, A.; González-Alfaro, V.; Orchillés, A.V. The role of pore topology on the behaviour of FCC zeolite additives. *Appl. Catal. A* **1999**, *187*, 245–254.

11. Haag, W.O.; Lago, R.M.; Weisz, P.B. The active site of acidic aluminosilicate catalysts. *Nature* **1984**, *309*, 589–591.

12. Corma, A.; Orchillés, A.V. Current views on the mechanism of catalytic cracking. *Microporous Mesoporous Mater.* **2000**, *35–36*, 21–30.

13. Müller, M.; Harvey, G.; Prins, R. Comparison of the dealumination of zeolites beta, mordenite, ZSM-5 and ferrierite by thermal treatment, leaching with oxalic acid and treatment with $SiCl_4$ by ^1H, ^{29}Si and ^{27}Al MAS NMR. *Microporous Mesoporous* **2000**, *34*, 135–147.

14. Baran, R.; Millot, Y.; Onfroy, T.; Krafft, J.M.; Dzwigaj, S. Influence of the nitric acid treatment on Al removal, framework composition and acidity of BEA zeolite investigated by XRD, FTIR and NMR. *Microporous Mesoporous Mater.* **2012**, *163*, 122–130.

15. Maier, S.M.; Jentys, A.; Lercher, J.A. Steaming of zeolite BEA and its effect on acidity: A comparative NMR and IR spectroscopic study. *J. Phys. Chem. C* **2011**, *115*, 8005–8013.

16. Parker, W.O.; Angelis, A.; Flego, C.; Millini, R.; Perego, C.; Zanardi, S. Unexpected destructive dealumination of zeolite Beta by silylation. *J. Phys. Chem. C* **2010**, *114*, 8459–8468.

17. Wang, Y.; Otomo, R.; Tatsumi, T.; Yokoi, T. Dealumination of organic structure-directing agent (OSDA) free betazeolite for enhancing its catalytic performance in *n*-hexane cracking. *Microporous Mesoporous Mater.* **2016**, *220*, 275–281.

18. Ogura, M.; Shinomiya, S.Y.; Tateno, J.; Nara, Y.; Nomura, M.; Kikuchi, E.; Matsukata, M. Alkali-treatment technique-new method for modification of structural and acid-catalyticproperties of ZSM-5 zeolites. *Appl. Catal. A* **2001**, *219*, 33–43.

19. Groen, J.C.; Moulijn, J.A.; Pérez-Ramırez, J. Desilication: On the controlled generation of mesoporosity in MFI zeolites. *J. Mater. Chem.* **2006**, *16*, 2121–2131.

20. Mochizuki, H.; Yokoi, T.; Imai, H.; Namba, S.; Kondo, J.N.; Tatsumi, T. Effect of desilication of H-ZSM-5 by alkali treatment on catalytic performance in hexane cracking. *Appl. Catal. A* **2012**, *449*, 188–197.

21. Bjørgen, M.; Joensen, F.; Holm, M.S.; Olsbye, U.; Lillerud, K.P.; Svelle, S. Methanol to gasoline over zeolite H-ZSM-5: Improved catalyst performance by treatment with NaOH. *Appl. Catal. A* **2008**, *345*, 43–50.

22. Jung, J.S.; Park, J.W.; Seo, G. Catalytic cracking of *n*-octane over alkali-treated MFI zeolites. *Appl. Catal. A* **2005**, *288*, 149–157.

23. Schmidt, F.; Lohe, M.R.; Büchner, B.; Giordanino, F.; Bonino, F.; Kaskel, S. Improved catalytic performance of hierarchical ZSM-5 synthesized by desilication with surfactants. *Microporous Mesoporous Mater.* **2013**, *165*, 148–157.

24. Groen, J.C.; Peffer, L.A.A.; Moulijn, J.A.; Pérez-Ramırez, J. On the introduction of intracrystalline mesoporosity in zeolites upon desilication in alkaline medium. *Microporous Mesoporous Mater.* **2004**, *69*, 29–34.

25. Wu, Y.H.; Tian, F.P.; Liu, J.; Song, D.; Jia, C.Y.; Chen, Y.Y. Enhanced catalytic isomerization of α-pinene over mesoporous zeolite beta of low Si/Al ratio by NaOH treatment. *Microporous Mesoporous Mater.* **2012**, *162*, 168–174.

26. Camblor, M.A.; Corma, A.; Valencia, S. Characterization of nanocrystalline zeolite Beta. *Microporous Mesoporous Mater.* **1998**, *25*, 59–74.

27. Niwa, M.; Iwamoto, M.; Segawa, K. Temperature-Programmed Desorption of Ammonia on Zeolites. Influence of the experimental conditions on the acidity measurement. *Bull. Chem. Soc. Jpn.* **1986**, *59*, 3735–3739.

28. Katada, N.; Igi, H.; Kim, J.H.; Niwa, M. Determination of the acidic properties of zeolite by theoretical analysis of Temperature-Programmed Desorption of ammonia based on adsorption equilibrium. *J. Phys. Chem. B* **1997**, *101*, 5969–5977.

29. Marques, J.P.; Gener, I.; Ayrault, P.; Bordado, J.C.; Lopes, J.M.; Ribeiroa, F.R.; Guisnet, M. Infrared spectroscopic study of the acid properties of dealuminated BEA zeolites. *Microporous Mesoporous Mater.* **2003**, *60*, 251–262.

30. Oumi, Y.; Mizuno, R.; Azuma, K.; Nawata, S.; Fukushima, T.; Uozumi, T.; Sano, T. Reversibility of dealumination-realumination process of BEA zeolite. *Appl. Catal. A* **2001**, *49*, 103–109.

31. Jiang, G.Y.; Zhang, L.; Zhao, Z.; Zhou, X.Y.; Duan, A.J.; Xu, C.M.; Gao, J.S. Highly effective P-modified HZSM-5 catalyst for the cracking of C_4 alkanes to produce light olefins. *Appl. Catal. A* **2008**, *340*, 176–182.

Switching off H_2O_2 Decomposition during TS-1 Catalysed Epoxidation via Post-Synthetic Active Site Modification

Ceri Hammond and Giulia Tarantino

Abstract: Despite its widespread use, the Lewis acidic zeolite, TS-1, still exhibits several unfavourable properties, such as excessive H_2O_2 decomposition, which decrease its overall performance. In this manuscript, we demonstrate that post-synthetic modification of TS-1 with aqueous NH_4HF_2 leads to modifications in epoxidation catalysis, which both improves the levels of epoxide selectivity obtained, and drastically minimises undesirable H_2O_2 decomposition. Through *in situ* spectroscopic study with UV-resonance enhanced Raman spectroscopy, we also observe a change in Ti site speciation, which occurs via the extraction of mononuclear $[Ti(OSi)_4]$ atoms, and which may be responsible for the changes in observed activity.

Reprinted from *Catalysts*. Cite as: Hammond, C.; Tarantino, G. Switching off H_2O_2 Decomposition during TS-1 Catalysed Epoxidation via Post-Synthetic Active Site Modification. *Catalysts* **2015**, *5*, 2309–2323.

1. Introduction

Over recent decades, remarkable achievements have been observed in the design of Lewis acidic zeolites [1–3]. The biggest breakthrough in this context is titanium silicalite-1 (TS-1), which is a crystalline, porous, MFI-type zeolite doped with a low quantity of framework Ti^{IV} atoms [4]. Possessing active Lewis acid sites, TS-1 has been shown to be an exceptional catalyst for a wide range of oxidation challenges, including olefin epoxidation, aromatic hydroxylation and ketone ammoximation, amongst others. The unique reactivity of TS-1 is attained when it is reacted at relatively mild conditions with hydrogen peroxide (H_2O_2), an oxidant with high potential in the area of sustainable chemistry. A pertinent example of this catalysis is the recently commercialised "hydrogen peroxide to propylene oxide" (HPPO) process [5,6]. Despite its industrial exploitation, conventional TS-1 catalyst still exhibits some unfavourable properties. Particularly when employed to perform epoxidation reactions, decreases in epoxide selectivity at high levels of conversion, and the undesirable decomposition of H_2O_2, yielding O_2, are two of its major disadvantages. The second is especially problematic, as unnecessary H_2O_2 decomposition leads both to decreased economic performance through loss of a relative expensive oxidant, and safety concerns associated with the generation of O_2 in the system.

By now, it is widely accepted that the reactivity of TS-1 arises from tetrahedral $[Ti^{IV}(OSi)_4]$ sites that are isomorphously substituted into the zeolitic framework, and which may—or may not—be partially hydrolysed [7]. The active sites are proposed to exist in *pseudo*-tetrahedral geometry, comparable to conventional framework atoms, but are believed to be able to expand their co-ordination sphere to six following interaction with solvent molecules and/or the oxidant, due to the flexibility of the MFI framework [7]. Despite this, debate remains regarding the speciation of the Ti active sites, and the nature of the reactive intermediates formed during catalysis, which remains given that *in situ* spectroscopic study is complicated by the presence of water (as by-product), the condensed phase nature of the reaction (solvent spectators), and the lability of Ti-(hydro)peroxo complexes.

Recent work has demonstrated that post-synthetic modification of TS-1 with NH_4HF_2 leads to exciting improvements for aromatic hydroxylation activity and selectivity [8–10]. Despite this, the impact of this treatment on: (i) the active site speciation; and (ii) the activity of TS-1 for other oxidation reactions, particularly epoxidation, remains almost unexplored. In this manuscript we demonstrate the positive impact of NH_4HF_2 treatment on the catalytic activity of TS-1 for olefin epoxidation. Kinetic measurements reveal that following post-synthetic modification, changes in epoxidation activity are observed, and H_2O_2 decomposition activity is almost entirely eliminated, leading to promising breakthroughs in H_2O_2-based selectivity. Through *in situ* spectroscopic study with resonance enhanced Raman spectroscopy, we also observe modifications to the Ti site speciation, which may be responsible for the changes in kinetic behaviour activity.

2. Results and Discussion

2.1. Synthesis of TS-1 and Post-Synthetic Modification with NH_4HF_2

TS-1 was prepared according to a hydrothermal synthesis protocol, and XRD, porosimetry, UV-Vis and FTIR analysis confirm its successful synthesis (ESI Figures S1–S3, Table S1). The post-synthetic modification of calcined TS-1 with NH_4HF_2 and H_2O_2 was performed in an aqueous solution under the conditions optimised by Bianchi and co-workers for 4 h (catalyst henceforth denoted TS-1B) [8–10]. The treatment does not lead to any substantial differences in crystallographic structure, porosity or Ti loading, with each of these parameters remaining constant following NH_4HF_2 treatment. The material is clearly still an MFI type zeolite, and a micropore volume of 0.138 $cm^3 \cdot g^{-1}$ indicates that any potential modification undergone by TS-1 does not change its overall topology or porosity. The Ti loading remains constant, within experimental error, and the Si/Ti molar ratio remains comparable at 43.5 (Table S1). Clearly, any changes occurring to the material do not involve changes to the crystalline structure, or a major loss of Ti atoms through leaching.

In line with previous research, treatment of TS-1 in NH_4HF_2 does lead to a decrease in the main absorbance band at 210 nm—generally attributed to isomorphously substituted Ti^{IV} atoms [4]—and an increase in absorbance at a wavelength of ±270 nm (ESI Figure S2). Changes are also observed in the FTIR data (ESI Figure S3). The 960 cm^{-1} band—generally attributed to the framework Ti^{IV} atoms [11]—is partially eroded by NH_4HF_2 treatment (spectral intensity normalised to the Si–O–Si overtones of the zeolite lattice), with intensity decreasing by approximately 30% following treatment. According to this provisional characterisation, it may tentatively be attributed that NH_4HF_2 treatment leads to a change to the framework Ti^{IV} atoms. We note that the spectral data presented in Part 1 is in good agreement to the scattered reports of TS-1B in the open literature [8–10].

2.2. Catalytic Studies of Allyl Alcohol Epoxidation

Whilst previous studies of TS-1B have focused upon aromatic hydroxylation [8–10], we chose to investigate the impact of NH_4HF_2 treatment on the epoxidation activity of TS-1. Allyl alcohol was chosen as a model HPPO substrate [6]. Both TS-1 (Figure 1) and TS-1B (Figure 2) were found to be very active catalysts for this reaction. Both catalysts are able to reach full conversion in 120 min or less at the relatively mild reaction temperature of 50 °C. In both cases, glycidol is the main reaction product obtained with >95% selectivity, although 3-methoxy-1,2-propanediol (3M12PD) was also detected, almost fully accounting for the remaining carbon balance. Carbon balances of 95%–98% were always obtained, thus indicated that any other potential by-products below the detectability limit of our analytical protocols were not formed to a major extent. The observation of 3M12PD indicates that ring opening methanolysis of glycidol occurs across the least hindered end of the epoxide. Under a conventional acid catalysed mechanism, 2M13PD would be expected according to Markovnikov's rules. This difference may be due to the active sites in TS-1 catalysing ring opening via a basic mechanism, or may alternatively be a consequence of shape selectivity favouring the linear methoxy-substituted diol over the branched. Although comparable in activity at 50 °C, further kinetic analysis at multiple temperatures reveals distinct differences in the activities of TS-1 and TS-1B.

Initial rate analysis (ESI Figures S4 and S5), calculated from the linear region of the rate plot, and the resulting Arrhenius plot (Figure 3) clearly shows a change to both the temperature dependence and the pre-exponential factor following NH_4HF_2 treatment. The activation energy between 30 and 50 °C is found to increase from 39.1 kJ· mol^{-1} for TS-1, to 53.7 kJ·mol^{-1} for TS-1B. The Arrhenius barrier obtained for TS-1 is found to be in good agreement to previous theoretical studies [12,13]. Although this data indicates that TS-1B is a less active epoxidation catalyst—being less able to reduce the kinetic barrier—this decrease in activity is partially compensated for by an increase in the pre-exponential factor, which increases from 1×10^3 for TS-1, to 2×10^5 for TS-1B. These changes may indicate a modification to the rate-limiting step, and potentially some modifications to the epoxidation pathway.

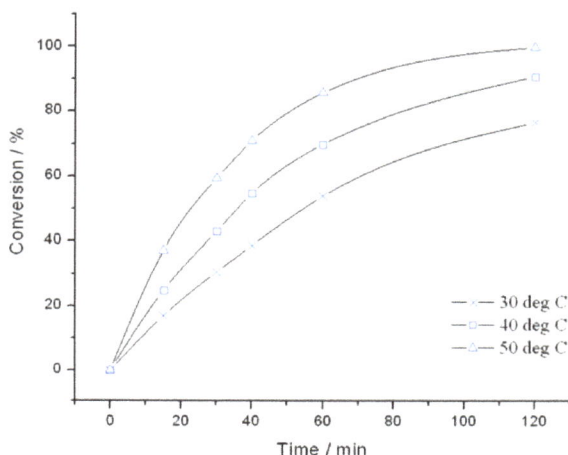

Figure 1. Time online analysis of allyl alcohol epoxidation with TS-1. Reaction conditions: [allyl alcohol] = 0.5 M, [H_2O_2] = 0.25 M, mass TS-1 = 50 mg, various temperatures.

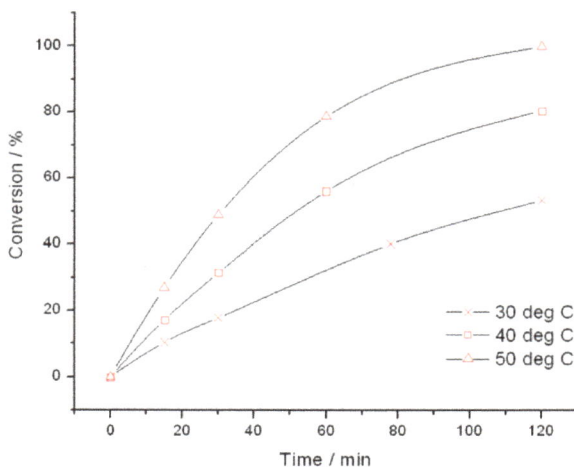

Figure 2. Time online analysis of allyl alcohol epoxidation with TS-1B. Reaction conditions: [allyl alcohol] = 0.5 M, [H_2O_2] = 0.25 M, mass TS-1B = 50 mg, various temperatures. Conversion is calculated according to the initial concentration of H_2O_2 (see experimental section).

103

Further changes in catalytic performance may also be observed from the conversion *versus* selectivity plots in Figure 4, determined at maximum levels of conversion. Although both catalysts are almost exclusively selective to the main product, glycidol, up to a peroxide conversion level of 80%, it is clear that TS-1B is able to maintain excellent glycidol selectivity even at substantially higher levels of peroxide conversion. Indeed, whilst glycidol selectivity decreases rapidly to $\pm85\%$ at 95% conversion with TS-1, glycidol selectivity remains well above 90% throughout the conversion range for TS-1B. This increase in selectivity is certainly non-trivial, and would allow the conversion to be held at higher levels. Not only does this avoid the formation of unwanted by-products, and hence decreases separation issues, but it also leads to higher space-time-yields and the avoidance of unnecessary recycles, which could improve the potential performance of the system on the whole. This improved selectivity may be related to a change in the acid/base character of the active sites, to a change in the way H_2O_2 is activated, or to the better availability of H_2O_2 at very high levels of conversion (See below), limiting the ability of the active sites to catalyse the consecutive reaction.

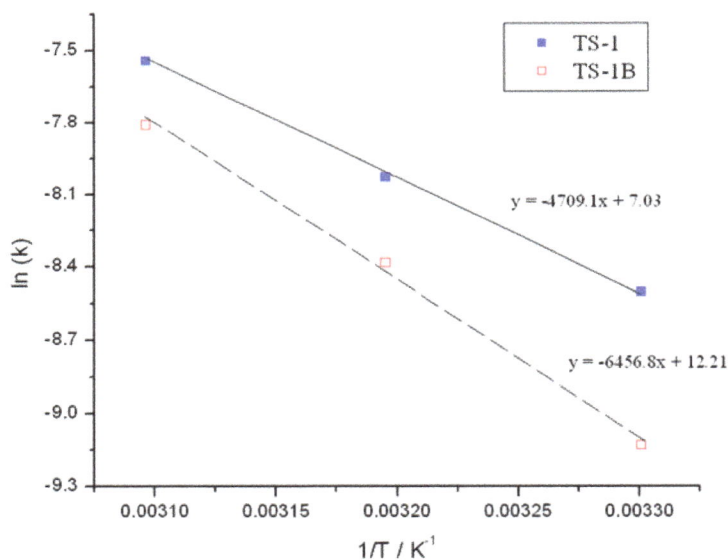

Figure 3. Arrhenius plots obtained for (blue squares) TS-1 and (hollow red squares) TS-1B.

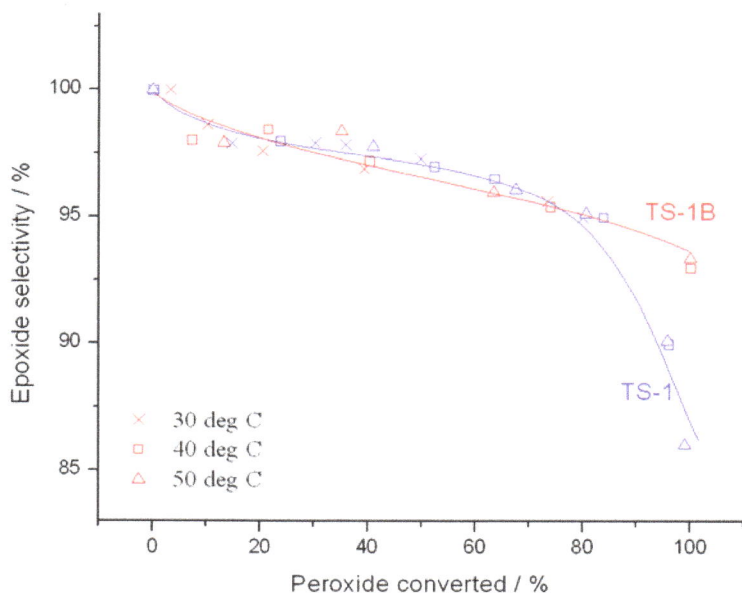

Figure 4. Comparison of selectivity as a function of conversion for (blue) TS-1 and (red) TS-1B. Reaction conditions: [allyl alcohol] = 0.5 M, [H_2O_2] = 0.25 M, mass of catalyst = 50 mg, various temperatures. Conversion is calculated according to the initial concentration of H_2O_2, as a 2:1 olefin:H_2O_2 ratio was employed.

2.3. Catalytic Studies of H_2O_2 Decomposition

To further understand the impact of NH_4HF_2 treatment, we also investigated the ability of TS-1 and TS-1B to catalyse H_2O_2 decomposition. As described above, unwanted H_2O_2 decomposition is clearly disadvantageous, leading to an unwanted loss of expensive oxidant, and potentially increasing safety issues associated. To evaluate the catalytic decomposition of H_2O_2, we performed the epoxidation reaction without the olefin being present, *i.e.*, solvent, catalyst and H_2O_2 concentrations were maintained, but the olefin substrate was absent. The loss of H_2O_2 with time could then be compared to control experiments in the absence of catalyst over the same time period, so as to rule out purely thermal losses (we note here that no thermal losses were observed at temperatures at or below 70 °C).

Both TS-1 and TS-1B are unsurprisingly less reactive for H_2O_2 decomposition than for epoxidation, and longer reaction times are required to induce substantial decomposition. Under reaction conditions identical to the epoxidation reaction, *i.e.*, at 50 °C, neither catalyst decomposes more than 6% of the initial H_2O_2 present in the first 90 min, even in the absence of the olefin. Despite this, clear differences between the decomposition activities of each catalyst can be observed following

105

NH$_4$HF$_2$ treatment. Between 40 and 60 °C, TS-1 displays considerable decomposition activity, with conversion reaching a maximum of 53% over TS-1 after 270 min at 60 °C (Figure 5). Whilst these values appear high, we note that decomposition in general is substantially retarded by the co-presence of the olefin, particularly under the reaction conditions employed for epoxidation. The presence of an epoxidisable double bond likely prohibits the Ti-(hydro)peroxo intermediate from liberating O$_2$ via decomposition or oxidation. Over TS-1, an activation barrier for H$_2$O$_2$ decomposition of 70.2 kJ·mol^{-1} was obtained, in excellent agreement to the values found in the open literature (69–72 kJ·mol^{-1}) [14–16].

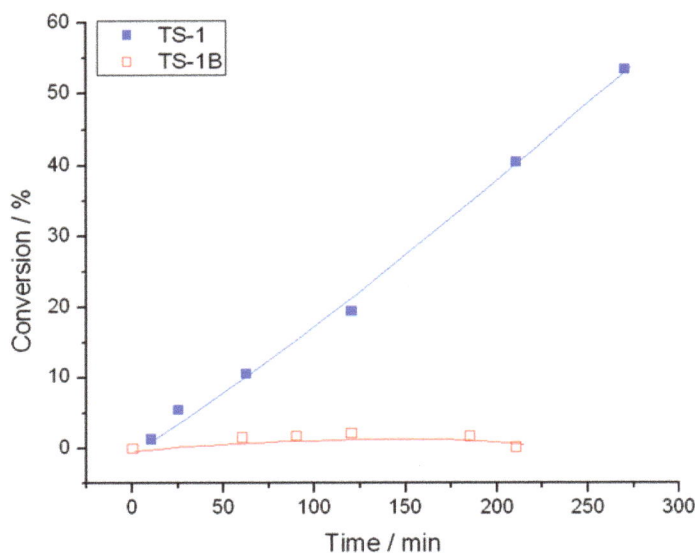

Figure 5. Conversion *versus* time for H$_2$O$_2$ decomposition at 60 °C over TS-1 and TS-1B.

In contrast, H$_2$O$_2$ decomposition activity is almost completely suppressed following NH$_4$HF$_2$ treatment. Indeed, even at 60 °C almost no decomposition is observed following the post-synthetic treatment, with a maximum conversion of 2.1% being observed under the same conditions. The suppression of H$_2$O$_2$ decomposition is further exemplified by the Arrhenius plot in Figure 6, following calculation of the initial rate constants over the linear region of the kinetic plot (ESI Figures S6 and S7). Due to the extremely low reactivity observed over TS-1B we were forced to work at slightly higher temperatures for experiments with this material. Nevertheless, an acceptable Arrhenius barrier of 140 kJ·mol^{-1} was obtained over TS-1B, a two-fold increase in kinetic barrier.

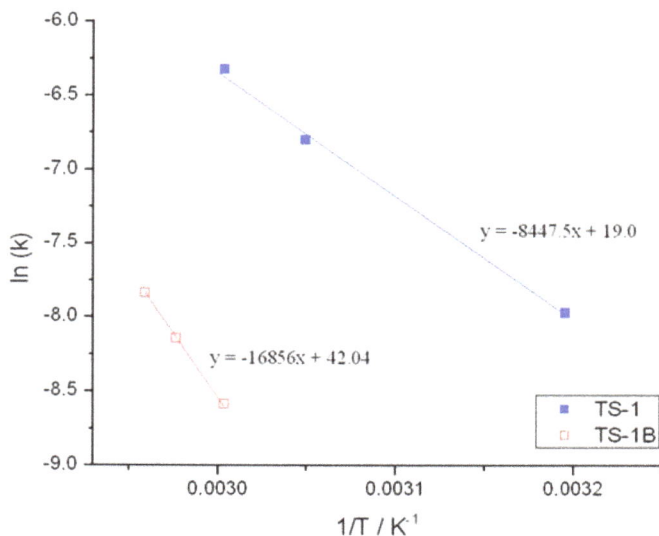

Figure 6. Arrhenius expression for H_2O_2 decomposition over (blue squares) TS-1 and (hollow red squares) TS-1B. Reaction conditions: $[H_2O_2]$ = 0.25 M, mass of catalyst = 50 mg, various temperatures.

2.4. Spectroscopic Studies with in Situ Raman

To account for these interesting effects, particularly the suppression of H_2O_2 decomposition, we considered several possibilities. Although the treatment of various zeolites with NH_4HF_2 has been found to induce some mesopores formation [17], the impact of this was ruled out given the negligible changes observed in the porosimetry data (Table S1). Similarly, changes to the potential acid/base character of TS-1—an important factor in H_2O_2 decomposition chemistry [18]—were also ruled out, based on previous observations that NH_4HF_2 treatment does not change surface acidity (neither Lewis nor Brønsted) to any extent [17]. Fluorine analysis of TS-1B revealed that no residual fluorine remained after treatment (Table S2), and thus changes to the hydrophilicity or polarity of the catalyst based on fluorine are unlikely. Further indication of this was provided through FTIR spectroscopy: Although additional defect sites are indeed observed in TS-1B (ESI Figure S8), these do not obviously change the observed dehydration/rehydration rate compared to TS-1. The potential role of defect sites can also be ruled out given that theoretical studies indicate these would decrease the activation barrier for epoxidation [13], in contrast to our observed kinetics. Whilst ICP analysis also revealed a decrease in trace Fe content following etching (Fe content decreased from 326 to 254 ppm), control measurements investigating the ability of Fe-silicalite-1 to decompose H_2O_2 revealed that the impurity Fe content was not responsible for H_2O_2

decomposition (ESI Table S3). The inability of this MFI material to catalyse H_2O_2 decomposition at a similar Fe loading to that found in TS-1 as impurities clearly indicates that Ti is required for decomposition to proceed, in good agreement with previous studies [18].

Accordingly, we re-focused our attention on spectroscopic study of the Ti active sites with more sensitive spectroscopic techniques. UV-resonance enhanced Raman (UVR) spectroscopy is a particularly promising technique, given its ability to directly probe the Ti active site speciation of each catalyst, improve sensitivity, minimise fluorescence, and selectively probe the different active sites present through the resonance Raman effect [19–21]. The tremendous insight offered by UVR is illustrated in Figure 7. Irradiating TS-1 with a progressively higher energy excitation wavelength allows the LMCT bands of the Ti^{IV} species to be probed, and leads to the identification of Ti^{IV}-specific vibrations more sensitively and selectively. Furthermore, vibrations that are totally undetectable by conventional Raman spectroscopy with IR (Figure 7a) and visible (Figure 7b) sources are also observed due to resonance enhancement effects, although it should be added that species that do not undergo resonance enhancement will not experience improved detectability.

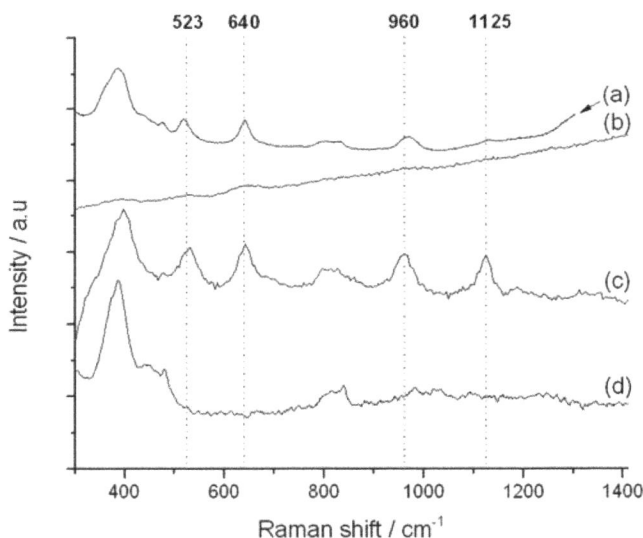

Figure 7. Impact of Raman excitation laser wavelength on Raman spectrum of TS-1. (**a**) TS-1 @785 nm excitation; (**b**) TS-1 @514 nm excitation; (**c**) TS-1 @325 nm excitation; (**d**) silicalite-1 @325 nm excitation.

The 325 nm UV-Raman spectrum of TS-1 is presented in Figure 7c. In addition to the two vibrations related to the zeolite framework at 380 and 800 cm^{-1} (also present in Ti-free silicalite-1, Figure 7d), four vibrations associated with the Ti^{IV} active sites are visible at 523, 640, 960 and 1125 cm^{-1}. The first band (532 cm^{-1}) has previously been assigned to the symmetric stretching vibration of framework Ti–O–Si species [22,23]. Both the 960 and 1125 cm^{-1} bands have also been attributed to the Ti^{IV} active sites of TS-1. These arise from the combination of three asymmetric stretching modes of a tetrahedral $[Ti(OSi)_4]$ unit (960 cm^{-1}), and a totally symmetric stretching mode, *i.e.*, breathing mode, of the same unit (1125 cm^{-1}), respectively [11]. Accordingly, these final two stretches are directly related to the tetrahedrally-coordinated Ti atom present in the $[Ti(OSi)_4]$ active sites. We note that 1125 cm^{-1} band is also known to undergo significant resonance enhancement as the laser is progressively tuned into the LMCT modes, in good agreement with our data (Figure 10) [11]. Accordingly, the 530, 960 and 1125 cm^{-1} UV-Raman bands provide a unique opportunity to monitor the Ti site speciation in TS-1 and TS-1B as a function of post-synthetic modification.

Figure 8 displays the UV-Raman (325 nm) spectra of TS-1 (blue) and TS-1B (red), with the right inset showing the relative intensities of the 960 and 1125 cm^{-1} bands. As can be seen, NH_4HF_2 treatment of TS-1 leads to significant changes in the UV Raman spectrum. The 1125 (inset), 960 (inset), and to a lesser extent the 530 cm^{-1} band (not shown) of TS-1 experience a significant decrease in relative intensity upon NH_4HF_2 treatment. Since the 960 and 1125 cm^{-1} bands have been assigned to stretching modes of a purely tetrahedral $[Ti(OSi)_4]$ unit [11], it can be proposed that a general decrease in tetrahedral, framework $[Ti(OSi)_4]$ atoms occurs upon post-synthetic modification, *i.e.*, extraction of framework Ti occurs. This is in line with the decrease in the intensity of the 210 nm band in the UV-Vis spectra (ESI Figure S2). Based on the relative intensities of the 960 and 1125 cm^{-1} bands before and after treatment, we estimate the concentration of the original $[Ti(OSi)_4]$ atoms have decreased by approximately 25%–50% following NH_4HF_2 treatment, in further agreement to the decreased intensity of the Si–O–Ti band in the FTIR spectra (ESI Figure S3).

Concurrently, a new Raman stretch at 695 cm^{-1} is observed. The generation of a new Raman band is in good agreement to the original UV-Vis data, which demonstrated that a new LMCT band was formed (270 nm) following NH_4HF_2 treatment. Given that this Raman stretch is in the framework, *i.e.*, fingerprint, region of the Raman spectrum, it is likely associated with the Ti^{IV} active sites. However, to verify if this new species was indeed responsible of the absorption band at 270 nm (ESI Figure S2), and is therefore related to the extraction of the $[Ti(OSi)_4]$ units, an additional Raman experiment at 266 nm excitation wavelength was performed (Figure 8, left inset). As can be seen, irradiating TS-1B with a 266 nm laser significantly increases the intensity of this stretch by several orders of magnitude. The 266 nm excitation laser overlaps perfectly with the 270 nm absorbance band, thus allowing true resonance conditions to be met, strongly indicating that this Raman stretch at 695 cm^{-1} is associated with the 270 nm UV-Vis band, and is hence directly proportional to the decrease in [Ti(OSi)$_4$] atoms. It appears, therefore, that treatment of TS-1 with NH$_4$HF$_2$ leads to a general extraction of framework [Ti(OSi)$_4$] atoms, and the concurrent generation of new Ti sites, which exhibit LMCT bands at ±270 nm, and a Raman stretch at 695 cm^{-1}.

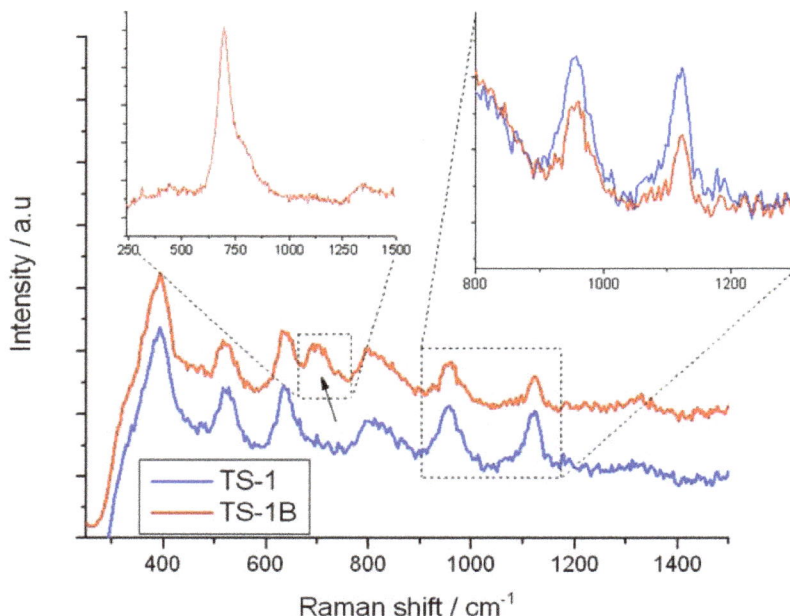

Figure 8. UV-Raman (325 nm) spectra of (blue/bottom) TS-1 and (red/top) TS-1B. The right inset shows the relative intensities of the 960 and 1125 cm^{-1} bands, whilst the left inset shows the intensity of the 695 cm^{-1} band when TS-1B is excited with a 266 nm laser.

To gain further insight into the active sites present in both TS-1 and TS-1B, we also performed *in situ* UV-Raman spectroscopy (Figure 9), where both TS-1 and TS-1B were treated with H$_2$O$_2$. In good agreement to previous *in situ* Raman studies of TS-1, complete elimination of the 960 and 1125 cm^{-1} bands are observed upon treatment of TS-1 with H$_2$O$_2$ [24,25]. Given that these bands are associated with various stretching and breathing modes of the [Ti(OSi)$_4$] tetrahedron, the total loss of these stretches can be attributed to the change in geometry (from tetrahedral to octahedral) that occurs upon the formation of the relevant Ti-(hydro)peroxo species. We note that the original spectrum can be restored by a heat treatment at 90 °C, and may even partially be restored by extended scanning under the Raman beam.

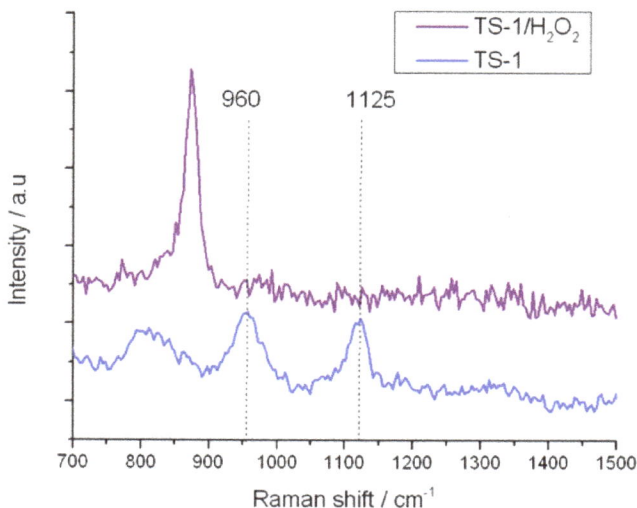

Figure 9. *In situ* UV-Raman (325 nm) spectra of (blue) TS-1 and (purple) TS-1 treated with H_2O_2.

Similarly, the residual 960 and 1125 cm^{-1} stretches in TS-1B are also eliminated upon treatment with H_2O_2, confirming that at least some of the original active site remains following NH_4HF_2 treatment. However, it is also notable that the 695 cm^{-1} band, which has been attributed to the new 270 nm UV-Vis absorbance band, is also eliminated from the Raman spectrum upon coordination of H_2O_2. The total loss of this Raman band confirms (i) that this Raman mode arises from a species whose vibrational signature is destroyed upon coordination of H_2O_2; and (ii) that the new active site formed following NH_4HF_2 treatment reacts with H_2O_2, indicating that this species also provides a route toward H_2O_2 activation. From the characterization data obtained thus far, we assign this change in Ti speciation to the extraction of framework $[Ti(OSi)_4]$ atoms, and to the concurrent generation of new active sites. Given that a redistribution of Ti from one T-site to another would not change the intensity of the breathing mode of the $[Ti(OSi)_4]$ atoms at 1125 cm^{-1}, it is likely that the new active sites are located in extra-framework positions of the zeolite.

At this time, we are not able to conclusively attribute the modified kinetic behaviour to the generation of extra-framework Ti species. Yet, the *in situ* UV-Raman data and the absence of any other contributing factors (see above) strongly suggest that these factors are related. We consider there to be two potential ways that the extraction of framework Ti could lead to such pronounced changes. The first possibility is that NH_4HF_2 treatment removes a particularly non-selective framework $[Ti(OSi)_4]$ site, responsible for both epoxidation and undesirable decomposition. Alternatively, the extraction of framework Ti could

111

result in the formation of active, extra-framework Ti complexes, which may provide an alternative pathway for H_2O_2 activation. In light of the *in situ* UV-Raman analysis (Figure 10), we favour the latter. Given the narrow FWHM observed in the UV-Vis spectrum of TS-1B (ESI Figure S2), and the generation of only one new Raman active mode, we hypothesise that any extra-framework complex must be either mononuclear, or of low nuclearity (*cf.* dimeric). Although reports suggest mononuclear, extra-framework Ti to be inactive for epoxidation [26], several recent reports have highlighted the potential for two concerted Ti sites to catalyse epoxidation reactions: Kortz *et al.* demonstrated that a polyoxometallate containing a unique di-titanium centre ([$Ti_2(OH)_2As_2W_{19}O_{67}(H_2O)$] [8]) is able to activate H_2O_2 for epoxidation catalysis [27], and the barrier calculated for H_2O_2 activation over a model Ti-dimer complex was found to be 53.5 kJ·mol^{-1} by Lundin *et al.* [28] This value is especially close to the experimental barrier of TS-1B. Definitive identification of the new active sites, and their potential role in the epoxidation/decomposition process, will require further (*in situ*) spectroscopic study with complimentary techniques, such as X-ray Absorption spectroscopy and computational methodologies. These additional spectroscopic, computational and kinetic studies remain the focus of our on-going work.

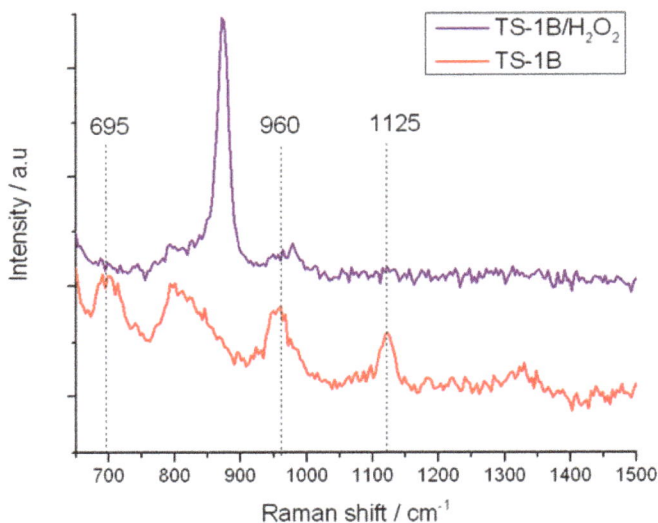

Figure 10. *In situ* UV-Raman (325 nm) spectra of (red) TS-1B and (purple) TS-1B treated with H_2O_2.

3. Experimental Section

3.1. Catalyst Synthesis and Pre-Treatment

TS-1 was prepared according to a method described elsewhere [29]. TS-1B was prepared as follows. TS-1 (4 g) was suspended in an aqueous solution (80 mL) containing NH_4HF_2 (120 mg) and H_2O_2 (50 wt. %, 3.2 mL). The slurry was stirred for 4 h at 80 °C, prior to filtration. The sample was finally dried overnight (110 °C, 16 h).

3.2. Catalyst Characterization

Powder X-ray diffraction was performed on a PANalytical X'PertPRO X-ray diffractometer (PANalytical, Almelo, the Netherlands), with a CuKα radiation source (40 kV and 30 mA). Diffraction patterns were recorded between 6° and 55°. FTIR spectroscopy was performed on a Bruker Tensor spectrometer (Bruker Corporation, MA, USA) over a range of 4000–650 cm^{-1} at a resolution of 2 cm^{-1}. UV-Vis analysis was performed on an Agilent Cary 4000 UV-Visible Spectrophotometer (Agilent Technologies, CA, USA) in Diffuse Reflectance mode. Samples were measured between 190 and 800 nm at 600 nm·min^{-1} scan rate. Ti content and Si/Ti molar ratios were determined by EDX. Specific surface area was determined from nitrogen adsorption using the BET equation, and microporous volume was determined from nitrogen adsorption isotherms using the t-plot method. Porosimetry measurements were performed on a Quantachrome Autosorb (Quantachrome, FL, USA), and samples were degassed prior to use (275 °C, 3h). Adsorption isotherms were obtained at 77 K. UV-Raman spectroscopy was performed on a Renishaw inVia spectrometer (Renishaw plc, Gloucestershire, UK) equipped with laser lines at 785, 514, 325 and 266 nm. Laser power was kept at or below 10 mW for solid powders, and at or below 2 mW for *in situ* measurements with H_2O_2.

3.3. Kinetic Evaluation and Analytical Methods

Allyl alcohol epoxidation was performed in a 100 mL round bottom flask equipped with a reflux condenser, thermostatically controlled by immersion in a silicon oil bath. The vessel was first charged with a 10 mL solution of allyl alcohol in methanol (0.5 M), which also contained an internal standard (biphenyl, 0.025M). The desired amount of catalyst (50 mg) was also added. The mixture was subsequently heated to the desired temperature (stated in the manuscript, all temperatures correspond to the oil temperature outside the reactor), and the reaction was initiated by addition of an appropriate amount of H_2O_2 (0.25 M, corresponding to a 1:2 H_2O_2:olefin ratio). The solution was stirred at ±750 rpm with an oval magnetic stirrer bar. Aliquots of reaction solution were taken periodically for analysis, and were centrifuged prior to injection into a GC (Agilent 7820 (Agilent Technologies, CA, USA), 25 m CP-Wax 52 CB column). Reactants were quantified against the biphenyl internal standard. Conversion was calculated as follows: Conversion = $\Sigma[\text{products, M}]/[\text{H}_2\text{O}_2]_0 \times 100$. All chemicals were purchased from Sigma-Aldrich (St. Louis, MO, USA).

4. Conclusions

Titanium silicalite-1 is a uniquely active catalyst for a range of commercial scale oxidation reactions. In this manuscript, we demonstrate that treatment of TS-1 with NH_4HF_2 leads to a change in Ti site speciation. Through UV-Vis, FTIR and UV-resonance enhanced Raman spectroscopy, we interpret this change as being related to the extraction of mononuclear $[\text{Ti(OSi)}_4]$ atoms from the zeolite framework, and the generation of new active Ti species that are present in extra-framework sites of the structure. These new active sites may be responsible both for the changes in epoxidation catalysis observed (increased selectivity, modified temperature dependence), and also for significant improvements in H_2O_2-based selectivity, as non-selective H_2O_2 decomposition is found to be drastically decreased following post-synthetic treatment of TS-1 with NH_4HF_2.

Acknowledgments: C.H. gratefully appreciates the support of The Royal Society for the provision of a University Research Fellowship (UF140207) and additional research funding (RG140754).

Author Contributions: C.H. Conceived and designed the experiments and guided the research. G.T. and C.H. performed characterization and catalytic studies. C.H. performed the Raman measurements. Both authors contributed to the writing of the manuscript.

Conflicts of Interest: The authors declare no conflict of interest.

References

1. Corma, A.; Garcia, H. Lewis Acids as Catalysts in Oxidation Reactions: From Homogeneous to Heterogeneous Systems. *Chem. Rev.* **2002**, *102*, 3837–3892.
2. Moliner, M. State of the Art of Lewis Acid-Containing Zeolites: Lessons From Fine Chemistry to New Biomass Transformation Processes. *Dalton Trans.* **2014**, *43*, 4197–4208.
3. Dapsens, P.Y.; Mondelli, C.; Perez-Ramirez, J. Design of Lewis-acid centres in zeolitic matrices for the conversion of renewables. *Chem. Soc. Rev.* **2015**, *44*, 7025–7043.
4. Ratnasamy, P.; Srinivas, D.; Knözinger, H. Active Sites and Reactive Intermediates in Titanium Silicate Molecular Sieves. *Adv. Catal.* **2004**, *48*, 1–169.
5. Cavani, F.; Teles, J.H. Sustainability in Catalytic Oxidation: An Alternative Approach or Structural Evolution? *ChemSusChem* **2009**, *2*, 508–534.
6. Cavani, F. Catalytic Selective Oxidation: The Forefront in the Challenge for a More Sustainable Chemical Industry. *Catal. Today* **2010**, *157*, 8–15.

7.   Bordiga, S.; Bonino, F.; Damin, A.; Lamberti, C. Reactivity of Ti(IV) species hosted in TS-1 towards H_2O_2-H_2O Solutions investigated by ab initio cluster and periodic approaches combined with experimental XANES and EXAFS data: a review and new highlights. *Phys. Chem. Chem. Phys.* **2007**, *9*, 4854–4878.

8.   Balducci, L.; Bianchi, D.; Bortolo, R.; D'Aloisio, R.; Ricci, M.; Tassinari, R.; Ungarelli, R. Direct Oxidation of Benzene to Phenol with Hydrogen Peroxide over a Modified Titanium Silicalite. *Angew. Chem. Int. Ed.* **2003**, *42*, 4937–4940.

9.   Bianchi, D.; D'Aloisio, R.; Bortolo, R.; Ricci, M. Oxidation of Mono- and Bicyclic Aromatic Compounds with Hydrogen Peroxide Catalyzed by Titanium Silicalites TS-1 and TS-1B. *Appl. Catal. A* **2007**, *327*, 295–299.

10. Bianchi, D.; Balducci, L.; Bortolo, R.; D'Aloisio, R.; Ricci, M.; Spanò, G.; Tassinari, R.; Tonini, C.; Ungarelli, R. Oxidation of Benzene to Phenol with Hydrogen Peroxide Catalyzed by a Modified Titanium Silicalite (TS-1B). *Adv. Synth. Catal.* **2007**, *349*, 979–986.

11. Ricchiardi, G.; Damin, A.; Bordiga, S.; Lamberti, C.; Spanò, G.; Rivetti, F.; Zecchina, A. Vibrational Structure of Titanium Silicate Catalysts. A Spectroscopic and Theoretical Study. *J. Am. Chem. Soc.* **2001**, *123*, 11409–11419.

12. Wells, D.H., Jr.; Joshi, A.M.; Delgass, N.; Thomson, K.T. A Quantum Chemical Study of Comparison of Various Propylene Epoxidation Mechanisms Using H_2O_2 and TS-1 Catalyst. *J. Phys. Chem. B* **2006**, *110*, 14627–14639.

13. Wells, D.H., Jr.; Delgass, N.; Thomson, K.T. Evidence of Defect-Promoted Reactivity for Epoxidation of Propylene in Titanosilicate (TS-1) Catalysts: A DFT Study. *J. Am. Chem. Soc.* **2004**, *126*, 2956–2962.

14. Wang, L.; Wang, Y.; Wu, G.; Feng, W.; Zhang, T.; Yang, R.; Jin, X.; Shi, H. A Novel Kinetics Study on H_2O_2 Decomposition in the Propylene Epoxidation System in a Fixed Bed Reactor. *Int. J. Chem. React. Eng.* **2013**, *11*, 265–269.

15. Potekhin, V.V.; Kulikova, V.A.; Kochina, E.G.; Potekhin, V.M. Decomposition of Hydrogen Peroxide in Protic and Polar Aprotic Solvents on TS-1 Heterogeneous Catalyst. *Russ. J. Appl. Chem.* **2011**, *84*, 1195–1200.

16. Shan, Z.C.; Lu, Z.D.; Wang, L.; Zhou, C.; Ren, L.M.; Zhang, L.; Meng, X.J.; Ma, S.J.; Xiao, F.S. Stable Bulky Particles Formed by TS-1 Zeolite Nanocrystals in the Presence of H_2O_2. *ChemCatChem* **2010**, *2*, 407–412.

17. Qin, Z.; Lakiss, L.; Gilson, J.-P.; Thomas, K.; Goupil, J.-M.; Fernandez, C.; Valtchev, V. Chemical Equilibrium Controlled Etching of MFI-Type Zeolite and Its Influence on Zeolite Structure, Acidity, and Catalytic Activity. *Chem. Mater.* **2013**, *25*, 2759–2766.

18. Yoon, C.W.; Hirsekorn, K.F.; Neidig, M.L.; Yang, X.; Tilley, T.D. Mechanism of the Decomposition of Aqueous Hydrogen Peroxide over Heterogeneous TiSBA15 and TS-1 Selective Oxidation Catalysts: Insights from Spectroscopic and Density Functional Theory Studies. *ACS Catal.* **2011**, *1*, 1665–1678.

19. Kim, H.; Kosuda, K.M.; van Duyne, R.P.; Stair, P.C. Resonance Raman and Surface- and Tip-enhanced Raman Spectroscopy Methods to Study Solid Catalysts and Heterogeneous Catalytic Reactions. *Chem. Soc. Rev.* **2010**, *39*, 4820–4844.

20. Fan, F.; Feng, Z.; Li, C. UV Raman Spectroscopic Study on the Synthesis Mechanism and Assembly of Molecular Sieves. *Chem. Soc. Rev.* **2010**, *39*, 4794–4801.

21. Fan, F.; Feng, Z.; Li, C. UV Raman Spectroscopic Studies on Active Sites and Synthesis Mechanisms of Transition Metal-Containing Microporous and Mesoporous Materials. *Acc. Chem. Res.* **2010**, *43*, 378–387.

22. Guo, Q.; Sun, K.; Feng, Z.; Li, G.; Guo, M.; Fan, F.; Li, C. A Thorough Investigation of the Active Titanium Species in TS-1 Zeolite by In Situ UV Resonance Raman Spectroscopy. *Chem. Eur. J.* **2012**, *18*, 13854–13860.

23. Li, C.; Xiong, G.; Xin, Q.; Liu, J.; Ying, P.; Feng, Z.; Li, J.; Yang, W.; Wang, Y.; Wang, G.; *et al.* UV Resonance Raman Spectroscopic Identification of Titanium Atoms in the Framework of TS-1 Zeolite. *Angew. Chem. Int. Ed.* **1999**, *38*, 2220–2221.

24. Bordiga, S.; Damin, A.; Bonino, F.; Ricchiardi, G.; Zecchina, A.; Tagliapietra, R.; Lamberti, C. Resonance Raman Effects in TS-1: the Structure of Ti(IV) Species and Reactivity Towards H_2O, NH_3 and H_2O_2: an in situ Study. *Phys. Chem. Chem. Phys.* **2003**, *5*, 4390–4393.

25. Bordiga, S.; Damin, A.; Bonino, F.; Ricchiardi, G.; Lamberti, C.; Zecchina, A. The Structure of the Peroxo Species in the TS-1 Catalyst as Investigated by Resonant Raman Spectroscopy. *Angew. Chem. Int. Ed.* **2002**, *41*, 4734–4737.

26. Li, G.; Wang, X.; Guo, X.; Liu, S.; Zhao, Z.; Bao, X.; Lin, L. Titanium Species in TS-1 Prepared by Hydrothermal Method. *Mat. Chem. Phys.* **2001**, *71*, 195–201.

27. Hussain, F.; Bassil, B.S.; Kortz, U.; Kholdeeva, O.A.; Timofeeva, M.N.; de Oliveira, P.; Keita, B.; Nadjo, L. Di-titanium-containing 19-tungstodiarsenate (III) $[Ti_2(OH)_2As_2W_{19}O_{67}(H_2O)]^{8-}$: Synthesis, structure, electrochemistry, and oxidation catalysis. *Chem. Eur. J.* **2007**, *13*, 4733–4742.

28. Lundin, A.; Panas, I.; Ahlberg, E. Quantum Chemical Modeling of Propene and Butene Epoxidation with Hydrogen Peroxide. *J. Phys. Chem. A* **2009**, *113*, 282–290.

29. Hammond, C.; Dimitratos, N.; Jenkins, R.L.; Lopez-Sanchez, J.A.; Kondrat, S.A.; Rahim, M.H.A.; Forde, M.M.; Thetford, A.; Taylor, S.; Hagen, H.; *et al.* Elucidation and Evolution of the Active Component within Cu/Fe/ZSM-5 for Catalytic Methane Oxidation: From Synthesis to Catalysis. *ACS Catal.* **2013**, *3*, 689–699.

Zeolite Catalysts for Phenol Benzoylation with Benzoic Acid: Exploring the Synthesis of Hydroxybenzophenones

Gherardo Gliozzi, Sauro Passeri, Francesca Bortolani, Mattia Ardizzi, Patrizia Mangifesta and Fabrizio Cavani

Abstract: In this paper, we report on the reaction of phenol benzoylation with benzoic acid, which was carried out in the absence of solvent. The aim of this reaction is the synthesis of hydroxybenzophenones, which are important intermediates for the chemical industry. H-beta zeolites offered superior performance compared to H-Y, with a remarkably high conversion of phenol and high yields to the desired compounds, when using a stoichiometric amount of benzoic acid. It was found that the reaction mechanism did not include the intramolecular Fries rearrangement of the primary product phenyl benzoate, but indeed, the bimolecular reaction between phenyl benzoate and phenol mainly contributed to the formation of hydroxybenzophenones. The product distribution was greatly affected by the presence of Lewis-type acid sites in H-beta; it was suggested that the interaction between the aromatic ring and the electrophilic Al^{3+} species led to the preferred formation of o-hydroxybenzophenone, because of the decreased charge density on the C atom at the para position of the phenolic ring. H-Y zeolites were efficient than H-beta in phenyl benzoate transformation into hydroxybenzophenones.

Reprinted from *Catalysts*. Cite as: Gliozzi, G.; Passeri, S.; Bortolani, F.; Ardizzi, M.; Mangifesta, P.; Cavani, F. Zeolite Catalysts for Phenol Benzoylation with Benzoic Acid: Exploring the Synthesis of Hydroxybenzophenones. *Catalysts* **2015**, *5*, 2223–2243.

1. Introduction

The acylation of phenolic compounds is an important class of reactions with several applications in the chemical industry [1,2]. In fact, aromatic hydroxy ketones are intermediates for the synthesis of various pharmaceuticals and fragrances. p-Hydroxyacetophenone and o-hydroxyacetophenone are used for the synthesis of paracetamol and aspirin, respectively; the ortho isomer is also an intermediate for the synthesis of 4-hydroxycoumarin and flavanones. Furthermore, the acylation of resorcinol is an important process for the synthesis of a precursor for the UV-light absorbent for polymers, 4-o-octyl-o-hydroxybenzophenone [3].

In general, these reactions are catalyzed by homogeneous Friedel-Crafts Lewis-type catalysts; however, due to the environmental problems associated with the use of these compounds, alternative heterogeneous systems have been sought and, in some cases, successfully implemented in industrial uses. Furthermore, zeolites have been widely studied as catalysts for the acylation of both aromatic hydrocarbons and phenolics; several reports describe the reaction of these substrates, mainly in the liquid phase, but also in the gas phase, with different acylating agents, such as acyl halides (typically benzoyl chloride and acetyl chloride), anhydrides and even aliphatic and aromatic carboxylic acids, which are typically considered poorly reactive and, thus, react preferably with activated substrates [1,4–39].

A summary of acylations where the catalyst is made up of a zeolite is shown in Table 1. Amongst the various zeolites tested, H-beta typically offers a superior performance in terms of reactant conversion and selectivity to the desired product; this was suggested to be due to its pore structure. In some cases, however, H-Y also showed excellent performances.

Table 1. A summary of the literature data on acylation with zeolites.

Substrate	Acylating Reagent	Desired Product	Catalyst	Ref.
Cyclohexene	Acetic anhydride, acetyl chloride	Acetylcyclohexenes	H-Y, H-beta, H-mordenite	[33]
Benzene	Benzoyl chloride	Benzophenone	Ga, In-H-beta	[26]
Toluene	Phthalic acid	2-methylanthraquinone	H-Y, H-beta, H-mordenite	[39]
Toluene	Benzoyl chloride	4-methylbenzophenone	H-beta	[4]
Toluene	Acetic anhydride	4-methylacetophenone	H-beta (nano)	[8]
o-Xylene	Benzoyl chloride	3,4-dimethylbenzophenone	H-beta	[6]
m-Xylene	Benzoyl chloride, benzoic anhydride	2,4-dimethylbenzophenone	H-Y	[40]
Naphthalene	Benzoyl chloride	2-benzoylnaphthalene	H-beta	[5]
2-Methoxynaphthalene	Acetic anhydride	2-acetyl-6-methoxynaphthalene	H-beta, H-Y, ITQ-7	[7,13,22,25,41]
Biphenyl	Acetic anhydride	4-acetylbiphenyl	H-Y, H-beta	[12]
Biphenyl	Benzoyl chloride	4-phenylbenzophenone	H-beta	[20]
Chlorobenzene	4-chlorobenozylchloride	4,4′-dichlorobenzophenone	H-beta	[14]
Phenol	Propionyl chloride	4- and 2-hydroxypropiophenone	H-beta	[10]
Phenol (gas-phase)	Acetic acid	Phenyl acetate, o-hydroxyacetophenone	H-Y, H-beta, H-ZSM-5	[11,30,31]
Phenol (gas-phase)	Acetic acid	p- and o-hydroxyacetophenone	Zn-exchanged NaY or ZSM-5	[42]
Phenol	Benzoic anhydride	p- and o-hydroxybenzophenone	H-beta	[19]
Phenol	Acetic anhydride	p- and o-hydroxyacetophenone	H-ZSM5 (Cu-, Co-doped)	[37]
Phenol	Acetic acid	p- and o-hydroxyacetophenone	HZSM-5, H-Y	[38,43]
Phenol	Phenylacetate	p- and o-hydroxyacetophenone	H-beta	[44]
p-Cresol	Acetic acid, propionic acid, butyric acid, etc.	Various o-hydroxy ketones	H-beta	[45]
Anisole	Octanoic acid	p-octanoyl anisole	H-beta	[24]
Anisole	Hexanoic, octanoic, decanoic acids	4-methoxyphenylalkylketone	H-beta, H-Y, H-mordenite	[29]
Anisole	Acetic anhydride	p- and o-methoxyacetophenone	H- beta, H-Y	[32,36]
Guaiacol	Acetic anhydride	2-methoxyphenyl acetate	H-Ferrierite	[16]
Veratrole	Benzoic anhydride, benzoyl chloride	Dimethoxybenzophenone	H-Y, H-beta	[15]
Veratrole	Propionyl chloride	3,4-dimethoxypropiophenone	H-beta	[17]
Dimethoxybenzenes	Various acyl chlorides	Various	H-Y	[18]

Table 1. *Cont.*

Substrate	Acylating Reagent	Desired Product	Catalyst	Ref.
Veratrole	Acetic anhydride	3-4-dimethoxyacetophenone	H-Y, H-beta	[9,21]
Phenylacetate	Fries rearrangement	*p*- and *o*-hydroxyacetophenone	H-beta	[23,27,46]
Phenylacetate (gas-phase)	Fries rearrangement	*p*- and *o*-hydroxyacetophenone	H₃PO₄/ZSM-5	[28]
Phenylacetate	Fries rearrangement	*p*- and *o*-hydroxyacetophenone	H-beta, H-Y, H-ZSM5	[34]
Phenylacetate, phenyl benzoate	Fries rearrangement	*p*- and *o*-hydroxyacetophenone	H-ZSM5, H-ZSM12	[35]

Other non-zeolitic, but heterogeneous catalysts have also been used, such as: (i) montmorillonites of the K series for the acetylation of various aliphatic and aromatic substrates [47,48] and of aliphatic and aromatic alcohols [49], for the acylation of resorcinol with phenylacetic chloride [50] and after an exchange with various metals for the reaction between aromatic aldehydes and acetic anhydride to produce the corresponding 1,1-diacetates [51]; (ii) silica-supported phosphotungstic acid for the acylation of aromatic hydrocarbons with acrylic and crotonic acid [52,53], for the Fries rearrangement of phenyl benzoate [54] and for the acylation of anisole with acetic anhydride [55,56]; (iii) clay-supported Cs-phosphotungstic acid for the benzoylation of anisole with benzoyl chloride [57], for the benzoylation of *p*-xylene [58,59] and of phenol with benzoic acid (BA) to obtain *p*-hydroxybenzophenone (HBP) [60]; (iv) giant P/W heteropolyacids for the acetylation of phenol with acetic anhydride [61] (the reactivity of heteropolyacids for Friedel-Crafts acylation was reviewed by Kozhevnikov [62]), and other P/W polyoxometalates for the acetylation of alcohols and phenols [61]; (v) NaGa-Mg hydrotalcite for the benzoylation of toluene with benzoyl chloride [63,64]; (vi) EPZG® for the acylation of 1-methoxynaphthalene and anisole with various acyl chlorides and anhydrides [65]; (vii) sulfated zirconia, even promoted with other metal oxides or supported over MCM-41, for the benzoylation of anisole with benzoic anhydride [66–70] or benzoyl chloride [71], for the acylation of phenol, anisole and chlorobenzene with acetic anhydride or benzoic anhydride [72–74], for the acylation of veratrole with acetic anhydride [75] and for the Fries rearrangement of 4-methylphenylbenzoate [76]; (viii) borate zirconia for the benzoylation of anisole with benzoyl chloride [77]; (ix) Zr hydroxide functionalized with trifluoromethanesulfonic acid for the benzoylation of biphenyl [78]; (x) Sn-doped sulfated zirconia for the benzoylation of anisole with benzoyl chloride [79]; (xi) triflic acid-functionalized Zr-TMS (zirconium oxide with a mesostructured framework; TMS, transition metal oxide mesoporous molecular sieves) for the benzoylation of biphenyl [80], of diphenyl ether to 4-phenoxybenzophenone [81] and of toluene to 4,4′-dimethylbenzophenone [82]; (xii) Fe-Zr phosphate for the benzoylation of different arenes [83]; (xiii) mesoporous UDCaT-5 (a sulphated tetragonal zirconia-based catalyst) for the acylation of anisole with propionic anhydride [84];

(xiv) KF/alumina for the benzoylation of phenolics with benzoyl chloride [85]; (xv) polystyrene-supported $GaCl_3$ for the benzoylation of phenols with benzoic anhydride [86]; (xvi) Fe_3O_4 nanoparticles for the benzoylation of phenols with aryl aldehydes to obtain the substituted o-HBPs (xanthones) [87]; (xvii) ZnO for the o-benzoylation of alcohols (including phenol) with benzoyl chloride [88,89] and $ZnAl_2O_4/SiO_2$ composites for the acetylation of alcohols and phenols with acetic anhydride [90]; (xviii) mesoporous Al-KIT-6 (a cubic mesoporous silica with Al incorporated) for the gas-phase acylation of phenol to phenyl acetate with acetic acid [91]; (xix) Al-MCM-41 for the acylation of phenol with acetic acid in the gas-phase [31]; (xx) yttria-zirconia for the acylation of alcohols, thiols and amines with carboxylic acids [92]; (xxi) nafion-in-silica composite for the Fries rearrangement of phenyl acetate [34] and for anisole acylation with acetic anhydride (also with expanded CO_2 as the solvent) [55]; (xxii) WO_3/ZrO_2 for the acylation of veratrole with acetic anhydride and of anisole with benzoic anhydride [93]; and (xxiii) Ni/SiO_2 for the acetylation of phenol, naphthol and other alcohols [94].

In regards to the synthesis of p- and o-HBP (or other substituted benzophenones) or benzyl benzoates via c- or o-benzoylation, respectively, papers in the literature describe the use of benzoyl chloride or benzyl trichloride also with various homogeneous catalysts: $AlCl_3$ for phenol (and phenol derivatives containing 2,2,6,6,-tetramethylpiperidine) [95], $Cu(OTf)_2$ and $Sn(OTf)_2$ for anisole and veratrole [96,97], Bi(III) salts for phenol [98] and $Hf(OTf)_4$ for phenol [99], even with carboxylic acids [100].

As shown in Table 1, there are no papers in the literature describing the use of zeolites for the benzoylation of phenol; there is only one paper reporting on phenol benzoylation with benzoic anhydride catalyzed by the H-beta zeolite [19] and one paper reporting about the benzoylation of phenol with BA using polyoxometalate catalyst [60]. In this latter case, Yadav et $al.$ [60] reported that the first step of the reaction, catalyzed by $Cs_{2.5}H_{0.5}PW_{12}O_{40}$ supported over K-10 clay, is the esterification, which is followed by Fries rearrangement towards hydroxybenzophenones (HBPs). The best selectivity to p-HBP was 32.5% at 70% BA conversion, with a phenol/BA ratio of 7/1. In the paper of Chaube et $al.$ [19], H-beta turned out to be the more active zeolite for the production of HBPs. The selectivity to p-HBP was found to increase with the increase of the strength of acid sites; HBPs formed by both Fries rearrangement and acylation of phenol by phenyl benzoate.

Here, we report on a study aimed at finding zeolitic catalysts and conditions for the liquid-phase benzoylation of phenol with BA; the reaction was carried out in the absence of any solvent. Typically, the aim of this reaction is the production of HBPs, which are intermediates for dyes, pharmaceuticals and fragrances. In the current industrial production of HBPs, benzotrichlorides are produced first; they are then reacted with benzene in the presence of $AlCl_3$, and afterwards, the adduct is hydrolysed into p-HBP. The direct reaction between phenol and BA is also carried out, using BF_3 as a Lewis-type catalyst [19].

2. Results and Discussion

2.1. Characterization of H-Beta Zeolites

Table 2 summarizes the main characteristics of the four H-beta zeolites used: the Si/Al ratio (as stated by the supplier), the overall amount of ammonia desorbed (as determined by NH_3-Thermal-Programmed -Desorption TPD), and main morphological features. As can be seen, the number of acid sites was not exactly in line with that expected based on the Si/Al ratio; in fact, sample HB-75 had a smaller number of acid sites than HB-150, while the opposite would normally be expected. This seems to be attributable to an overestimation of the number of acid sites in HB-150.

Table 2. Main characteristics of the H-beta (HB) and H-Y (HY) zeolites used as catalysts for phenol benzoylation.

Sample	Si/Al, Atomic Ratio	Overall Amount of NH_3 Desorbed (mmole NH_3/g)	Micropore Volume and Area (cm^3/g, m^2/g)	Mesopore Volume (cm^3/g)	Total Pore Volume (cm^3/g)	Total Surface Area (m^2/g)	Crystallite Size (nm; from XRD)
HB-13	13	0.39	0.13, 384	0.55	0.70	575	18
HB-38	38	0.24	0.15, 421	0.94	1.10	636	16
HB-75	75	0.11	0.15, 429	0.99	1.16	645	16
HB-150	150	0.17	0.16, 458	0.17	0.38	641	27
HY-3	3	1.0 *	Nd	Nd	Nd	584	Nd
HY-7	7.5	0.1 *	Nd	Nd	Nd	550 *	Nd
HY-100	100	<0.1 *	Nd	Nd	Nd	814	Nd

Nd: not determined; * as from the TOSOH website.

In regard to the strength of these sites, Figure 1 plots the relative concentration of the Lewis and Brønsted sites, as determined by means of FTIR spectra after pyridine adsorption, taking as reference bands those falling at 1455 and 1545 cm^{-1}, respectively, and using the formula reported in the literature [101]. As can be seen, the sample with the greater Al content showed the higher fraction of strong Lewis sites, which is an expected result based on the fact that the latter is attributable to extra-framework Al species. The strength and relative amount of these Lewis sites decreased when the Si/Al ratio was increased. Conversely, the zeolite with the stronger Brønsted sites was sample H-beta (HB)-150.

We also determined the degree of hydrophilicity of samples by means of H_2O-TPD; Figure 2 shows that, as expected, the increased Si/Al ratio led to samples with a lower affinity for water; however, the molar ratio between adsorbed H_2O

121

molecules and Al atoms increased, suggesting that a higher number of water molecules interacted with more isolated Al sites.

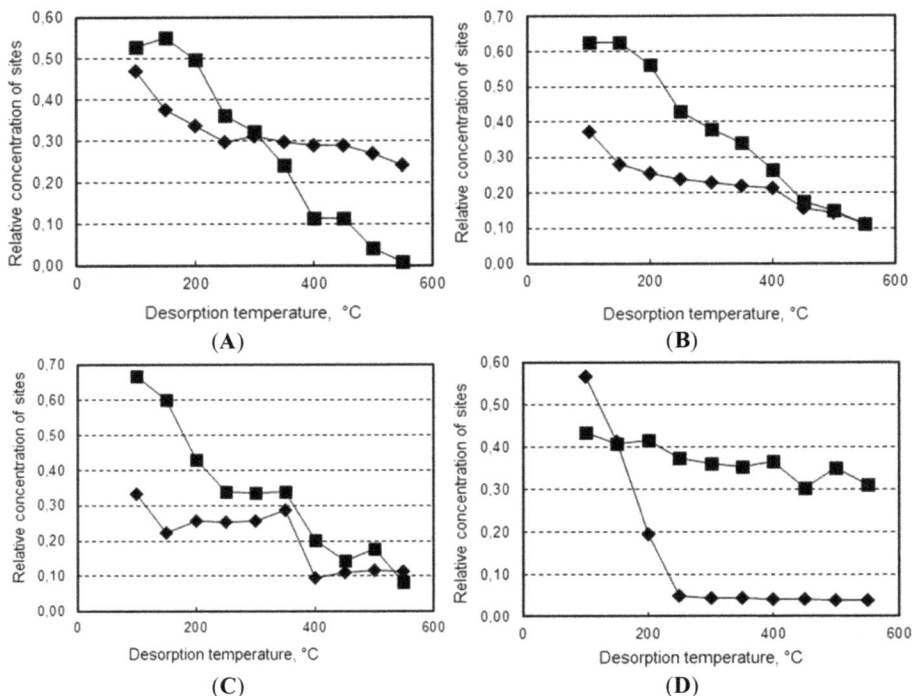

Figure 1. Relative concentration of Lewis (♦) and Brønsted (■) sites based on the pyridine desorption temperature, as determined by the intensity of the band associated with the pyridine adsorbed over the two types of sites. Samples: HB-13 (**A**), HB-38 (**B**), HB-75 (**C**) and HB-150 (**D**).

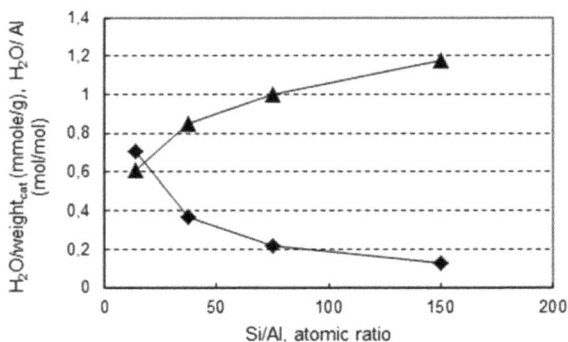

Figure 2. Ratio between amount of H_2O adsorbed and catalyst weight (♦) or the number of Al moles (▲), based on the Si/Al atomic ratio. Samples: H-beta zeolites.

2.2. Reactivity of H-Beta Zeolites

Figure 3 plots phenol conversion based on reaction time and product selectivity based on phenol conversion for the four H-beta zeolites studied. There was no obvious relationship between the Al content in zeolites and the catalyst activity. Indeed, comparing phenol conversion at short reaction times, it was shown to increase when the Silica-to-Alumina Ratio SAR ratio decreased from 150 to 38, but then, a further increase in Al content (in sample HB-13) led to its decrease; thus, the most active sample was HB-38. This may mean that the activity was not only related to the number of active acid sites, but also to their strength. It is also important to note that, as reported in the literature, during the liquid-phase acylation of aromatics, the accumulation of heavy compounds may cause catalyst deactivation and a strong inhibition of reactant conversion [27,32,44]. Under our reaction conditions, however, the data shown in Figure 2 seem to rule out important deactivation phenomena; indeed, significant coking phenomena also could be ruled out, because spent catalysts did not appear as being covered by carbonaceous residua. However, HB-13 exhibited a slower rate of conversion increase during time; this might be due to a partial deactivation of Lewis sites (see below for the discussion on the role of acid sites); in fact, HB-13 was the sample showing the greater strength for Lewis acid sites.

Products were phenyl benzoate (PB), o- and p-hydroxybenzophenone (HBP) and o- and p-benzoylphenyl benzoate (BPB); the latter formed with selectivity no higher than 10% and were clearly more prominent with samples having the greatest SAR ratio; in fact, with HB-13, the formation of these compounds was negligible, despite the significant selectivity to HBP, which is a precursor for BPB formation.

With all catalysts, PB was clearly the only primary product, since its selectivity extrapolated at low conversion was 100%. The formation of HBP occurred by consecutive transformations of the ester; it is also interesting to note that the molar ratio between the two HBP isomers (also shown in Figure 3) was very different depending on the sample characteristics. The formation of the para isomer was more facilitated with the catalyst that had the highest Al content (HB-150), with a p-/o-HBP selectivity ratio much greater than one, while it was less facilitated over the sample with the lowest Al content (HB-13), with a selectivity ratio much lower than one. Furthermore, in all samples, with the exception of HB-13, the para/ortho ratio increased with the increase in phenol conversion, up to a maximum value shown at 20%–40% phenol conversion, after which it decreased; lastly, all samples reached a p-/o-HBP selectivity ratio close to one.

These trends may have two different interpretations. The first is that two different mechanisms take place for PB transformation into HBPs:

(a)   An intramolecular Fries rearrangement of PB, which does not involve phenol, and leads preferentially to the formation of o-HBP. The greatest contribution of

123

this reaction is registered with samples having the highest Al content, leading to a *p-/o*-HBP selectivity ratio lower than one with HB-13.

(b) An intermolecular reaction between PB and phenol, which leads to both of the two HBP isomers. The greatest contribution of this reaction is registered with samples having the lowest Al content, leading to a *p-/o*-HBP selectivity ratio greater than one with HB-150 and HB-75.

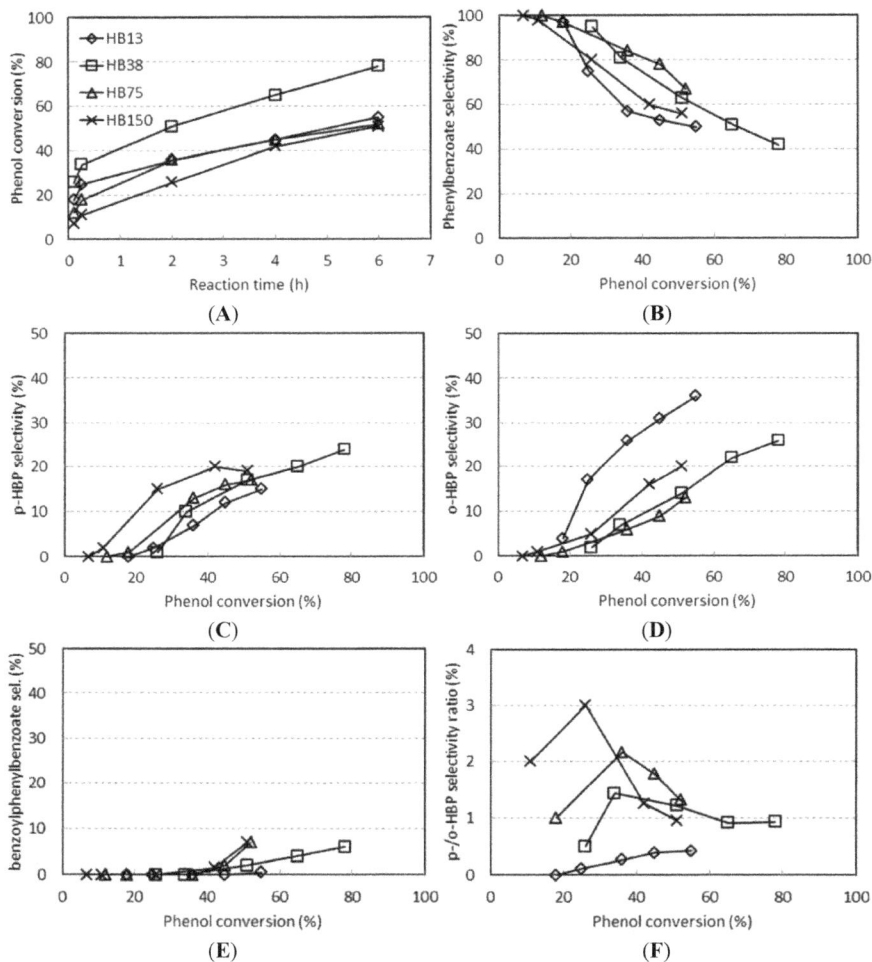

Figure 3. Phenol benzoylation with benzoic acid (BA): phenol conversion based on reaction time (**A**) and selectivity to phenylbenzoate (**B**), *p*-hydroxybenzophenone (**C**), *o*-hydroxybenzophenone (**D**) and benzoylphenylbenzoates (**E**) based on phenol conversion at T =190 °C. (**F**): selectivity ratio between *p*- and *o*-hydroxybenzophenone. Catalysts: H-beta zeolites.

An important role in the reaction might be played by the different zeolite properties; with Al-rich zeolites, those showing a greater affinity for more polar molecules (Figure 2), it may be expected that the phenol/BA molar ratio inside pores is greater than with Al-poor zeolites. This might be the reason for the effects shown, *i.e.*,: (i) a maximum value for the *p-/o*-HBP selectivity ratio obtained at about 30%–40% phenol conversion; and (ii) a different trend in the selectivity ratio shown by samples, both effects being attributable to a different concentration of the two reactants inside pores and, thus, to a change in the relative contribution of intra- and inter-molecular reactions.

An alternative hypothesis is that the formation of HBPs occurs only by means of the bimolecular mechanism between PB and phenol. In this case, the different acidity of zeolites might be the reason for the different *p-/o*-HBP selectivity ratio shown. In particular, the stronger Lewis-type acidity shown by the H-beta with higher Al content might be responsible for an enhanced interaction with the phenol aromatic ring, which in turn might entail a lower delocalization of the negative charge over the ring itself; thus, the regio-selectivity of the reaction might be strongly affected by the decreased contribution of the mesomeric effect. Therefore, with catalysts showing the strongest Lewis-type acid sites, the ring-acylation at the ortho position might be more probable over the acylation at the para position, due to the prevailing contribution of the inductive effect, which exerts its effect on the ortho position only.

The trend observed for the *p-/o*-HBP selectivity ratio based on phenol conversion, with a maximum value registered at about 30%–40% conversion, may be due to the overlapping of two opposite effects. On the one hand, a progressive deactivation of the Lewis-acid sites, due to either the strong interaction of phenolics with the aromatic ring or the accumulation of water (the co-product in PB formation), causes the *in situ* transformation of the Lewis sites into Brønsted acid sites. On the other hand, under conditions of phenol scarcity (*i.e.*, after 30%–40% phenol conversion), either the intramolecular rearrangement of PB into *o*-HBP might become kinetically preferred over the bimolecular reaction or an equilibration between the two isomers might lead the *p-/o*-HBP selectivity ratio to approach a value close to the equilibrium one.

In the literature, the role of Lewis sites in acylation reactions is also highlighted; it is reported that P/W heteropoly salts or metal-exchanged clays are efficient catalysts for the acetylation of alcohols or of aromatic substrates, in which the reaction rate increases with the cation electronegativity. The reaction is initiated by the coordination of acetic anhydride to the metal cation, with polarization of the C–O bond in the coordinated species [102]. In our case, instead, the main effect of the Lewis acidity is on the selectivity ratio between the two HBP isomers.

Additional experiments carried out with PB as the reactant will help clarify the aspects related to the reaction network and to the role of acid sites.

2.3. Reactivity Experiments with PB

Figure 4 plots the selectivity to products obtained by reacting PB over the HB-38 zeolite, based on phenol conversion. Unexpectedly, HBPs appeared to not be primary products of PB transformation, because their selectivity extrapolated at nil conversion was close to zero. This implies that the contribution of the intramolecular Fries rearrangement is negligible and that during phenol benzoylation, the formation of HBPs only occurs by means of the bimolecular reaction between PB and phenol. In the case of the reaction starting from PB, the intermolecular reaction may only occur between two PB molecules, to produce mainly p-BPB (selectivity to the ortho isomer was very low) and phenol. A further consecutive reaction between phenol and BPB or phenol and PB leads to the production of HBP isomers, with the coproduction of PB or phenol, respectively (however, in the latter case, the decrease in phenol selectivity shown in Figure 4 would not be justified). A rather similar trend was observed during PB reaction with HB-150.

The overall reaction network, as inferred from the experiments carried out starting from phenol and from PB, is shown in Scheme 1. Results suggest that differences shown among the H-beta samples are not due to the presence of various mechanisms for the transformation of phenyl benzoate into HBPs, but are instead attributable to the acidity features of the H-beta zeolites tested.

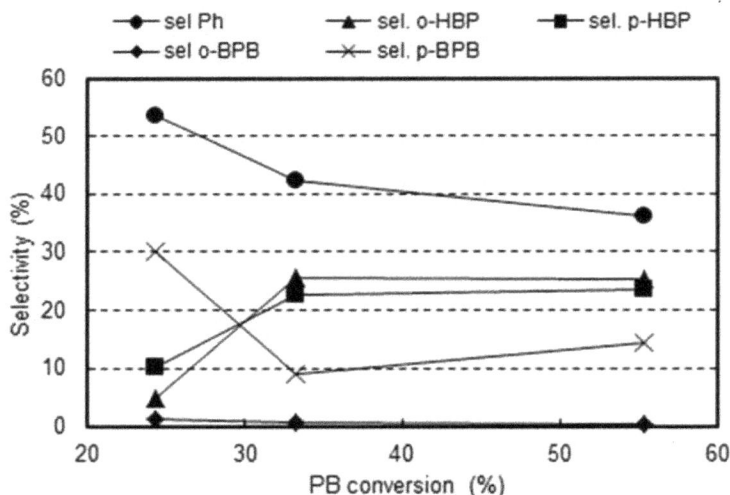

Figure 4. Phenyl benzoate (PB) reactivity: selectivity to main products based on PB conversion. Temperature: 190 °C.

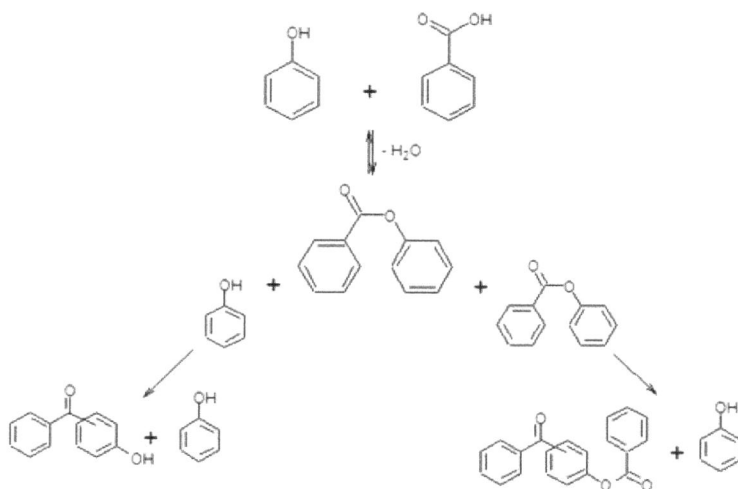

Scheme 1. Reaction scheme for phenol benzoylation with BA over H-beta zeolites.

2.4. The Reactivity of H-Y Zeolites

We tested the reactivity of three different commercial H-Y zeolites (Table 2). In regard to the acidity of H-Y, it is reported in the literature that Lewis acidity may also play an important role, even for samples having low Al content (see, for instance [103], and the references therein).

The results of catalytic experiments are shown in Figure 5. As can be seen, in all cases, phenol conversion was lower than that obtained with H-beta zeolites. If the initial conversion is taken into account, e.g., at a 0.25 h reaction time, the scale of activity was HY-7 > HY-3 > HY-100; thus, in this case also, as with H-beta zeolites, the greatest activity was obtained with the intermediate SAR ratio. Overall, however, the activity was not much affected by the Si/Al ratio.

With these catalysts, also, PB was the only primary product, and the formation of HBPs occurred by the consecutive transformation of the ester. However, the selectivity to HBPs was lower than that shown with H-beta zeolites, even with the more active H-Y sample. Therefore, H-Y zeolites were also poorly efficient in the consecutive transformation of PB into HBPs; this is particularly apparent with the sample HY-100, which showed a negligible consecutive transformation of PB into HBP, even at 30% phenol conversion. With all of the H-Y zeolites tested, no BPB formation was shown.

The lowest *p-/o*-HBP selectivity ratio was achieved with HY-100, the highest with H-Y samples having the greater Al content, which is the opposite of what was observed with H-beta zeolites; however, in this case also, the selectivity ratio seemed to approach a value close to one when the phenol conversion was increased.

127

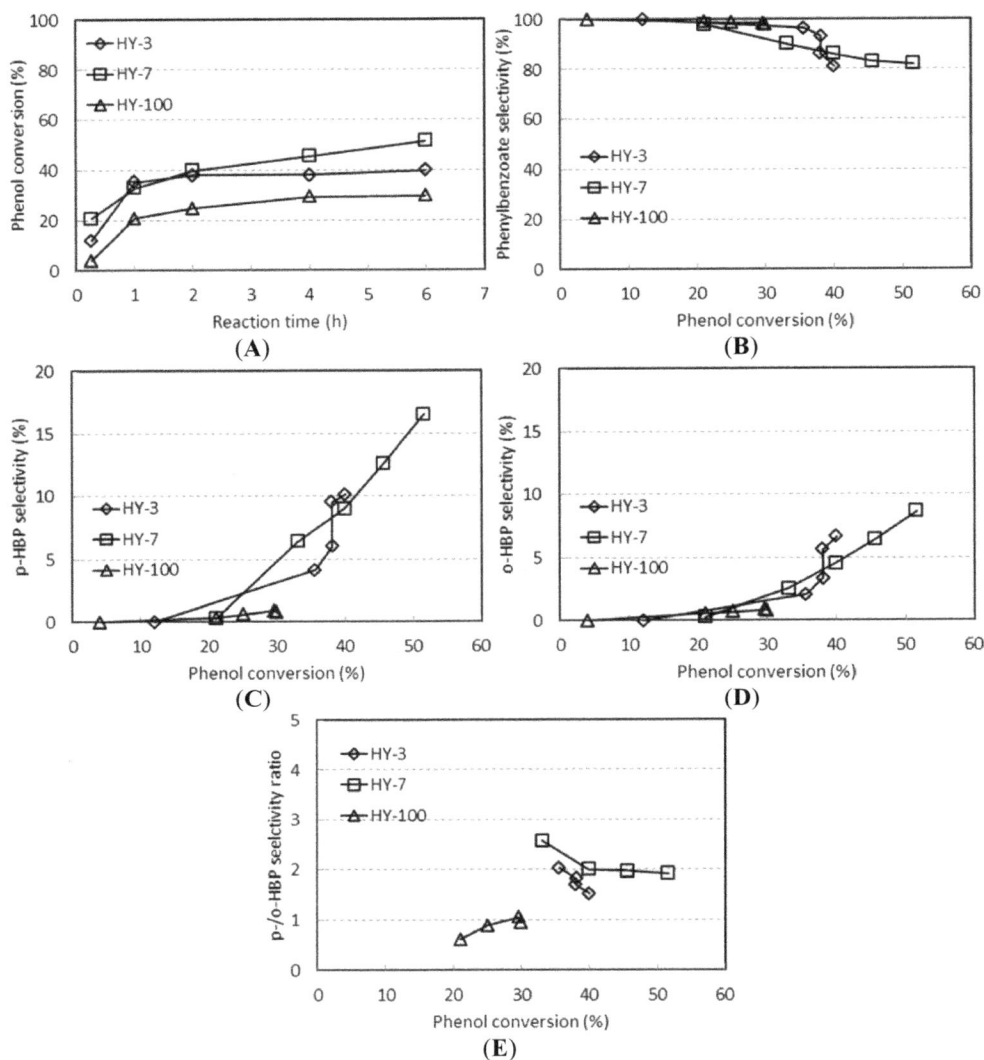

Figure 5. Phenol benzoylation with BA: phenol conversion based on reaction time (**A**) and selectivity to phenylbenzoate (**B**), to *p*-hydroxybenzophenone (**C**), and *o*-hydroxybenzophenone (**D**) based on phenol conversion at $T = 190\ °C$. (**E**): selectivity ratio between *p*- and *o*-hydroxybenzophenone. Catalysts: H-Y zeolites.

Even when we changed the BA/phenol molar ratio (Figure 6), the highest overall selectivity to HBPs achieved was not higher than 32%. An excess of phenol, in experiments carried out at a BA/phenol molar ratio equal to 0.5 and 0.16, should be kinetically unconducive to an intermolecular transformation of PB to HBPs compared to a Fries-type rearrangement; therefore, we might expect a change in the p-/o-HBP ratio in the presence of both the inter- and intra-molecular mechanisms of PB transformation into HBPs. With experiments carried out using a BA/phenol molar ratio ≤ 1, the p-/o-HBP selectivity ratio was between 1.5 and 2.0, whereas in the case of the BA/phenol ratio of two, it was lower than 1.0.

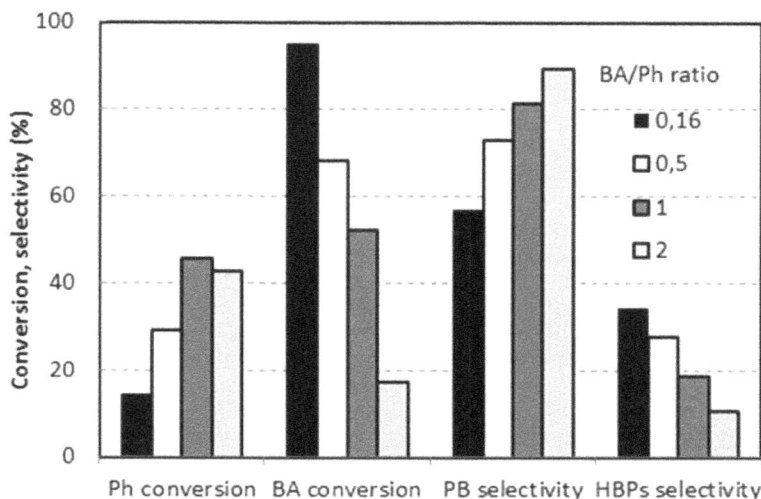

Figure 6. Effect of the BA/Ph molar ratio on phenol (Ph) and benzoic acid (BA) conversion and selectivity to phenylbenzoate (PB) and hydroxybenzophenones (HBPs). Catalyst: HY-7, reaction time 4 h, $T = 190\,°C$.

In order to gain more information on the reaction network, we conducted some experiments in which we reacted PB; we observed that the peak conversion of PB was reached with HY-3 (43%), the lowest with HY-100 (6%) and the intermediate value of 18% for HY-7. These results were in line with the PB selectivity decrease registered in phenol benzoylation experiments, during which PB underwent almost no consecutive transformation with HY-100. Surprisingly, however, no HBP was formed in these experiments, with the exception of some traces observed with the most active HY-3 compound. The only products found were phenol and BA. On the one hand, this indicates the presence of water in the zeolites (the greater amount being obviously shown with the more hydrophilic zeolite), which are responsible for the hydrolysis of the ester, but on the other hand, this suggests that both the intramolecular Fries rearrangement and the intermolecular reaction between two molecules of PB are slowed greatly under these conditions with H-Y catalysts.

It can be concluded that with H-Y zeolites, the presence of water (which is generated *in situ* during phenol benzoylation tests) has a pronounced slowing effect on the rate of consecutive PB transformation into HPBs, thus explaining the lower selectivity to HBPs recorded with these catalysts. It is possible that Lewis acid sites, which play an important role in the reaction mechanism with H-beta, are strongly inhibited by the presence of water.

When we carried out the benzoylation of phenol with benzoic anhydride, a reaction that does not lead to the co-production of water, we obtained a greater phenol conversion, which is an expected result, because BA (the co-product in benzoylation with benzoic anhydride) is a more efficient leaving group than water (the co-product in benzoylation with BA). However, despite the greater phenol conversion achieved, the selectivity to HBPs still remained low (Figure 7); this indicates that the difference observed between H-beta and H-Y zeolites in phenol benzoylation with BA was not due to water and to its interaction with Lewis acid sites.

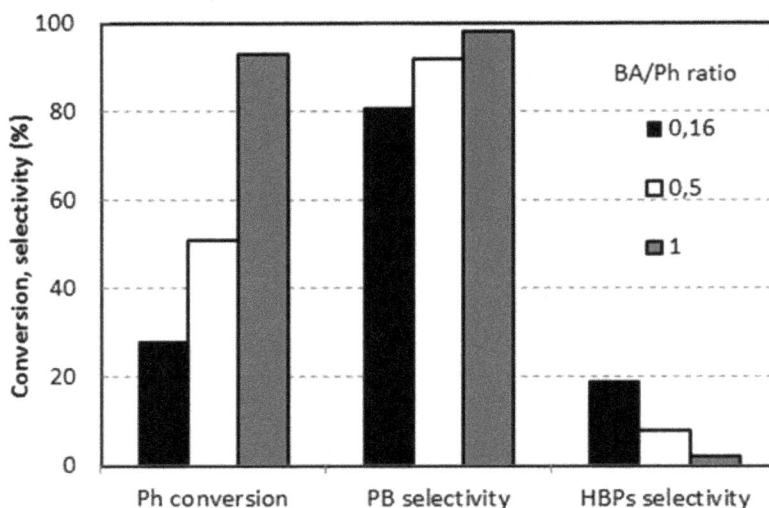

Figure 7. Effect of the benzoic anhydride/Ph molar ratio on phenol (Ph) conversion and selectivity to phenylbenzoate (PB) and hydroxybenzophenones (HBPs). Catalyst: HY-7, reaction time 4 h, $T = 190\ °C$.

Excellent performances in the benzoylation of phenol with benzoic anhydride were reported by Chaube *et al.* [19] with the use of an H-beta zeolite; the authors observed a *p-/o*-HBP selectivity ratio close to two, similar to that one also observed in our experiments with HY-7. However, a selectivity to HBPs as high as almost 70% at over 95% benzoic anhydride conversion could be obtained with the H-beta. Hoefnagel and Van Bekkum studied the acylation of resorcinol with benzoic acid, with H-beta in *p*-chlorotoluene solvent; a 70% yield to 2,4-dihydroxybenzophenone was reported, with 20% residual resorcinol monobenzoate [104]. Indeed, the better performance of H-beta in acylation of phenolics is well known and widely described in the literature, for example in the acylation of anisole with acetic anhydride (a process developed by Rhodia); however, acylation of veratrole is preferably carried out with H-Y zeolites [2].

Overall, we can state that despite both the large number of papers reporting on the use of zeolites for the acylation of phenolics (see Table 1) and the various interpretations advanced to explain the superior performance of one zeolite type over others, a clear picture as to the possible reasons for the differences in the catalytic behavior for this reaction has not yet been produced. Furthermore, in our case, the better performance of H-beta zeolites over H-Y for HBPs' production, notwithstanding the well-known analogies of H-BEA and H-FAU zeolite types, is difficult to explain, even though it is clearly attributable to a remarkably less efficient transformation rate of the intermediately-formed PB with H-Y. This difference is greater between samples having the lowest Al content, *i.e.*, HB-150 and HY-100; in fact, the latter catalyst was substantially inactive in PB conversion into HBPs. Further studies are needed in order to develop a better understanding of the main zeolite characteristics that affect the catalytic properties in phenolic acylation.

3. Experimental Section

H-beta zeolites: commercial samples purchased from Zeolyst International (Conshohocken, PA, USA) (HB-38, HB-75 and HB-150) and from ZEOCHEM (Uetikon am See, Switzerland) (HB-13) were used. H-Y zeolites: commercial samples were purchased from TOSOH (Tokyo, Japan) (HY-3, HY-7 and HY-100).

Porosimetry was determined using a Micromeritics (Norcross, GA, USA) ASAP 2020 instrument.

Thermal-programmed-desorption of ammonia (10% in He) was carried out using a TPDRO 1000 Thermo Finnigan Italia (Rodano, Italia) instrument. Samples were first pre-treated at 550 °C for 1 h; then, pulses of ammonia were fed onto the sample maintained at a temperature of 180 °C. The temperature was then raised (10 °C/min) up to 550 °C, and the sample was left at this temperature for 1 h, until the complete desorption of the ammonia.

FTIR spectra in the transmission mode were recorded after the adsorption of pyridine vapors under a vacuum on a self-supported thin film of the sample. The sample was initially pre-treated at 550 °C for 2 h; then, the cell was saturated with pyridine vapors at 100 °C, for 30 min. Lastly, the sample was heated at increasing temperatures, and the IR spectrum was recorded at defined temperature intervals.

In a typical benzoylation reaction, phenol (Sigma-Aldrich, St. Louis, MO, USA) (0.1 g), benzoic acid (Sigma-Aldrich, St. Louis, MO, USA) (0.13 g) (molar ratio phenol/benzoic acid = 1) and the catalyst (10 wt. % with respect to the reactants) were loaded without solvent in a sealed test tube equipped with a magnetic stirrer.

The test tube was then placed in an oil bath and heated up to the temperature desired (190–200 °C). After the reaction, the crude was cooled, diluted with HPLC acetone, filtered and analyzed. The reaction mixture was analyzed with a Thermo Focus from Thermo Fisher (Waltham, MA, USA) GC gas-chromatograph equipped with an FID detector and Agilent (Santa Clara, CA, USA) HP-5 column, using *n*-decane as the internal standard. The analysis conditions were: 60 °C for 0 min, 40 °C/min up to 280 °C, 2 min at 280 °C.

4. Conclusions

Here, we report on the unprecedented reaction of phenol benzoylation with benzoic acid catalyzed by zeolites, which was carried out in the absence of solvent; this reaction is aimed at the production of hydroxybenzophenones, compounds of great interest for the chemical industry. H-beta zeolites offered a superior performance, in terms not only of activity, but also of selectivity to the desired compounds. The reaction network was found to consist of the primary reaction of phenol esterification with the formation of phenyl benzoate, which is then consecutively transformed into hydroxybenzophenones by means of an intermolecular reaction between the ester and phenol; conversely, the contribution of the Fries rearrangement seemed to be negligible. With H-beta zeolites, the Si/Al ratio significantly affected performance, especially in terms of the selectivity ratio between the two hydroxybenzophenone isomers. An important role was played by Lewis-type acid sites, which were most abundant with Al-richer zeolites. Conversely, in the case of H-Y zeolites, the transformation of the ester into the desired ketones was greatly inhibited.

Author Contributions: G.G., S.P., F.B., P.M. and M.A. designed and performed experiments, and analyzed data. F.C. analyzed data and wrote the paper.

Conflicts of Interest: The authors declare no conflict of interest.

References

1. Sartori, G.; Maggi, R. Use of solid catalysts in Friedel-Crafts acylation reactions. *Chem. Rev.* **2006**, *106*, 1077–1104.
2. Bejblová, M.; Procházková, D.; Čejka, J. Acylation Reactions over Zeolites and Mesoporous Catalysts. *ChemSusChem* **2009**, *2*, 486–499.
3. Bolognini, M.; Cavani, F.; Cimini, M.; Pozzo, L.D.; Maselli, L.; Venerito, D.; Pizzoli, F.; Veronesi, G. An environmentally friendly synthesis of 2,4-dihydroxybenzophenone by the single-step *o*-mono-benzoylation of 1,3-dihydroxybenzene (resorcinol) and Fries rearrangement of intermediate resorcinol monobenzoate: the activity of acid-treated montmorillonite. *Comptes Rendus Chim.* **2004**, *7*, 143–150.
4. Singh, A.P.; Bhattacharya, D.; Sharma, S. Benzoylation of toluene with benzoyl chloride over zeolite catalysts. *J. Mol. Catal. A* **1995**, *102*, 139–145.

5. Bhattacharya, D.; Sharma, S.; Singh, A.P. Selective benzoylation of naphthalene to 2-benzoylnaphthalene using zeolite H-beta catalysts. *Appl. Catal. A* **1997**, *150*, 53–62.

6. Jacob, B.; Sugunan, S.; Singh, A.P. Selective benzoylation of *o*-xylene to 3,4-dimethylbenzophenone using various zeolite catalysts. *J. Mol. Catal. A* **1999**, *139*, 43–53.

7. Andy, P.; Garcia-Martinez, J.; Lee, G.; Gonzalez, H.; Jones, C.W.; Davis, M.E. Acylation of 2-Methoxynaphthalene and Isobutylbenzene over Zeolite Beta. *J. Catal.* **2000**, *192*, 215–223.

8. Botella, P.; Corma, A.; López-Nieto, J.M.; Valencia, S.; Jacquot, R. Acylation of Toluene with Acetic Anhydride over Beta Zeolites: Influence of Reaction Conditions and Physicochemical Properties of the Catalyst. *J. Catal.* **2000**, *195*, 161–168.

9. Moreau, P.; Finiels, A.; Meric, P. Acetylation of dimethoxybenzenes with acetic anhydride in the presence of acidic zeolites. *J. Mol. Catal. A* **2000**, *154*, 185–192.

10. Chaube, V.D.; Moreau, P.; Finiels, A.; Ramaswamy, A.V.; Singh, A.P. Propionylation of phenol to 4-hydroxypropiophenone over zeolite H-beta. *J. Mol. Catal. A* **2001**, *174*, 255–264.

11. Neves, I.; Jayat, F.; Magnoux, P.; Perot, G.; Ribeiro, F.R.; Gubelmann, M.; Guisnet, M. Acylation of phenol with acetic acid over a HZSM-5 zeolite: the reaction scheme. *J. Mol. Catal.* **1994**, *93*, 169–179.

12. Escola, J.M.; Davis, M.E. Acylation of biphenyl with acetic anhydride and carboxylic acids over zeolite catalysts. *Appl. Catal. A* **2001**, *214*, 111–120.

13. Botella, P.; Corma, A.; Sastre, G. Al-ITQ-7, a Shape-Selective Zeolite for Acylation of 2-Methoxynaphthalene. *J. Catal.* **2001**, *197*, 81–90.

14. Venkatesan, C.; Jaimol, T.; Moreau, P.; Finiels, A.; Ramaswamy, A.V.; Singh, A.P. Liquid phase selective benzoylation of chlorobenzene. *Catal. Lett.* **2001**, *75*, 119–123.

15. Raja, T.; Singh, A.P.; Ramaswamy, A.V.; Finiels, A.; Moreau, P. Benzoylation of 1,2-dimethoxybenzene with benzoic anhydride and substituted benzoyl chlorides over large pore zeolites. *Appl. Catal. A* **2001**, *211*, 31–39.

16. Chavan, S.; Anand, R.; Pasupathy, K.; Rao, B. Catalytic acetylation of alcohols, phenols, thiols and amines with zeolite H-FER under solventless conditions. *Green Chem.* **2001**, 320–322.

17. Jaimol, T.; Moreau, P.; Finiels, A.; Ramaswamy, A.V.; Singh, A.P. Selective propionylation of veratrole to 3,4-dimethoxypropiophenone using zeolite H-beta catalysts. *Appl. Catal. A* **2001**, *214*, 1–10.

18. Bigi, F.; Carloni, S.; Flego, C.; Maggi, R.; Mazzacani, A.; Rastelli, M.; Sartori, G. HY zeolite-promoted electrophilic acylation of methoxyarenes with linear acid chlorides. *J. Mol. Catal. A* **2002**, *178*, 139–146.

19. Chaube, V.D.; Moreau, P.; Finiels, A.; Ramaswamy, A.V.; Singh, A.P. A novel single step selective synthesis of 4-hydroxybenzophenone (4-HBP) using zeolite H-beta. *Catal. Lett.* **2002**, *79*, 89–94.

20. Chidambaram, M.; Venkatesan, C.; Moreau, P.; Finiels, A.; Ramaswamy, A.V.; Singh, A.P. Selective benzoylation of biphenyl to 4-phenylbenzophenone over zeolite H-beta. *Appl. Catal. A* **2002**, *224*, 129–140.

21. Guignard, C.; Pédron, V.; Richard, F.; Jacquot, R.; Spagnol, M.; Coustard, J.M.; Pérot, G. Acylation of veratrole by acetic anhydride over Hβ and HY zeolites: Possible role of di- and triketone by-products in the deactivation process. *Appl. Catal. A* **2002**, *234*, 79–90.

22. Meric, P.; Finiels, A.; Moreau, P. Kinetics of 2-methoxynaphthalene acetylation with acetic anhydride over dealuminated HY zeolites. *J. Mol. Catal. A* **2002**, *189*, 251–262.

23. Wang, H.; Zou, Y. Modified beta zeolite as catalyst for Fries rearrangement reaction. *Catal. Lett.* **2003**, *86*, 163–167.

24. Beers, A.E.W.; Van Bokhoven, J.A.; De Lathouder, K.M.; Kapteijn, F.; Moulijn, J.A. Optimization of zeolite Beta by steaming and acid leaching for the acylation of anisole with octanoic acid: A structure-activity relation. *J. Catal.* **2003**, *218*, 239–248.

25. Botella, P.; Corma, A.; Navarro, M.T.; Rey, F.; Sastre, G. On the shape selective acylation of 2-methoxynaphthalene over polymorph C of Beta (ITQ-17). *J. Catal.* **2003**, *217*, 406–416.

26. Choudhary, V.R.; Jana, S.K.; Patil, N.S.; Bhargava, S.K. Friedel-Crafts type benzylation and benzoylation of aromatic compounds over HY zeolite modified by oxides or chlorides of gallium and indium. *Microporous Mesoporous Mater.* **2003**, *57*, 21–35.

27. Heitling, E.; Roessner, F.; Van Steen, E. Origin of catalyst deactivation in Fries rearrangement of phenyl acetate over zeolite H-Beta. *J. Mol. Catal. A* **2004**, *216*, 61–65.

28. Ghiaci, M.; Abbaspur, A.; Kalbasi, R.J. Internal versus external surface active sites in ZSM-5 zeolite: Part 1. Fries rearrangement catalyzed by modified and unmodified H_3PO_4/ZSM-5. *Appl. Catal. A* **2006**, *298*, 32–39.

29. Wagholikar, S.G.; Niphadkar, P.S.; Mayadevi, S.; Sivasanker, S. Acylation of anisole with long-chain carboxylic acids over wide pore zeolites. *Appl. Catal. A* **2007**, *317*, 250–257.

30. Padró, C.L.; Sad, M.E.; Apesteguía, C.R. Acid site requirements for the synthesis of *o*-hydroxyacetophenone by acylation of phenol with acetic acid. *Catal. Today* **2006**, *116*, 184–190.

31. Padró, C.L.; Apesteguía, C.R. Acylation of phenol on solid acids: Study of the deactivation mechanism. *Catal. Today* **2005**, *107–108*, 258–265.

32. Freese, U.; Heinrich, F.; Roessner, F. Acylation of aromatic compounds on H-Beta zeolites. *Catal. Today* **1999**, *49*, 237–244.

33. Armengol, E.; Corma, A.; Fernández, L.; García, H.; Primo, J. Acid zeolites as catalysts in organic reactions. Acetylation of cyclohexene and 1-methylcyclohexene. *Appl. Catal. A* **1997**, *158*, 323–335.

34. Heidekum, A.; Harmer, M.A.; Hoelderich, W.F. Highly Selective Fries Rearrangement over Zeolites and Nafion in Silica Composite Catalysts: A Comparison. *J. Catal.* **1998**, *176*, 260–263.

35. Vogt, A.; Kouwenhoven, H.W.; Prins, R. Fries rearrangement over zeolitic catalysts. *Appl. Catal. A* **1995**, *123*, 37–49.

36. Wei, H.; Liu, K.; Xie, S.; Xin, W.; Li, X.; Liu, S.; Xu, L. Determination of different acid sites in Beta zeolite for anisole acylation with acetic anhydride. *J. Catal.* **2013**, *307*, 103–110.

37. Subba Rao, Y.V.; Kulkarni, S.J.; Subrahmanyam, M.; Rama Rao, A.V. An improved acylation of phenol over modified ZSM-5 catalysts. *Appl. Catal. A* **1995**, *133*, L1–L6.

38. Padró, C.L.; Apesteguía, C.R. Gas-phase synthesis of hydroxyacetophenones by acylation of phenol with acetic acid. *J. Catal.* **2004**, *226*, 308–320.

39. Hou, Q.; Zheng, B.; Bi, C.; Luan, J.; Zhao, Z.; Guo, H.; Wang, G.; Li, Z. Liquid-phase cascade acylation/dehydration over various zeolite catalysts to synthesize 2-methylanthraquinone through an efficient one-pot strategy. *J. Catal.* **2009**, *268*, 376–383.

40. Fang, R.; Harvey, G.; Kouwenhoven, H.W.; Prins, R. Effects of non-framework alumina in the acylation of xylene over USY. *Appl. Catal. A* **1995**, *130*, 67–77.

41. Patil, S.P.; Yadav, G.D. Selective acylation of 2 methoxynaphthalene by large pore zeolites: Catalyst selection through molecular modeling. *Comput. Biol. Chem.* **2003**, *27*, 393–404.

42. Padró, C.L.; Rey, E.A.; González Peña, L.F.; Apesteguía, C.R. Activity, selectivity and stability of Zn-exchanged NaY and ZSM5 zeolites for the synthesis of *o*-hydroxyacetophenone by phenol acylation. *Microporous Mesoporous Mater.* **2011**, *143*, 236–242.

43. Guisnet, M.; Lukyanov, D.B.; Jayat, F.; Magnoux, P.; Hap, H. Kinetic modeling of phenol acylation with acetic acid on HZSM5. *Ind. Eng. Chem. Res.* **1995**, *34*, 1624–1629.

44. Rohan, D.; Canaff, C.; Magnoux, P.; Guisnet, M. Origin of the deactivation of HBEA zeolites during the acylation of phenol with phenylacetate. *J. Mol. Catal. A* **1998**, *129*, 69–78.

45. Chandra Shekara, B.M.; Jai Prakash, B.S.; Bhat, Y.S. Microwave-induced deactivation-free catalytic activity of BEA zeolite in acylation reactions. *J. Catal.* **2012**, *290*, 101–107.

46. Jayat, F.; Picot, M.J.S.; Guisnet, M. Solvent effects in liquid phase Fries rearrangement of phenyl acetate over a HBEA zeolite. *Catal. Lett.* **1996**, *41*, 181–187.

47. Li, A.-X.; Li, T.-S.; Ding, T.-H. Montmorillonite K-10 and KSF as remarkable acetylation catalysts. *Chem. Commun.* **1997**, 1389–1390.

48. Li, T.; Li, A. Acylation of alcohols, phenols, thiols and amines. *J. Chem. Soc. Perkin Trans. 1* **1998**, 1913–1917.

49. Choudary, B.M.; Bhaskar, V.; Lakshmi Kantam, M.; Koteswara Rao, K.; Raghavan, K.V. Acylation of alcohols with carboxylic acids; the evolution of compatible acidic sites in montmorillonites. *Green Chem.* **2000**, *2*, 67–70.

50. Békássy, S.; Farkas, J.; Agai, B.; Figueras, F. Selectivity of *C*-versus *O*-acylation of diphenols by clay catalysts. I. Acylation of resorcinol with phenylacetyl chloride. *Top. Catal.* **2000**, *13*, 287–290.

51. Janković, L.; Komadel, P. Metal cation-exchanged montmorillonite catalyzed protection of aromatic aldehydes with Ac_2O. *J. Catal.* **2003**, *218*, 227–233.

52. De Castro, C.; Primo, J.; Corma, A. Heteropolyacids and large-pore zeolites as catalysts in acylation reactions using α, β-unsaturated organic acids as acylating agents. *J. Mol. Catal. A* **1998**, *134*, 215–222.

53. Castro, C.; Corma, A.; Primo, J. On the acylation reactions of anisole using α,β-unsaturated organic acids as acylating agents and solid acids as catalysts: A mechanistic overview. *J. Mol. Catal. A* **2002**, *177*, 273–280.

135

54. Kozhevnikova, E.F.; Rafiee, E.; Kozhevnikov, I.V. Fries rearrangement of aryl esters catalysed by heteropoly acid: Catalyst regeneration and reuse. *Appl. Catal. A* **2004**, *260*, 25–34.

55. Sarsani, V.S.R.; Lyon, C.J.; Hutchenson, K.W.; Harmer, M.A.; Subramaniam, B. Continuous acylation of anisole by acetic anhydride in mesoporous solid acid catalysts: Reaction media effects on catalyst deactivation. *J. Catal.* **2007**, *245*, 184–190.

56. Bachiller-Baeza, B.; Anderson, J.A. FTIR and reaction studies of the acylation of anisole with acetic anhydride over supported HPA catalysts. *J. Catal.* **2004**, *228*, 225–233.

57. Yadav, G.D.; Asthana, N.S.; Kamble, V.S. Cesium-substituted dodecatungstophosphoric acid on K-10 clay for benzoylation of anisole with benzoyl chloride. *J. Catal.* **2003**, *217*, 88–99.

58. Yadav, G.D.; Asthana, N.S.; Kamble, V.S. Friedel-crafts benzoylation of *p*-xylene over clay supported catalysts: Novelty of cesium substituted dodecatungstophosphoric acid on K-10 clay. *Appl. Catal. A* **2003**, *240*, 53–69.

59. Yadav, G.D.; Asthana, N.S.; Salgaonkar, S.S. Regio-selective benzoylation of xylenes over caesium modified heteropolyacid supported on K-10 clay. *Clean Technol. Environ. Policy* **2004**, *6*, 105–113.

60. Yadav, G.D.; George, G. Single step synthesis of 4-hydroxybenzophenone via esterification and Fries rearrangement: Novelty of cesium substituted heteropoly acid supported on clay. *J. Mol. Catal. A* **2008**, *292*, 54–61.

61. Heravi, M.M.; Behbahani, F.K.; Bamoharram, F.F. $H_{14}[NaP_5W_{30}O_{110}]$: A heteropoly acid catalyzed acetylation of alcohols and phenols in acetic anhydride. *J. Mol. Catal. A* **2006**, *253*, 16–19.

62. Kozhevnikov, I.V. Friedel-Crafts acylation and related reactions catalysed by heteropoly acids. *Appl. Catal. A* **2003**, *256*, 3–18.

63. Choudhary, V.R.; Jana, S.K.; Mandale, A.B. Highly active, reusable and moisture insensitive catalyst obtained from basic Ga-Mg-hydrotalcite anionic clay for Friedel-Crafts type benzylation and acylation reactions. *Catal. Lett.* **2001**, *74*, 95–98.

64. Choudhary, V.R.; Jana, S.K.; Narkhede, V.S. Benzylation and benzoylation of substituted benzenes over solid catalysts containing Ga- and Mg-oxides and/or chlorides derived from Ga-Mg-hydrotalcite by its HCl pre-treatment or calcination. *Appl. Catal. A* **2002**, *235*, 207–215.

65. Veverkova, E.; Meciarova, M.; Gotov, B.; Toma, S. Microwave assisted acylation of methoxyarenes catalyzed by EPZG catalyst. *Green Chem.* **2002**, *4*, 361–365.

66. Trunschke, A.; Deutsch, J.; Müller, D.; Lieske, H.; Quaschning, V.; Kemnitz, E. Nature of surface deposits on sulfated zirconia used as catalyst in the benzoylation of anisole. *Catal. Lett.* **2002**, *83*, 271–279.

67. Zane, F.; Melada, S.; Signoretto, M.; Pinna, F. Active and recyclable sulphated zirconia catalysts for the acylation of aromatic compounds. *Appl. Catal. A* **2006**, *299*, 137–144.

68. Signoretto, M.; Breda, A.; Somma, F.; Pinna, F.; Cruciani, G. Mesoporous sulphated zirconia by liquid-crystal templating method. *Microporous Mesoporous Mater.* **2006**, *91*, 23–32.

69. Ghedini, E.; Signoretto, M.; Pinna, F.; Cerrato, G.; Morterra, C. Gas and liquid phase reactions on MCM-41/SZ catalysts. *Appl. Catal. B* **2006**, *67*, 24–33.

70. Ghedini, E.; Signoretto, M.; Pinna, F.; Cruciani, G. Mesoporous silica-zirconia systems for catalytic applications. *Catal. Lett.* **2008**, *125*, 359–370.

71. Melada, S.; Signoretto, M.; Somma, F.; Pinna, F.; Cerrato, G.; Meligrana, G.; Morterra, C. Gas-and liquid-phase reactions on sulphated zirconia prepared by precipitation. *Catal. Lett.* **2004**, *94*, 193–198.

72. Deutsch, J.; Trunschke, A.; Mu, D. Different acylating agents in the synthesis of aromatic ketones on sulfated zirconia. *Catal. Lett.* **2003**, *88*, 9–15.

73. Signoretto, M.; Torchiaro, A.; Breda, A.; Pinna, F.; Cerrato, G.; Morterra, C. Study on reuse of metal oxide-promoted sulphated zirconia in acylation reactions. *Appl. Catal. B* **2008**, *84*, 363–371.

74. Ratnam, K.J.; Reddy, R.S.; Sekhar, N.S.; Kantam, M.L.; Figueras, F. Sulphated zirconia catalyzed acylation of phenols, alcohols and amines under solvent free conditions. *J. Mol. Catal. A* **2007**, *276*, 230–234.

75. Breda, A.; Signoretto, M.; Ghedini, E.; Pinna, F.; Cruciani, G. Acylation of veratrole over promoted SZ/MCM-41 catalysts: Influence of metal promotion. *Appl. Catal. A* **2006**, *308*, 216–222.

76. Clark, J.H.; Dekamin, M.G.; Moghaddam, F.M. Genuinely catalytic Fries rearrangement using sulfated zirconia. *Green Chem.* **2002**, *4*, 366–368.

77. Patil, P.T.; Malshe, K.M.; Kumar, P.; Dongare, M.K.; Kemnitz, E. Benzoylation of anisole over borate zirconia solid acid catalyst. *Catal. Commun.* **2002**, *3*, 411–416.

78. Chidambaram, M.; Venkatesan, C.; Rajamohanan, P.R.; Singh, A.P. Synthesis of acid functionalized mesoporous Zr-O-SO$_2$-CF$_3$ catalysts; heterogenization of CF$_3$SO$_3$H over mesoporous Zr(OH)$_4$. *Appl. Catal. A* **2003**, *244*, 27–37.

79. Sakthivel, R.; Prescott, H.A.; Deutsch, J.; Lieske, H.; Kemnitz, E. Synthesis, characterization, and catalytic activity of SO$_4$/Zr$_{1-x}$Sn$_x$O$_2$. *Appl. Catal. A* **2003**, *253*, 237–247.

80. Chidambaram, M.; Curulla-Ferre, D.; Singh, A.P.; Anderson, B.G. Synthesis and characterization of triflic acid-functionalized mesoporous Zr-TMS catalysts: Heterogenization of CF$_3$SO$_3$H over Zr-TMS and its catalytic activity. *J. Catal.* **2003**, *220*, 442–456.

81. Parambadath, S.; Chidambaram, M.; Singh, A.P. Synthesis, characterization and catalytic properties of benzyl sulphonic acid functionalized Zr-TMS catalysts. *Catal. Today* **2004**, *97*, 233–240.

82. Landge, S.M.; Chidambaram, M.; Singh, A.P. Benzoylation of toluene with p-toluoyl chloride over triflic acid functionalized mesoporous Zr-TMS catalyst. *J. Mol. Catal. A* **2004**, *213*, 257–266.

83. Gawande, M.B.; Deshpande, S.S.; Sonavane, S.U.; Jayaram, R.V. A novel sol-gel synthesized catalyst for Friedel-Crafts benzoylation reaction under solvent-free conditions. *J. Mol. Catal. A* **2005**, *241*, 151–155.

84. Yadav, G.D.; George, G. Friedel-Crafts acylation of anisole with propionic anhydride over mesoporous superacid catalyst UDCaT-5. *Microporous Mesoporous Mater.* **2006**, *96*, 36–43.

85. Jain, S.K.; Meena, S.; Singh, V.P.; Bharate, J.B.; Joshi, P.; Singh, V.P.; Vishwakarma, R.A.; Bharate, S.B. KF/alumina catalyzed regioselective benzylation and benzoylation using solvent-free grind-stone chemistry. *RSC Adv.* **2012**, *2*, 8929–8933.

86. Rahmatpour, A. Polystyrene-supported GaCl$_3$: A new, highly efficient and recyclable heterogeneous Lewis acid catalyst for acetylation and benzoylation of alcohols and phenols. *Comptes Rendus Chim.* **2012**, *15*, 1048–1054.

87. Ramani, T.; Umadevi, P.; Prasanth, K.L.; Sreedhar, B. Synthesis of ortho -Hydroxybenzophenones Catalyzed by Magnetically Retrievable Fe$_3$O$_4$ Nanoparticles under Ligand-Free Conditions. *European J. Org. Chem.* **2013**, 6021–6026.

88. Tamaddon, F.; Amrollahi, M.A.; Sharafat, L. A green protocol for chemoselective O-acylation in the presence of zinc oxide as a heterogeneous, reusable and eco-friendly catalyst. *Tetrahedron Lett.* **2005**, *46*, 7841–7844.

89. Moghaddam, F.M.; Saeidian, H. Controlled microwave-assisted synthesis of ZnO nanopowder and its catalytic activity for O-acylation of alcohol and phenol. *Mater. Sci. Eng. B.* **2007**, *139*, 265–269.

90. Farhadi, S.; Jahanara, K. ZnAl$_2$O$_4$@SiO$_2$ nanocomposite catalyst for the acetylation of alcohols, phenols and amines with acetic anhydride under solvent-free conditions. *Chin. J. Catal.* **2014**, *35*, 368–375.

91. Prabhu, A.; Kumaresan, L.; Palanichamy, M.; Murugesan, V. Synthesis and characterization of aluminium incorporated mesoporous KIT-6: Efficient catalyst for acylation of phenol. *Appl. Catal. A* **2009**, *360*, 59–65.

92. Kumar, P.; Pandey, R.K.; Bodas, M.S.; Dagade, S.P.; Dongare, M.K.; Ramaswamy, A.V. Acylation of alcohols, thiols and amines with carboxylic acids catalyzed by yttria-zirconia-based Lewis acid. *J. Mol. Catal. A* **2002**, *181*, 207–213.

93. Signoretto, M.; Ghedini, E.; Menegazzo, F.; Cerrato, G.; Crocellà, V.; Bianchi, C.L. Aerogel and xerogel WO3/ZrO2 samples for fine chemicals production. *Microporous Mesoporous Mater.* **2013**, *165*, 134–141.

94. Alam, M.; Rahman, A.; Alandis, N.M.; Shaik, M.R. Ni/Silica catalyzed acetylation of phenols and naphthols: An eco-friendly approach. *Arab. J. Chem.* **2014**, *7*, 53–56.

95. Zakrzewski, J.; Szymanowski, J. 2-Hydroxybenzophenone UV-absorbers containing 2,2,6,6-tetramethylpiperidine (HALS) group—benzoylation of corresponding phenol derivatives. *Polym. Degrad. Stab.* **2000**, *67*, 279–283.

96. Singh, R.P.; Kamble, R.M.; Chandra, K.L.; Saravanan, P.; Singh, V.K. An efficient method for aromatic Friedel-Crafts alkylation, acylation, benzoylation, and sulfonylation reactions. *Tetrahedron* **2001**, *57*, 241–247.

97. Saravanan, P.; Singh, V.K. An efficient method for acylation reactions. *Tetrahedron Lett.* **1999**, *40*, 2611–2614.

98. Mohammadpoor-baltork, I.; Aliyan, H.; Khosropour, A.R. Bismuth (III) salts as convenient and efficient catalysts for the selective acetylation and benzoylation of alcohols and phenols. *Tetrahedron* **2001**, *57*, 5851–5854.

99. Kobayashi, S.; Moriwaki, M.; Hachiya, I. Hafnium trifluoromethanesulfonate (Hf(OTf)$_4$) as an efficient catalyst in the Fries rearrangement and direct acylation of phenol and naphthol derivatives. *Tetrahedron Lett.* **1996**, *37*, 2053–2056.

100. Kobayashi, S.; Moriwaki, M.; Hachiya, I. Catalytic direct *C*-acylation of phenol and naphthol derivatives using carboxylic acids as acylating reagents. *Tetrahedron Lett.* **1996**, *37*, 4183–4186.

101. Emeis, C.A. Determination of Integrated Molar Extinction Coefficients for Infrared Absorption Bands of Pyridine Adsorbed on Solid Acid Catalysts. *J. Catal.* **1993**, *141*, 347–354.

102. Shimizu, K.; Satsuma, A. Toward a rational control of solid acid catalysis for green synthesis and biomass conversion. *Energy Environ. Sci.* **2011**, *4*, 3140–3153.

103. Phung, T.K.; Busca, G. On the Lewis acidity of protonic zeolites. *Appl. Catal. A* **2015**, *504*, 151–157.

104. Hoefnagel, A.J.; Van Bekkum, H. Direct Fries reaction of resorcinol with benzoic acids catalyzed by zeolite H-beta. *Appl. Catal. A* **1993**, *97*, 87–102.

Facile Synthesis of Yolk/Core-Shell Structured TS-1@Mesosilica Composites for Enhanced Hydroxylation of Phenol

Houbing Zou, Qingli Sun, Dongyu Fan, Weiwei Fu, Lijia Liu and Runwei Wang

Abstract: In the current work, we developed a facile synthesis of yolk/core-shell structured TS-1@mesosilica composites and studied their catalytic performances in the hydroxylation of phenol with H_2O_2 as the oxidant. The core-shell TS-1@mesosilica composites were prepared via a uniform coating process, while the yolk-shell TS-1@mesosilica composite was prepared using a resorcinol-formaldehyde resin (RF) middle-layer as the sacrificial template. The obtained materials were characterized by X-ray diffraction (XRD), N_2 sorption, Fourier transform infrared spectoscopy (FT-IR) UV-Visible spectroscopy, scanning electron microscopy (SEM) and transmission electron microscopy (TEM). The characterization results showed that these samples possessed highly uniform yolk/core-shell structures, high surface area (560–700 $m^2\ g^{-1}$) and hierarchical pore structures from oriented mesochannels to zeolite micropores. Importantly, owing to their unique structural properties, these composites exhibited enhanced activity, and also selectivity in the phenol hydroxylation reaction.

Reprinted from *Catalysts*. Cite as: Zou, H.; Sun, Q.; Fan, D.; Fu, W.; Liu, L.; Wang, R. Facile Synthesis of Yolk/Core-Shell Structured TS-1@Mesosilica Composites for Enhanced Hydroxylation of Phenol. *Catalysts* **2015**, *5*, 2134–2146.

1. Introduction

Hydroxylation of phenol with hydrogen peroxide (H_2O_2) as oxidant has been widely applied for the production of two useful compounds, *viz.* catechol and hydroquinone [1,2]. Catechol is used as agrichemical, and in the synthesis of aromatic compounds, whereas hydroquinone is used as an antioxidant in elastics pigment. Various homogeneous and heterogeneous catalysts have been developed to achieve hydroxylation, including Fenton's reagent [3], Fe-containing molecular sieves [4–7], Cu-containing materials [8], and Ti-containing porous materials [9–14]. Among these functional materials, TS-1 zeolite is one of the most efficient solid catalysts with high activity and selectivity coupled with environmentally benign catalytic performance [13,14]. However, the small pores in this zeolite make the diffusion of substrate and products through the channel difficult, thereby leading to limited catalytic performance. Therefore, many studies were directed toward improving the accessibility of the active sites in TS-1 catalysts. To this end, introduction of a second mesopore into TS-1 zeolite crystals to form hierarchical structured TS-1 is an efficient strategy [15–21]. While significant achievements have been realized in this approach, the methods employed usually involved either cost-intensive mesostructured templates [15,19–21] or complex synthetic procedures [17,18]. In this regard, synthesis of TS-1 zeolite-based composites is an alternative approach.

Recently, hierarchical core-shell structured composites with a microporous zeolite crystal core and a mesoporous silica/carbon shell have received much attention [22–27] and have been widely applied in the some important catalytic reactions, such as epoxidation of alkene [23], cyclohexanone ammoximation [24], methanol to propylene (MTP) conversion [22] and catalytic cracking of large molecules [25]. Owing to the unique structural properties and the bimodal pore structure with open junction and retainable diffusion efficiency, these zeolite-based core-shell composite materials exhibit superior catalytic activity and selectivity. Moreover, multifunctional core-shell composites with various catalytic sites have been also designed to catalyze one pot cascade reactions. For-example, $Rh(OH)_3$ species were supported on the core-shell structured TS-1@KCC-1 composite for one-pot synthesis of benzamide from benzaldehyde, ammonia and hydrogen peroxide [28]. However, TS-1 zeolite-based yolk/core-shell composites have been rarely reported for application in hydroxylation of phenol. In the current work, we devoted our efforts to study the catalytic properties of TS-1 zeolite-based yolk/core-shell composites in the hydroxylation of phenol and determine the relationship between the mircostructure of the material and its catalytic performance.

To study the catalyzed hydroxylation of phenol, specifically we prepared yolk/core-shell structured TS-1@mesosilica composites and used H_2O_2 as the oxidant. The core-shell structured TS-1@mesosilica composite was prepared via a uniform coating process, whereas the yolk-shell structured TS-1@mesosilica composite was prepared, for the first time, using a resorcinol-formaldehyde resin (RF) middle-layer as the sacrificial template. The obtained materials were characterized by X-ray diffraction (XRD), N_2 sorption, FT-IR, UV-Visible spectroscopy, scanning electron microscopy (SEM) and transmission electron microscopy (TEM).

2. Results and Discussion

2.1. Synthesis and Characterization of CS-TS-1@mSiO₂

The micro-sized TS-1 single crystal was prepared by hydrothermal method. The SEM image (Figure 1a) and TEM image (Figure 2a,b) indicate that the TS-1 sample consist of uniform monodisperse nanocrystals with a particle size of 290–350 nm. Core-shell structured TS-1@mesosilica composites (designated as CS-TS-1@mSiO$_{2-x}$, x is the thickness of mSiO$_2$ layer) were obtained through a uniform coating process using water-ethanol (2:1), ammonia aqueous, CTAB (hexadecyltrimethylammonium bromide) and TEOS (tetraethylorthosilicate) as solvent, basic catalyst, surfactant and precursor, respectively [29,30]. The SEM micrograph (Figure 1b) shows that the particles are oval in shape with a large diameter (350–400 nm), suggesting that an mSiO$_2$ layer of approximately 30 nm was formed on the surface of the TS-1 crystals. From the TEM image in Figure 2d, it can be clearly seen that every TS-1 crystal is encapsulated with a 30 nm mSiO$_2$ layer. Furthermore, the HRTEM images in Figure 2e,f reveal that the mesopores are nearly perpendicular to the microporous frameworks (directed by red arrows) and there are close connections between the mesoporous silica shells and the zeolite cores, facilitating efficient transportation of reactants into active microporous framework through the mesochannels. It is worth mentioning that the current coating process did not involve any surface modifiers or additions that were often necessary for obtaining uniform core-shell structured zeolite@mSiO$_2$ composites with oriented mesochannels in mSiO$_2$ shell [22,23], which can be attributed to the flexibility in current synthesis system [29,30]. In addition, the thickness of mesoporous silica shells could be tailored by adjusting the amount of TEOS precursor. For example, when the amount of TEOS was increased to 0.20 mL, we obtained CS-TS-1@mSiO$_2$ with a uniform mSiO$_2$ shell of 50 nm (Figure 2g–i). This control on thickness of the mSiO$_2$ shell is significant in regulating the catalytic properties of materials.

Figure 1. (a) SEM images of pristine TS-1; (b) CS-TS-1@mSiO$_2$-30 and (c) CS-TS-1@mSiO$_2$-50.

The nanoporous structure of CS-TS-1@mSiO$_2$ composites was further characterized by XRD and nitrogen sorption analysis. Apparently, the small-angle XRD pattern of CS-TS-1@mSiO$_2$-30 (Figure 3A) shows a broad diffraction peak at a 2θ value of approximately 2.4° whereas pristine TS-1 shows no diffraction peaks. This suggests an unordered mesostructure in mSiO$_2$ shell. Moreover, a broad diffraction peak with higher intensity and a weak diffraction peak are simultaneously present at 2θ values of 2.4° and 4.8° in the small-angle XRD pattern of CS-TS-1@mSiO$_2$-50 (Figure 3A), revealing that the mesochannels in mSiO$_2$ shell are partly ordered. These results indicate that the order degree of the mesochannels increased with increase in the thickness of them SiO$_2$ shell, which is in accordance with the TEM results and was likely caused due to the increase in the domain-size of the ordered mesostructure [25]. The wide-angle XRD patterns of both the composites (Figure 3B) exhibit the same characteristic diffraction peaks as those of the pristine TS-1, assigned to the typical MFI structure. The lower diffraction intensity is probably due to the shielding effects of the mesosilica shell.

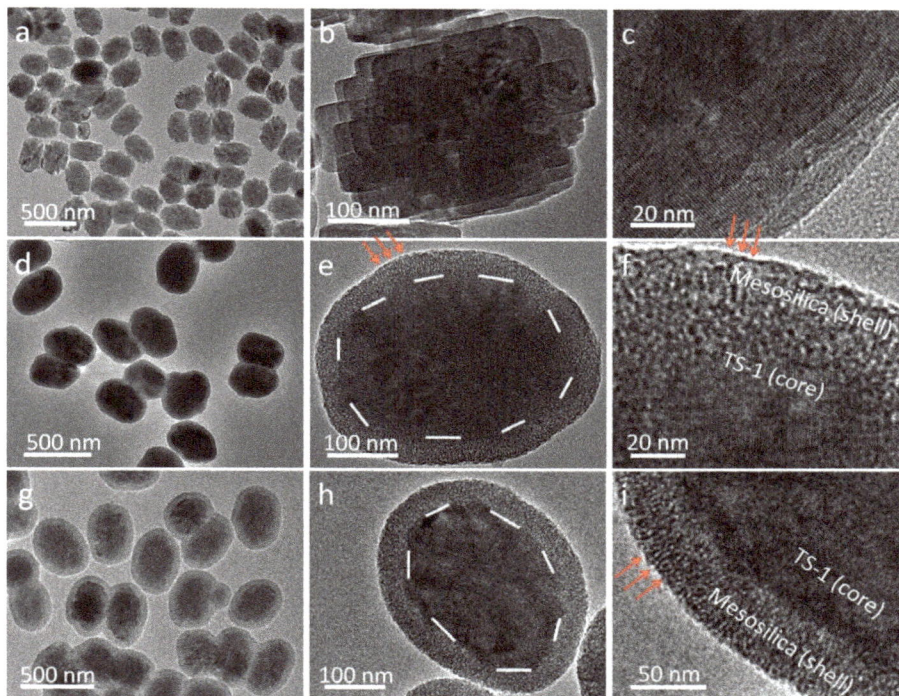

Figure 2. Low- moderate- and high-magnification TEM images of (**a**–**c**) pristine TS-1; (**d**–**f**) CS-TS-1@mSiO$_2$-30 and (**g**–**i**) CS-TS-1@mSiO$_2$-50.

In the N$_2$ sorption analysis, the pristine TS-1 shows a type-I isotherm according to the IUPAC (International Union of Pure and Applied Chemistry) classification with a sharp uptake in the P/P_0 range 0–0.01, a characteristic behavior of completely microporous materials (Figure 3C). On the other hand, the CS-TS-1@mSiO$_2$ composites with different shell thickness (30 nm or 50 nm) exhibit a similar uptake at low P/P_0 (type-I curves); however, an additional uptake following typical type-IV curves with a capillary condensation step appears at moderate P/P_0 (0.2–0.6). A distinct hysteresis loop of H2 type is also clearly observed for both the composites. These phenomena reveal the bimodal-pore properties of the CS-TS-1@mSiO$_2$ composites from micropores to mesopores as well as the accessible microporous ZSM-5 cores covered with a mesoporous silica shell. Moreover, in comparison with CS-TS-1@mSiO$_2$-30, a higher uptake in the moderate P/P_0 of 0.2–0.6 and a more obvious H2 type hysteresis loop in the P/P_0 of 0.4–0.6 are apparent for the CS-TS-1@mSiO$_2$-50, further suggesting the intrinsic bimodal-pore properties and the tunable mesopore shells. The pore size distribution (PSD) of pristine TS-1 reflects undetectable mesopores and a distinct micropore with a diameter of 0.55 nm. By contrast, both the CS-TS-1@mSiO$_2$ show two distinct PSD models corresponding to micropores with a similar size to pristine TS-1 (around 0.55 nm) and mesopores with a Barrett-Joyner-Halenda (BJH) diameter centered at 2.6 nm, respectively (Figure 3D), which indicates the formation of uniform hierarchical pores. Both the CS-TS-1@mSiO$_2$ composites exhibit a high Brunauer-Emmett-Teller (BET) surface area of $\geq$ 550 m^2 g^{-1} of which the microporous surface area was $\geq$50 m^2 g^{-1} (Table 1), and a high total pore volume of $\geq$0.45 m^3 g^{-1}. The lower micropore surface area was possibly due to presence of a low portion of TS-1 in the composites. All the above results indicate that a highly uniform core-shell structured TS-1@mSiO$_2$ composite with tunable shell thickness was successfully synthesized via a simple coating process.

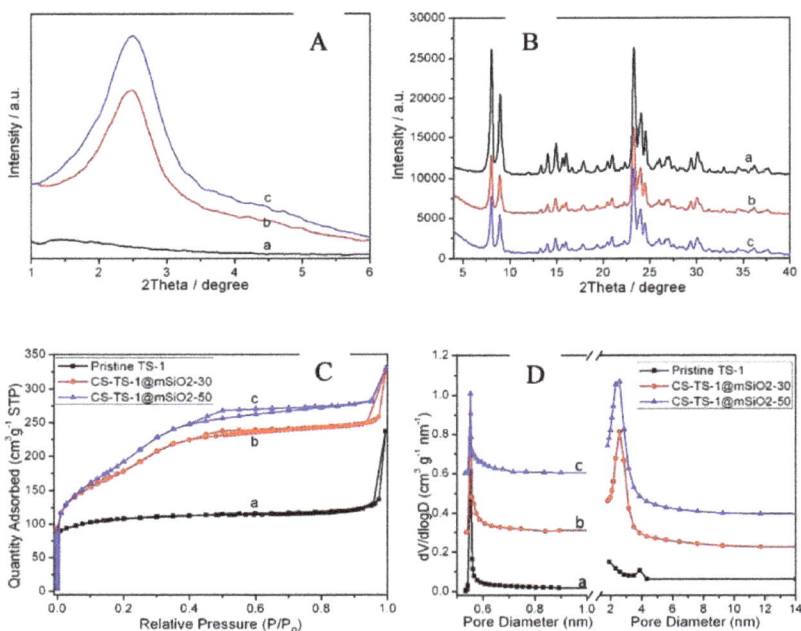

Figure 3. (A) Small-angle XRD patterns of (a) pristine TS-1, (b) CS-TS-1@mSiO$_2$-30 and (c) CS-TS-1@mSiO$_2$-50; **(B)** wide-angle XRD patterns of (a) pristine TS-1, (b) CS-TS-1@mSiO$_2$-30 and (c) CS-TS-1@mSiO$_2$-50; **(C)** N$_2$ sorption isotherms; **(D)** HK and BJH pore size distribution curves.

Table 1. Textural properties of pristine TS-1, core-shell structured TS-1@mesosilica composite (CS-TS-1@mSiO2) and yolk-shell structured TS-1@mesosilica composite (YS-TS-1@mSiO2) [a].

Sample	S_{BET} [m^2 g^{-1}]	S_{micro} [m^2 g^{-1}]	S_{ext} [m^2 g^{-1}]	V_{micro} [m^3 g^{-1}]	V_{ext} [m^3 g^{-1}]	D_{micro} [nm]	D_{meso} [nm]
pristine TS-1	368.5	249.3	119.2	0.12	0.26	0.55	-
CS-TS-1@mSiO$_2$-30	564.6	71.8	492.8	0.031	0.412	0.55	2.6
CS-TS-1@mSiO$_2$-50	635.9	53.2	582.7	0.027	0.454	0.55	2.6
YS-TS-1@mSiO$_2$	695.2	57.2	638.0	0.028	0.484	0.55	2.8

[a] S_{BET}: specific BET surface area; S_{micro}: micropore area; S_{ext}: external surface area; V_{micro}: micropore volume; V_{ext}: mesopore volume; D_{micro}: micropore diameter; D_{meso}: mesopore diameter.

The nature and coordination of Ti species in silica network was characterized by FT-IR and UV-Vis spectroscopy. The FT-IR spectrum of pristine TS-1 shows three different bands at 550 cm^{-1}, 804 cm^{-1} and 960 cm^{-1} (Figure 4A). The band at 550 and 804 cm^{-1} are assigned to δ(Si-O-Si) and ν(Si-O-Si), respectively. The peak at 960 cm^{-1} is often referred to as Ti-O-Si band and can be considered as a good indication of titanium substitution into the zeolite framework. In the UV-vis spectrum (Figure 4B), the maximum absorption peak appears around 210 nm, which further confirms the existence of the Ti framework. Furthermore, the presence of a weak absorption at 330 nm suggests the presence of a low amount of anatase phase. Importantly, both the CS-TS-1@mSiO$_2$ exhibit identical FT-IR and UV-Vis spectra as those of the pristine TS-1, revealing that the Ti active species was retained in the TS-1 core during the coating process. Additionally, the absence of the bands at 250–280 nm in these UV-Vis spectra reveals that amorphous Ti species is neither present in the TS-1 core nor in the mSiO$_2$ shell. These results indicate that the process of coating mSiO$_2$ did not have any negative effect on the catalytic behavior of TS-1.

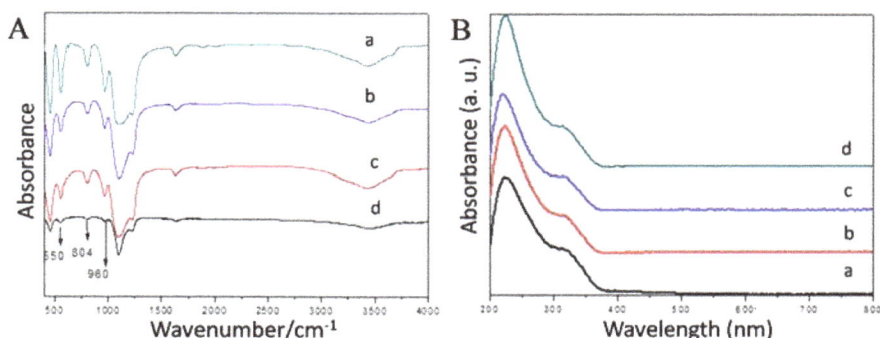

Figure 4. (**A**) FT-IR spectra and (**B**) UV-Vis spectra of (a) pristine TS-1, (b) CS-TS-1@mSiO$_2$-30, (c) CS-TS-1@mSiO$_2$-50 and (d) YS-TS-1@mSiO$_2$.

2.2. Synthesis and Characterization of YS-TS-1@mSiO₂

The yolk-shell structured TS-1@mesosilica composite (YS-TS-1@mSiO$_2$) was prepared using a resorcinol-formaldehyde resin (RF) middle-layer as the sacrificial template. Following a published report [31], the TS-1 crystals were first coated with the RF layer via simple cooperative template-directed coating method to form core-shell structured TS-1@RF. A successful coating was confirmed from the large decrease in the nitrogen uptake at low P/P_0 (0–0.01) (Figure 5e), and the microporous surface area from 249.3 m^2 g^{-1} to 36 m^2 g^{-1} of the synthesized TS-1@RF. A similar coating for CS-TS-1@mSiO$_2$-30 was also done to obtain sandwich type TS-1@RF@mSiO$_2$. This was followed by the removal of the middle-layer RF, surfactant CTAB and structure direct agent TPA$^+$ by calcination at 550 °C in air for 6 h, which yielded YS-TS-1@mSiO$_2$ with an open hierarchical pore structure from the inherent micropore in TS-1 core and the mesopore in mSiO$_2$ shell to nanocapsule. The sandwich type TS-1@RF@mSiO$_2$ displays a similar oval sharp as that of CS-TS-1@mSiO$_2$-30 and a larger particle diameter (410–450 nm) (Figure 5a). By contrast, the obtained YS-TS-1@mSiO$_2$ exhibits a distinct yolk-shell structure after the simple calcination process in air (Figure 5b,c). One mSiO$_2$ shell encapsulates one TS-1 crystal and a void space of around 15 nm between the TS-1 core and mSiO$_2$ shell is clearly observed from the TEM images (Figure 5b,c). Moreover, the SEM image (Figure 5d) shows some broken particles indicated by white arrows, further confirming the yolk-shell structure. In addition, identical with the CS-TS-1@mSiO$_2$ composites, the YS-TS-1@mSiO$_2$ also exhibit a retained uptake at low P/P_0 (type-I curves) and typical type-IV curves with a capillary condensation step at moderate P/P_0(0.2–0.6) as well as a distinct hysteresis loop of H2 type. On the other hand, the corresponding DFT PSD not only shows two sharp peaks centered at 0.55 nm and 2.8 nm that was ascribed to the micropores in TS-1 core and mesopores in mSiO$_2$ shell, respectively, but also a broad small peak in the range of 10–20 nm, which resulted from the void space between the TS-1 core and the mSiO$_2$ shell. This phenomenon is in good agreement with the TEM result. YS-TS-1@mSiO$_2$ shows a high BET surface area of 695 m^2 g^{-1} (Table 1). All these results demonstrated that we were able to successfully prepare a TS-1-based nanocomposite consisted of a TS-1 core and a mSiO$_2$ shell as well as a void space of around 15 nm between the core and the shell. The FT-IR and UV-Vis spectra (Figure 4A,B) also reveal that the nature and coordination of Ti species keeps identical with pristine TS-1.

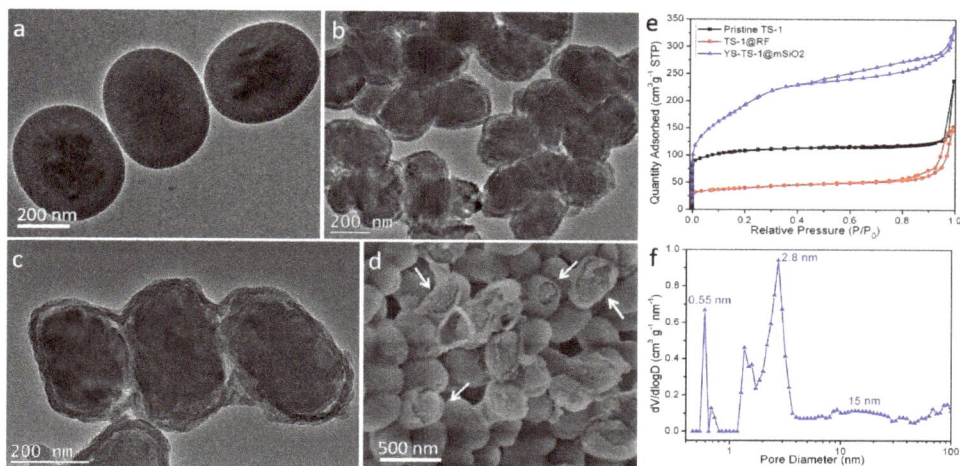

Figure 5. (a) TEM image of TS-1@RF@mSiO$_2$; (**b–d**) TEM images and SEM image of YS-TS-1@mSiO$_2$; (**e**) N$_2$ sorption isotherms of pristine TS-1 (●), TS-1@RF (●) and YS-TS-1@mSiO$_2$ (▲); (**f**) Relevant DFT pore size distribution curve of YS-TS-1@mSiO$_2$.

2.3. Catalytic Performance of CS/YS-TS-1@mSiO$_2$ in Hydroxylation of Phenol

Table 2 summarizes the catalytic results of various TS-1 catalysts in the hydroxylation of phenol using H$_2$O$_2$ as the oxidant and water as the solvent. The pristine TS-1 material (Entry 3) showed similar activity and selectivity with previous reports [12,13]. Notably, our prepared composites CS-TS-1@mSiO$_2$ (Entry 6) and YS-TS-1@mSiO$_2$ (Entry 9) gave a higher phenol conversion of 25.2% and 27.6% than that of pristine TS-1 (23.7%), respectively. The turnover frequency (TOF) of YS-TS-1@mSiO$_2$ (34.4) was even higher than some other hierarchical nanoporous TS-1 zeolite materials [18]. The enhanced activities should be attributed to the effect of external mesosilica layer that can effectively capture the large phenol molecules like a pump from the reaction solution due to the large specific surface area of mesopores. This "capture" effect was also reported for enhanced catalytic cracking reactions in previous publication [25]. Moreover, the void space between the mSiO$_2$ shell and the TS-1 core reinforced this positive "capture" effect for further enhancing the catalytic activity of material. In addition, we were aware that the selectivity of side product benzoquinone was slightly increased from 3.9% to 5.5% using our composites as the catalysts, which indicated that the external mesosilica layer and the void space also adsorbed products catechol and hydroquinone when capturing substrate phenol from reaction solution, and thus led to further oxidation. Moreover, it is worth mentioning that more product hydroquinone was produced in the cases of using composites CS-TS-1@mSiO$_2$ and YS-TS-1@mSiO$_2$ as catalysts. These catalytic

147

results clearly revealed that the external $mSiO_2$ shell and the void space between the $mSiO_2$ shell and the TS-1 core not only can capture substrate molecules like a pump for enhanced activity, but also adsorb product molecules, and thus posed diffusion resistance for negative selectivity.

Table 2. Catalytic activities of various TS-1 catalysts in hydroxylation of phenol [a].

Entry	Catalyst	Reaction Time	Conv. /%	Product Sel. [b] /%			HQ/CAT	TOF [c] /h^{-1}
				CAT	HQ	BQ		
1		1 h	6.2	10.3	1.5	88.2	0.15	
2	Pristine TS-1	3 h	16.6	60.7	25.1	14.1	0.41	29.6
3		5 h	23.7	55.1	41.0	3.9	0.74	
4		1 h	18.7	55.8	24.5	19.7	0.44	
5	CS-TS-1@mSiO$_2$	3 h	24.3	51.2	38.1	10.7	0.74	31.5
6		5 h	25.2	45.7	49.8	4.5	1.09	
7		1 h	20.7	49.7	24.1	26.2	0.48	
8	YS-TS-1@mSiO$_2$	3 h	24.8	50.8	38.6	10.6	0.76	34.4
9		5 h	27.6	47.4	47.1	5.5	0.99	

[a] Reaction conditions: water as solvent, 10 mL; phenol, 1.0 g (10.63 mmol); phenol/H_2O_2 molar ratio, 3.0; catalyst mass (containing TS-1), 40 mg (0.017 mmol Ti); temperature, 80 °C; reaction time, 5 h. [b] CAT: catechol, HQ: hydroquinone, BQ: benzoquinone. [c] Turnover frequency (TOF): moles of phenol converted per mole of titanium per hour.

In order to obtain more details, we performed a kinetic study for the above three catalysts at different reaction times. Obviously, the reaction rate was YS-TS-1@mSiO$_2$ > CS-TS-1@mSiO$_2$ > pristine TS-1 (Entries 1–9). At the initial stage of reaction, because the $mSiO_2$ shell and the void space could capture substrate molecules, the concentration of phenol around active Ti sites was YS-TS-1@mSiO$_2$ > CS-TS-1@mSiO$_2$ > pristine TS-1. Therefore, after reacting for 1 h, the conversions of composite YS-TS-1@mSiO$_2$ and CS-TS-1@mSiO$_2$ were much higher than that of pristine TS-1 (20.7% *vs.* 18.7% *vs.* 6.2%). Moreover, the decreasing rate of benzoquinone's selectivity of pristine TS-1 was faster than that of YS-TS-1@mSiO$_2$ and CS-TS-1@mSiO$_2$. This controlled experiment further confirmed aforementioned diffusion/adsorption effect. This will help us to design other multifunctional zeolites-based nanocomposites for various oxidation reactions.

3. Experimental Section

3.1. Chemicals

Tetraethylorthosilicate (TEOS, 98%), aqueous ammonia (NH$_3$·H$_2$O, 28%) and cetyltrimethyl-ammonium bromide (CTAB, 99.0%) were purchased from Sinopharm Chemical Reagent Co. Ltd., Shanghai, China. Tetrapropylammonium hydroxide (20%–25%) was obtained from TCI, Shanghai, China. Hydrogen peroxide (H$_2$O$_2$, 30%), phenol (AR) and tetrabutylorthotitanate (TBOT) were obtained from Beijing chemical works, Beijing, China. All chemicals were used as received without any further purification.

3.2. Synthesis of Pristine TS-1 Zeolite

In a typical synthesis, 8.0 mL of TPAOH, 5.0 mL of TEOS and 5.0 mL of water were mixed and stirred at room temperature for 24 h to form solution A; 0.2 mL of TBOT was added into 10.0 mL of 30% H_2O_2 aqueous to form solution B. The solution B was added dropwise into solution A under stirring, and a clear sol was obtained after 24 h. The obtained sol was taken in a Teflon container and placed in a stainless steel autoclave. The sol was hydrothermally treated at 100 °C for 24 h. The TS-1 zeolites were collected through centrifugation and then washed several times with water and ethanol.

3.3. Synthesis of Core-Shell TS-1@mesosilica Composites

The core-shell TS-1@mesosilica composite was prepared via a uniform coating process, according to our previous reports [29,30]. In a typical synthesis, 100 mg of the as-prepared TS-1 zeolite was homogeneously dispersed in a mixture containing 22 mL of water and 11 mL of ethanol by ultrasonication for 30 min. The resultant mixture was stirred at room temperature for 30 min followed by sequential addition of 0.12 g of CTAB, 0.25 mL of ammonium aqueous solution (25%–28%), and 0.15 mL of TEOS. The as-obtained white mixture was further stirred at room temperature for 6 h. The core-shell structured TS-1@mesosilica composites were collected by centrifugation and then washed with water and ethanol several times. Finally, the organic structure directing agent TPA^+ and the surfactant CTAB were removed by calcination at 550 °C in air for 6 h.

3.4. Synthesis of Yolk-Shell TS-1@Mesosilica Composites

The yolk-shell structured TS-1@mesosilica composite was prepared using an RF middle-layer as the sacrificial template. In a typical synthesis, 200 mg of the as-prepared TS-1 zeolite was homogeneously dispersed in a mixture containing 17.5 mL of water and 11 mL of ethanol by ultrasonication for 30 min. To the above solution were added 0.575 g of CTAB, 0.025 mL of ammonium aqueous solution (25%–28%) and 0.12 g of resorcinol sequentially and the resultant mixture was stirred at 35 °C for 30 min. Then, 0.3 mL of formalin solution was added to the dispersion under stirring maintaining 35 °C for 6 h. The mixture was cooled to room temperature after 6 h, and then aged at room temperature overnight without stirring. The product (designated as TS-1@RF) was collected by centrifugation and then washed with water and ethanol several times. Next, a uniform mesosilica layer was coated on the surface of TS-1@RF similarly like TS-1@mesosilica to form TS-1@RF@mesosilica. Finally, the yolk-shell structured TS-1@mesosilica composite was obtained after removal of the middle-layer RF, surfactant CTAB and structure direct agent TPA$^+$ by calcination at 550 °C in air for 6 h.

3.5. Catalytic Tests

In a typical catalytic reaction, desired amount of catalyst (containing 40 mg of TS-1 zeolite), 1.0 g of phenol, 0.36 mL of H_2O_2 and 10 mL of H_2O were mixed in a 50 mL round-bottomed flask equipped with a reflux condenser and a magnetic stirrer. After reaction at 80 °C for 5 h, the products were collected and analyzed by GC.

3.6. Materials Characterization

SEM images were taken with a JSM-6700F field-emission electron microscope (JEOL, Tokyo, Japan). TEM images were obtained from an Tecnai G^2 F20s-twin D573 field emission transmission electron microscope (FEI, Hillsboro, FL, USA) at an accelerating voltage of 200kV. Powder XRD patterns were obtained by using a 2550 diffractometer (Rigaku, Tokyo, Japan) with Cu Ka radiation (λ = 1.5418 Å). N_2 adsorption-desorption isotherms were obtained at −196 °C on a ASAP 2010 sorptometer (Micromeritics, Shanghai, China). Samples were degassed at 120 °C for a minimum of 12 h prior to analysis. BET surface areas were calculated from the linear part of the BET plot. Pore size distribution was estimated from the adsorption branch of the isotherm by the BJH method. The total pore volume was estimated from the adsorbed amount of nitrogen at P/P_0 = 0.995.

4. Conclusions

In summary, we have prepared two kinds of TS-1-based nanocomposites, namely, core-shell structured TS-1@mesosilica and yolk-shell structured TS-1@mesosilica. These nanocomposites showed highly uniform morphology and structure, high BET surface area, and hierarchical pore structures from oriented mesochannels to zeolite micropores. When used as catalysts in the hydroxylation of phenol with H_2O_2 as the oxidant, they exhibited enhanced activity and unique selectivity. We found that the external mesosilica shell and the void space between the mesosilica shell and the TS-1 core not only captured phenol molecules like a pump for enhanced activity but also posed diffusion resistance for negative selectivity. Moreover, these nanocomposites presented higher selectivity of the para-product. The current work is expected to offer some insights into designing of multifunctional zeolites-based nanocomposites for enhanced oxidation reactions.

Acknowledgments: This work was supported by National Natural Science Foundation of China (21390394), the National Basic Research Program of China (2012CB821700, 2011CB808703), NSFC (21261130584, 91022030), "111" project (B07016), Award Project of KAUST (CRG-1-2012-LAI-009) and Ministry of Education, Science and Technology Development Center Project (20120061130012).

150

Author Contributions: Houbing Zou designed the experiments, analyzed the experimental data and wrote the manuscript. Qingli Sun prepared and characterized the samples. Lijia Liu tested the catalytic performances of samples. Weiwei Fu supported the experiments. Dongyu Fan and Runwei Wang revised the final version of paper.

Conflicts of Interest: The authors declare no conflict of interest.

References

1.   Thangaraj, A.; Kumar, R.; Ratnasamy, P. Catalytic Properties of Crystalline Titanium Silicalites II. Hydroxylation of Phenol with Hydrogen Peroxide over TS-1 Zeolites. *J. Catal.* **1991**, *131*, 294–297.

2.   Martens, J.A.; Buskens, P.; Jacobs, P.A.; van der Pol, A.; van Hooff, J.H.C.; Ferrini, C.; Kouwenhoven, H.W.; Kooyman, P.J.; van Bekkum, H. Hydroxylation of Phenol with Hydrogen Peroxide on EUROTS-1 Catalyst. *Appl. Catal. A* **1993**, *99*, 71–84.

3.   Hamilton, G.; Friedman, J.; Campbell, P. The Hydroxylation of Anisole by Hydrogen Peroxide in the Presence of Catalytic Amounts of Ferric Ion and Catechol. Scope, Requirements, and Kinetic Studies. *J. Am. Chem. Soc.* **1966**, *88*, 5266–5268.

4.   Villa, A.; Caro, C.; Correa, C. Cu- and Fe-ZSM-5 as Catalysts for Phenol Hydroxylation. *J. Mol. Catal. A* **2005**, *228*, 233–240.

5.   Li, B.; Xu, J.; Liu, J.; Zuo, S.; Pan, Z.; Wu, Z. Preparation of Mesoporous Ferrisilicate with High Content of Framework Iron by pH-Modification Method and Its Catalytic Performance. *J. Colloid Interface Sci.* **2012**, *366*, 114–119.

6.   Choi, J.-S.; Yoon, S.-S.; Jang, S.-H.; Ahn, W.-S. Phenol Hydroxylation Using Fe-MCM-41 Catalysts. *Catal. Today* **2006**, *111*, 280–287.

7.   Liu, H.; Lu, G.; Guo, Y.; Guo, Y.; Wang, J. Study on the Synthesis and the Catalytic Properties of Fe-HMS Materials in the Hydroxylation of Phenol. *Microporous Mesoporous Mater.* **2008**, *108*, 56–64.

8.   Fu, W.; Wang, R.; Wu, L.; Wang, H.; Wang, X.; Wang, A.; Zhang, Z.; Qiu, S. Synthesis of $Cu_2(OH)PO_4$ Crystals with Various Morphologies and Their Catalytic Activity in Hydroxylation of Phenol. *Chem. Lett.* **2013**, *42*, 772–774.

9.   Clerici, M.G. Oxidation of Saturated Hydrocarbons with Hydrogen Peroxide Catalysed by Titanium Silicalite. *Appl. Catal.* **1991**, *68*, 249–261.

10. Xiao, F.; Han, Y.; Yu, Y.; Meng, X.; Yang, M.; Wu, S. Hydrothermally Stable Ordered Mesoporous Titanosilicates with Highly Active Catalytic Sites. *J. Am. Chem. Soc.* **2002**, *133*, 888–889.

11. Lin, K.; Wang, L.; Meng, F.; Sun, Z.; Yang, Q.; Cui, Y.; Jiang, D.; Xiao, F.S. Formation of Better Catalytically Active Titanium Species in Ti-MCM-41 by Vapor-phase Silylation. *J. Catal.* **2005**, *235*, 423–427.

12. Tsai, S.-T.; Chao, P.-Y.; Tsai, T.-C.; Wang, I.; Liu, X.; Guo, X.-W. Effects of Pore Structure of Post-treated TS-1 on Phenol Hydroxylation. *Catal. Today* **2009**, *148*, 174–178.

13. Zhu, G.; Ni, L.; Qi, W.; Ding, S.; Li, X.; Wang, R. Synthesis and Morphology Research of Framework Ti-rich TS-1 Containing No Extraframework Ti Species in the Presence of CO_2. *Inorg. Chem. Commun.* **2014**, *40*, 129–132.

14. Shi, C.; Chu, B.; Lin, M.; Long, J.; Wang, R. Cyclohexane Mild Oxidation Catalyzed by New Titanosilicate with Hollow Structure. *Catal. Today* **2011**, *175*, 398–403.

15. Cheneviere, Y.; Chieux, F.; Caps, V.; Tuel, A. Synthesis and Catalytic Properties of TS-1 with Mesoporous/microporous Hierarchical Structures Obtained in the Presence of Amphiphilic Organosilanes. *J. Catal.* **2010**, *269*, 161–168.

16. Moliner, M.; Corma, A. Advances in the synthesis of titanosilicates: From the medium pore TS-1 zeolite to highly-accessible ordered materials. *Micropor. Mesopor. Mater.* **2014**, *189*, 31–40.

17. Chen, L.; Li, X.; Tian, G.; Li, Y.; Rooke, J.; Zhu, G.; Qiu, S.; Yang, X.; Su, B. Highly Stable and Reusable Multimodal Zeolite TS-1 Based Catalysts with Hierarchically Interconnected Three-Level Micro-Meso-Macroporous Structure. *Angew. Chem. Int. Ed.* **2011**, *50*, 11156–11161.

18. Xin, H.; Zhao, J.; Xu, S.; Li, J.; Zhang, W.; Guo, X.; Hensen, E.; Li, C. Enhanced Catalytic Oxidation by Hierarchically Structured TS-1 Zeolite. *J. Phys. Chem. C* **2010**, *114*, 6553–6559.

19. Kang, Z.; Fang, G.; Ke, Q.; Hu, J.; Tang, T. Superior Catalytic Performance of Mesoporous Zeolite TS-1 for the Oxidation of Bulky Organic Sulfides. *ChemCatChem* **2013**, *5*, 2191–2194.

20. Na, K.; Jo, C.; Kim, J.; Ahn, W.; Ryoo, R. MFI Titanosilicate Nanosheets with Single-unit-cell Thickness as An Oxidation Catalyst Using Peroxides. *ACS Catal.* **2011**, *1*, 901–907.

21. Wang, R.; Liu, W.; Ding, S.; Zhang, Z.; Li, J.; Qiu, S. Mesoporous MFI Zeolites with Self-stacked Morphology Templated by Cationic Polymer by Cationic Polymer. *Chem. Commun.* **2010**, *46*, 7418–7420.

22. Qian, X.; Du, J.; Li, B.; Si, M.; Yang, Y.; Hu, Y.; Niu, G.; Zhang, Y.; Xu, H.; Tu, B.; *et al.* Controllable Fabrication of Uniform Core-Shell Structured Zeolite@SBA-15 Composites. *Chem. Sci.* **2011**, *2*, 2006–2016.

23. Xu, L.; Ren, Y.; Wu, H.; Liu, Y.; Wang, Z.; Zhang, Y.; Xu, J.; Peng, H.; Wu, P. Core/shell-Structured TS-1@mesoporous Silica-supported Au Nanoparticles for Selective Epoxidation of Propylene with H_2 and O_2. *J. Mater. Chem.* **2011**, *21*, 10852–10858.

24. Xu, L.; Peng, H.; Zhang, K.; Wu, H.; Chen, L.; Liu, Y.; Wu, P. Core-Shell-Structured Titanosilicate As A Robust Catalyst for Cyclohexanone Ammoximation. *ACS Catal.* **2012**, *3*, 103–110.

25. Qian, X.; Li, B.; Hu, Y.; Niu, G.; Zhang, D.; Che, R.; Tang, Y.; Su, D.; Asiri, A.; Zhao, D. Exploring Meso-/Microporous Composite Molecular Sieves with Core-Shell Structures. *Chem. Eur. J.* **2012**, *18*, 931–939.

26. Lv, Y.; Qian, X.; Tu, B.; Zhao, D. Generalized Synthesis of Core-Shell Structured Nano-zeolite@ordered mesoporous Silica Composites. *Catal. Today* **2013**, *204*, 2–7.

27. Li, C.; Lu, Y.; Wu, H.; Wu, P.; He, M. A Hierarchically Core/Shell-Structured Titanosilicate with Multiple Mesopore Systems for Highly Efficient Epoxidation of Alkenes. *Chem. Commun.* **2015**, *51*, 14905–14908.

28. Peng, H.; Xu, L.; Wu, H.; Zhang, K.; Wu, P. One-pot Synthesis of Benzamide Over A Robust Tandem Catalyst Based on Center Radially Fibrous Silica Encapsulated TS-1. *Chem. Commun.* **2013**, *49*, 2709–2711.

29. Zou, H.; Wang, R.; Li, X.; Wang, X.; Zeng, S.; Ding, S.; Li, L.; Zhang, Z.; Qiu, S. An Organosilane-directed Growth-induced Etching Strategy for Preparing Hollow/Yolk-shell Mesoporous Organosilica Nanospheres with Perpendicular Mesochannels and Amphiphilic Frameworks. *J. Mater. Chem. A* **2014**, *2*, 12403–12412.

30. Zou, H.; Wang, R.; Dai, J.; Wang, Y.; Wang, X.; Zhang, Z.; Qiu, S. Amphiphilic Hollow Porous Shell Encapsulated Au@Pd Bimetal Nanoparticles for Aerobic Oxidation of Alcohols in Water. *Chem. Commun.* **2015**, *51*, 14601–14604.

31. Guan, B.; Wang, X.; Xiao, Y.; Liu, Y.; Huo, Q. A Versatile Cooperative Template-directed Coating Method to Construct Uniform Microporous Carbon Shells for Multifunctional Core-Shell Nanocomposites. *Nanoscale* **2013**, *5*, 2469–2475.

Photocatalytic Oxidation of NO over Composites of Titanium Dioxide and Zeolite ZSM-5

Akram Tawari, Wolf-Dietrich Einicke and Roger Gläser

Abstract: Composites of TiO_2 (Hombikat, P25, sol-gel synthesis) and zeolite ZSM-5 (n_{Si}/n_{Al} = 55) with mass fractions from 25/75 to 75/25 were prepared by mechanical mixing, solid-state dispersion and sol-gel synthesis. Characterization of the composites by X-ray diffraction (XRD), N_2-sorption, scanning electron microscopy (SEM), and UV-Vis spectroscopy show that mechanical mixing and solid-state dispersion lead to comparable textural properties of the composites. A homogeneous distribution and intimate contact of small TiO_2 particles on the crystal surface of zeolite ZSM-5 were achieved by sol-gel synthesis. The composites were studied in the photocatalytic oxidation (PCO) of NO in a flatbed reactor under continuous flow according to ISO 22197-1. The highest NO conversion of 41% at an NO_2 selectivity as low as 19% stable for 24 h on-stream was reached over the TiO_2/ZSM-5 composite from sol-gel synthesis with equal amounts of the two components after calcination at 523 K. The higher activity and stability for complete NO oxidation than for pure TiO_2 from sol-gel synthesis, Hombikat, or P25 is attributed to the adsorptive properties of the zeolite ZSM-5 in the composite catalyst. Increasing the calcination temperature up to 823 K leads to larger TiO_2 particles and a lower photocatalytic activity.

Reprinted from *Catalysts*. Cite as: Tawari, A.; Einicke, W.-D.; Gläser, R. Photocatalytic Oxidation of NO over Composites of Titanium Dioxide and Zeolite ZSM-5. *Catalysts* **2016**, *6*, 31.

1. Introduction

The increasing consumption of fossil fuels and the worldwide industrial growth has resulted in increasing emissions and air pollution. Nitrogen oxides NO_x are among the major pollutants that strongly contribute to the air emissions [1,2]. NO_x emissions mainly consist of nitric oxide NO and nitrogen dioxide NO_2 [3]. In particular, NO_2 can cause severe impacts on the environment and on human health, e.g., by the formation of acid rain and photochemical smog or by damaging the lung tissue of human beings [1,2]. To reduce NO_x emissions from the atmosphere, several technologies are applied including selective catalytic reduction (SCR) [4,5]. Most of these technologies, however, are cost-intensive and require elevated operating temperatures and continuous maintenance. Hence, alternative approaches are needed for more environmentally friendly and cost-effective NO_x reduction [6]. One promising approach is the photocatalytic oxidation (PCO) for NO_x removal. It is particularly attractive, since it can be applied at ambient temperature and pressure and since solar illumination can be utilized as a readily available energy source [7]. Although Fujishima and Honda [8] already promoted the field of photocatalysis in 1972, the first report on PCO for the removal of NO_x over TiO_2 was first reported by Ibusuki and Takeuchi [9] only in 1994. TiO_2 is considered one of the most suitable semiconductor photocatalysts due to its high stability towards photocorrosion, biological and chemical inertness as well as low cost and low toxicity [10].

The commercialization of TiO_2-based photocatalysts commenced in the mid-1990s. Applications are found, e.g., in materials for construction, furnishing, or road-construction, as well as purification facilities and household goods [11]. PCO of NO_x over TiO_2 was studied using active building materials, e.g., paints, pavement stones and other concrete-based products [12–17]. The PCO approach includes oxidation of NO_x to NO_3^-, which can be washed off the surface by water, e.g., from rain [15–18]. By varying the calcination temperature of TiO_2 prepared by the hydrolysis of titanium alkoxide, Hashimoto et al. [19] have shown that a higher photocatalytic activity in the PCO of NO_x can be achieved over smaller TiO_2 crystals in comparison to a commercial TiO_2. Most importantly, it was proven by Nakamura et al. [20] using in situ Fourier-Transform infrared (FTIR) spectroscopy and by Dalton et al. [21] using surface-specific techniques such as X-ray photoelectron and Raman spectroscopy that the PCO of NO over TiO_2 proceeds consecutively from NO to NO_3^- via NO_2^-. The amounts of NO_3^- and NO_2^- formed over TiO_2 sheets can be influenced by added metal compounds [22]. Furthermore, NO_x can be oxidized over TiO_2 under UV irradiation in air to nitric acid (HNO_3) [9]. The formed HNO_3 molecules adsorbed on the TiO_2 surface act as a barrier reducing the photocatalytic activity. One approach to avoid this reduction in photocatalytic activity is to add porous materials such as activated carbon, alumina, silica or zeolites to the catalytic system to adsorb the PCO products and maintain high photocatalytic activity [23–25]. The combination of an adsorptive with a catalytic functionality was also applied, e.g., in the decomposition of 2-chloroethylethylsulfide over copper oxide on a spherical activated carbon [26]. The advantage of using an adsorbent during photocatalytic oxidation of H_2S was reported by Portela et al. [27]. A reduced selectivity for SO_2 was obtained employing a hybrid material consisting of TiO_2 with sepiolite as an adsorbent.

Zeolites are especially suitable as adsorbents for this purpose due to their high specific surface area, tunable hydrophobic/hydrophilic surface properties, availability on an industrial scale and low environmental impact [28–30]. Guo et al. [31] prepared composites of ZSM-5 with different n_{Si}/n_{Al} ratios of 500, 200 and 25 with TiO_2 (P25) using a solid-state dispersion method. An increasing hydrophobic character of zeolites increased the photocatalytic activity for NO removal by decreasing the amount of water adsorbed in the zeolite pores. Furthermore, the PCO of NO over TiO_2 mixed mechanically with zeolite A in a mass ratio of 7:3 showed a higher photocatalytic activity compared to bare TiO_2 [32].

The combination of TiO_2 with zeolites in sheets using paper-making techniques leads to effective photocatalysts for the removal of volatile organic compounds [33,34]. Takeuchi et al. [35] reported a high activity for the photooxidation of gaseous acetaldehyde using TiO_2/ZSM-5 catalysts synthesized by a simple impregnation. Jansson et al. [36] studied the photodegradation of formaldehyde and trichloroethylene over composites of TiO_2 and the zeolites Y and ZSM-5 with different n_{Si}/n_{Al} ratios prepared using incipient wetness impregnation of the zeolite with an acidic TiO_2 sol. Applying a composite of zeolite ZSM-5 loaded with as little as 8 wt.% of TiO_2 resulted in a significant increase of photodegradation of formaldehyde and trichloroethylene among the other composites and pure TiO_2. In addition, the role of zeolite H-beta in composites with TiO_2 in the photocatalytic degradation of an aqueous propoxur solution was shown to rely on concentrating the reactants and intermediates within its cavities and a subsequent and continuous supply to the TiO_2 surface [28].

It was, therefore, the aim of this work to study composites of TiO_2 and zeolite ZSM-5 over a broad range of mass ratios for the photocatalytic oxidation of NO with respect to the activity of pure TiO_2 and pure zeolite ZSM-5. In order to establish application relevant conditions, humid air streams containing NO and a flatbed reactor following a standard testing procedure according to ISO 22197-1 [37] were used. A particular goal of this study was to understand the role of the local proximity of the zeolite and the TiO_2 components within the composites at different mass ratios of zeolite and TiO_2. Therefore, the composites were prepared by different methods, i.e., solid-state dispersion, sol-gel synthesis of TiO_2 in the presence of zeolite powder and by mechanical mixing. The stability of the composite photocatalysts over an extended operation time of 24 h was also in the focus of this study. Furthermore, it was investigated if the photocatalytic activity can be improved by varying the calcination temperature of the composites and, thus, the crystallite size of the TiO_2 and the zeolite surface hydrophobicity, respectively.

2. Results and Discussion

2.1. Characterization of the Composite Photocatalysts

The powder X-ray diffraction (XRD) patterns of the TiO_2 materials P25, from sol-gel synthesis and Hombikat are shown in Figure 1a. For TiO_2 from sol-gel synthesis and Hombikat, the reflexes of anatase as the predominant crystalline phase are observed at $2\theta = 25.4°$, $37.8°$, $48.1°$, $55.1°$ and $62.7°$ [38,39]. In addition to the reflexes for anatase, a reflex at 2θ = 27.5° for the rutile phase is found for P25 and a reflex for the brookite phase at 2θ = 30° for TiO$_2$ from sol-gel synthesis [40].

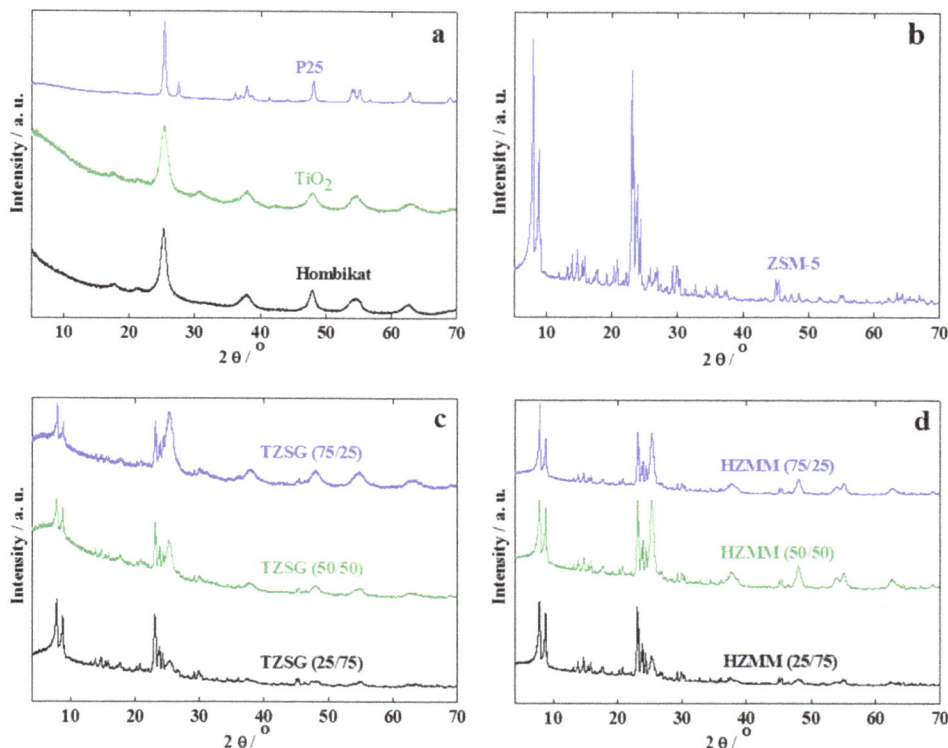

Figure 1. XRD patterns of (**a**) Hombikat, TiO$_2$ from sol-gel synthesis and P25; (**b**) zeolite ZSM-5; (**c**) composites from sol-gel synthesis (TZSG); and (**d**) composites from mechanical mixture (HZMM).

It is preferable, however, to keep the rutile fraction low for a high photocatalytic activity [41–43]. The XRD reflexes for P25 are sharper than for the other two TiO$_2$ materials indicative of larger crystallite sizes (determined by the Scherrer equation for the most intense (101) reflex at 2θ = 25.4°) of 21 nm, with respect to Hombikat (10 nm) and TiO$_2$ from sol-gel synthesis (8 nm), respectively.

For comparison, the XRD pattern of zeolite ZSM-5 is also shown (Figure 1b). It exhibits the characteristics typical for the MFI framework topology [44].

In the TiO$_2$/ZSM-5 composites from sol-gel synthesis, the reflexes from both anatase and the zeolite are detected in the XRD patterns (Figure 1c). As expected, the reflex intensities of the zeolite increase with increasing mass fraction in the composite. This is also the case for the composites obtained from mechanically mixing Hombikat and zeolite ZSM-5 (Figure 1d). At low TiO_2 contents in both the materials from sol-gel synthesis and mechanical mixing, *i.e.*, TZSG (25/75) and HZMM (25/75), the reflexes for anatase are scarcely observed and the zeolite reflexes dominate the XRD patterns (Figure 1c,d).

Although the anatase reflexes are similarly broad for pure Hombikat and TiO_2 from sol-gel synthesis (Figure 1a), they are much broader in the composite prepared by the sol-gel route when compared to the mechanical mixture (Figure 1d). This points at a higher dispersion of small TiO_2 particles over the crystallites of zeolite ZSM-5 in the sol-gel-derived composites. It is also consistent with earlier literature reports showing that zeolite ZSM-5 can suppress the crystal growth of anatase during sol-gel synthesis [45]. Note also that the XRD patterns of the composites from solid-state dispersion of Hombikat and zeolite ZSM-5 are identical to those of the mechanical mixture (Figure S1). The particle size distribution in the composite from solid-state dispersion can, therefore, be assumed to be comparable to that in the mechanical mixture. Due to the overlap of the most intense anatase reflex at $2\theta = 25.4°$ (101), the crystal size of TiO_2 in the composites could, however, not be accurately determined.

The differences in the textural properties of the composites from solid-state dispersion, mechanical mixing and sol-gel synthesis become clear from the N_2 sorption isotherms shown in Figure 2a–c. Only minor differences are observed in the sorption isotherms for the composites from mechanical mixing and solid-state dispersion (Figure 2a,b). The composite isotherms can be described as a superposition of the isotherms of the individual components, *i.e.*, Hombikat and zeolite ZSM-5, taking into account the corresponding mass fractions. Thus, all textural properties show additivity. This means that in all cases of these material series, the two components do not affect each other within the composite with respect to the accessibility of their pore systems. This is in accordance with the similar XRD patterns found for the HZMM and HZSSD materials with the same mass fractions.

A different behavior is observed for the materials from sol-gel synthesis (TZSG-series, Figure 2c). Distinct hysteresis loops in the desorption isotherms of the H1 type [46] occur at relative pressures of $P/P_0 = 0.4$ to 0.8 caused by the formation of mesopores during the sol-gel process. With increasing content of TiO_2 in the material, the hysteresis loop shifts to higher relative pressures. Correspondingly larger mesopores are formed with increasing mass fraction of TiO_2 in the TZSG-composites.

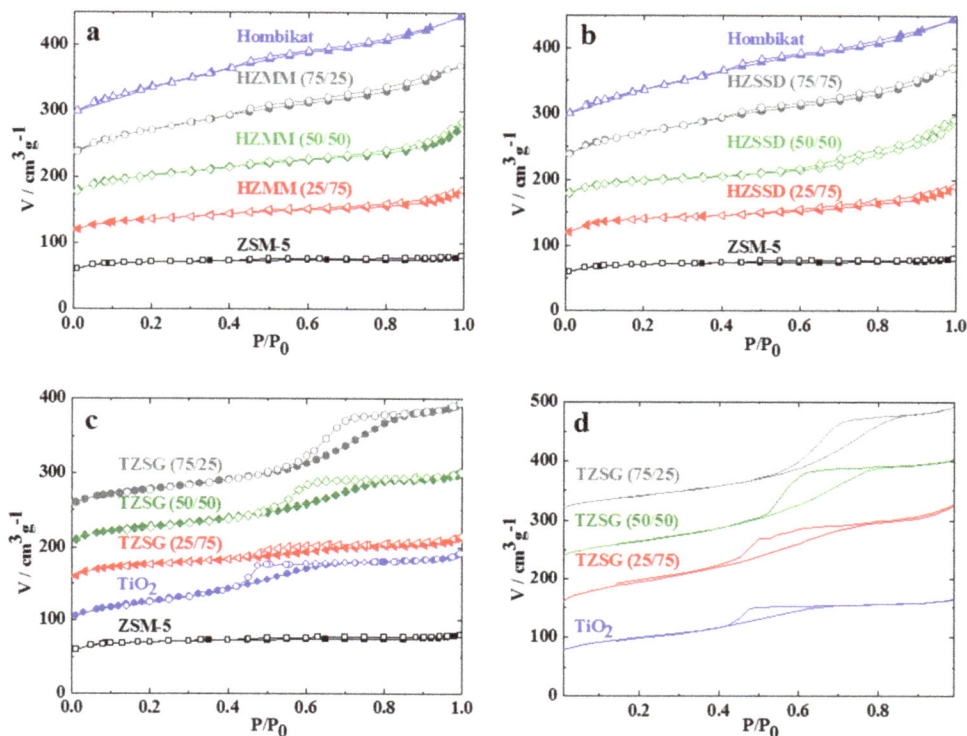

Figure 2. N$_2$ sorption isotherms of (**a**) Hombikat, composites from mechanical mixture (HZMM) and zeolite ZSM-5; (**b**) Hombikat, composites from solid-state dispersion (HZSSD) and zeolite ZSM-5; (**c**) composites from sol-gel synthesis (TZSG) and TiO$_2$; (**d**) calculated isotherms for the TiO$_2$ fraction of the TZSG composites. For clarity, all isotherms in parts a to c were vertically displaced by 60 cm$^3 \cdot$ g^{-1} except for the isotherms in part d, which were displaced by 80 cm$^3 \cdot$ g^{-1}.

In contrast to the materials from mechanical mixing and solid-state dispersion, the isotherms and, thus, the textural properties of the composites from sol-gel synthesis show no additivity based on the isotherms of the pure components. From the analysis of the isotherms in the low pressure range with the t-plot method (not shown), it becomes clear that the micropore volume of the zeolite ZSM-5 is not affected by the presence of the TiO$_2$ from sol-gel synthesis. This means that the TiO$_2$ particles exclusively reside outside the micropores of the zeolite. The deviations from additivity are, thus, caused by the changes in the mesopore range of the isotherms. In order to estimate the textural properties for the TiO$_2$ component in the TZSG composites, isotherms were calculated from the ones shown in Figure 2c by subtracting the contribution of the zeolite ZSM-5 taking into account its mass fraction.

159

The resulting isotherms for the TiO_2 fraction of the composite clearly display the hysteresis loops characteristic for a disordered mesopore system with a broad width distribution (Figure 2d). The pore size range of the TiO_2 material increases with increasing mass fraction of TiO_2 in the composite pointing at pore space between TiO_2 particles formed during the sol-gel synthesis. Clearly, however, the presence of the zeolite affects the textural properties of the TiO_2 formed, since the mesopores in the TiO_2 prepared in the absence of the zeolites are much smaller, *i.e.*, the hysteresis loop occurs at lower relative pressures (0.4–0.6, Figure 2d). It is, therefore, obvious that the TiO_2 from sol-gel synthesis strongly interacts with the outer surface of the ZSM-5 zeolite crystals, although this interaction is expectedly reduced at higher TiO_2 mass fractions. Details of the textural properties (specific surface area, pore volume, and average pore diameter) are shown in Table S1.

The morphology and the proximity of the TiO_2 particles and the crystals of zeolite ZSM-5 were further characterized by scanning electron microscopy (SEM) (Figure 3). The SEM images of the pure TiO_2 from sol-gel synthesis and Hombikat show a broad particle size distribution, while Hombikat possesses somewhat more regularly shaped particles with sizes below 1 μm (Figure 3a,b). The crystallites of zeolite ZSM-5 are between 3 and 8 μm in diameter and partly intertwined (Figure 3d). These crystals are clearly visible in the composites from mechanical mixing (HZMM (50/50), Figure 3d) and from solid-state dispersion (HZSSD (50/50), Figure 3e). The TiO_2 particles are aggregated and loosely packed around the ZSM-5 crystallites. In complete agreement with results from XRD and N_2 sorption, there is no apparent difference in the composites obtained by mechanically mixing or by solid-state dispersion. In sharp contrast, however, greater surface of the zeolite crystals is covered with small TiO_2 particles when prepared by sol-gel synthesis (TZSG, Figure 3f). Although larger TiO_2 particles and agglomerates of around 2–5 μm diameter are also seen in the SEM image, the TiO_2 particles are more evenly distributed over the zeolite surface and support the more intense and direct interaction between both components of the composites as already concluded from N_2 sorption analysis (*vide supra*). Nevertheless, the contact between the TiO_2 particles and the zeolite does not lead to a blockage of access to the zeolite pores as shown by the results from N_2 sorption discussed above.

Diffuse reflectance UV-Vis absorption spectra of TiO_2 from sol-gel synthesis, Hombikat, and the composites TZSG (50/50), HZSSD (50/50), and HZMM (50/50) are shown in Figure 4. It can be clearly seen that all materials absorb light in the UV region with only small differences between the pure TiO_2 samples and the TiO_2/ZSM-5 composites as well the different preparation methods. This indicates that the preparation techniques and the presence of ZSM-5 have only a minor effect on the optical absorption of the TiO_2 component in the composites. Nevertheless, a measurable blue shift of the absorption band edge of TiO_2 and TZSG (50/50) from sol-gel synthesis and from Hombikat and the composites of HZSSD (50/50) and HZMM (50/50) is observed. This is probably due to the smaller particle size of TiO_2 derived from sol-gel synthesis resulting in a quantum confinement effect [47].

Figure 3. SEM images of (**a**) TiO_2 from sol-gel synthesis; (**b**) Hombikat; (**c**) zeolite ZSM-5; (**d**) composite HZMM (50/50); (**e**) composite HZSSD (50/50); and (**f**) composite TZSG (50/50).

The smaller difference in the absorption spectra between TiO_2 and TZSG (50/50) might indicate the differences of the TiO_2 particle size formed on the zeolite surface as confirmed by XRD. The slight shift in the absorption spectra of TZSG (50/50) from HZSSD (50/50) and HZMM (50/50) suggests that TiO_2 small particles were well-dispersed on zeolite surface, while Hombikat particles were agglomerated over zeolite surface as confirmed by SEM images.

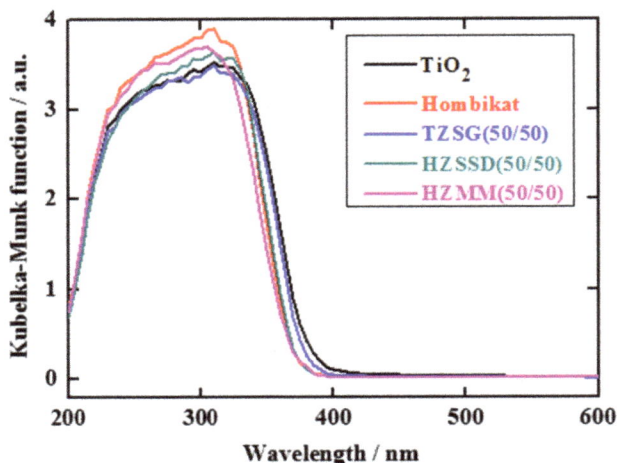

Figure 4. Diffuse reflectance UV-Vis absorption spectra of TiO_2 from sol-gel synthesis, Hombikat, and the composites TZSG (50/50), HZSSD (50/50) and HZMM (50/50).

2.2. Photocatalytic Activity

The conversion of NO (X_{NO}) and of NO_x (X_{NO_x}), the selectivity for NO_2 (S_{NO_2}), the amount of NO removed (Z_{NO}), the amount of NO_2 formed (Z_{NO_2}), and the net amount NO_x removed (Z_{NO_x}) in the photooxidation of NO over the TiO_2 materials and the composites are shown in Table 1. In the presence of the TiO_2-containing photocatalysts, *i.e.*, P25, TiO_2 from sol-gel synthesis or Hombikat, a higher NO conversion, but also a higher NO_2 selectivity than, for most of the composites are observed. Note that in the presence of only zeolite ZSM-5, no NO is converted. In addition, the dilution of the TiO_2 with sea sand results in only a slightly reduced NO conversion, while the selectivity for NO_2 remains high.

During the photocatalytic experiment, the concentration of NO increases slightly and approaches a steady state for the TiO_2-based catalysts (Figure 5a–c). However, the NO_2 concentration increases continuously in parallel with the NO_x concentration, especially pronounced for the TiO_2 from sol-gel synthesis (Figure 5c). This point at a deactivation of the catalysts, most likely caused by deposition of NO_3^- in the form of nitric acid and a blocking of photocatalytically active sites [9,22]. This is also consistent with the NO_3^- found on the catalyst after the photocatalytic conversion (Table 2). With ongoing deactivation, NO is, then, not oxidized to the desired NO_3^- or NO_2^-, but rather preferably to NO_2.

Table 1. Conversion of NO (X_{NO}) and of NO_x (X_{NO_x}), selectivity for NO_2 (S_{NO_2}), amount of NO removed (Z_{NO}), amount of NO_2 formed (Z_{NO_2}), and net amount NO_x removed (Z_{NO_x}) in the photocatalytic oxidation of NO over different TiO_2 and TiO_2/ZSM-5 composite catalysts after 24 h UV irradiation (5 g TiO_2 or composite, 1000 ppb NO, 50% relative humidity, 3 $dm^3 \cdot min^{-1}$ total volumetric flow rate, 10 $W \cdot m^{-2}$ UV irradiance, 0.005 m^2 irradiated area).

Catalysts	X_{NO} /%	X_{NO_x}/%	Z_{NO}/ $\mu mol \cdot m^{-2} \cdot h^{-1}$	Z_{NO_2}/ $\mu mol \cdot m^{-2} \cdot h^{-1}$	Z_{NO_x}/ $\mu mol \cdot m^{-2} \cdot h^{-1}$	S_{NO_2} / %
P25	45	16	704	455	249	65
TiO_2	32	19	601	291	310	48
Hombikat	35	16	518	271	247	52
TSSSG	31	12	502	305	197	62
HZMM (25/75)	16	9	257	108	149	33
HZMM (50/50)	21	14	327	103	224	31
HZMM (75/25)	24	16	373	132	241	38
HZSSD (25/75)	28	17	435	163	271	37
HZSSD (50/50)	30	22	553	169	384	30
HZSSD (75/25)	29	19	450	160	290	35
TZSG (25/75)	31	23	483	123	360	25
TZSG (50/50)	41	33	655	131	524	19
TZSG (75/25)	36	27	572	166	406	29

Table 2. Content of NO_3^- present in the catalysts after the photocatalytic experiment.

Samples	NO_3^-/$mg \cdot g^{-1}$
Hombikat	0.30
P25	0.26
TSSSG	0.22
TZSG (50/50)	0.66

If zeolite ZSM-5 is present in the photocatalysts, a different behavior is observed. Generally, the selectivity for NO_2 over the composites is lower than for the pure TiO_2-based analogues, although the NO conversion is lower. During the photocatalytic NO oxidation over the composites, the NO_2 concentration is much lower, and it decreases less strongly than in the case of the zeolite-free TiO_2 materials (Figure 5d,e). This indicates that NO_2 was effectively adsorbed by the zeolite ZSM-5 and is further oxidized to NO_3^- after diffusion to and conversion over the photocatalytically active TiO_2 particles.

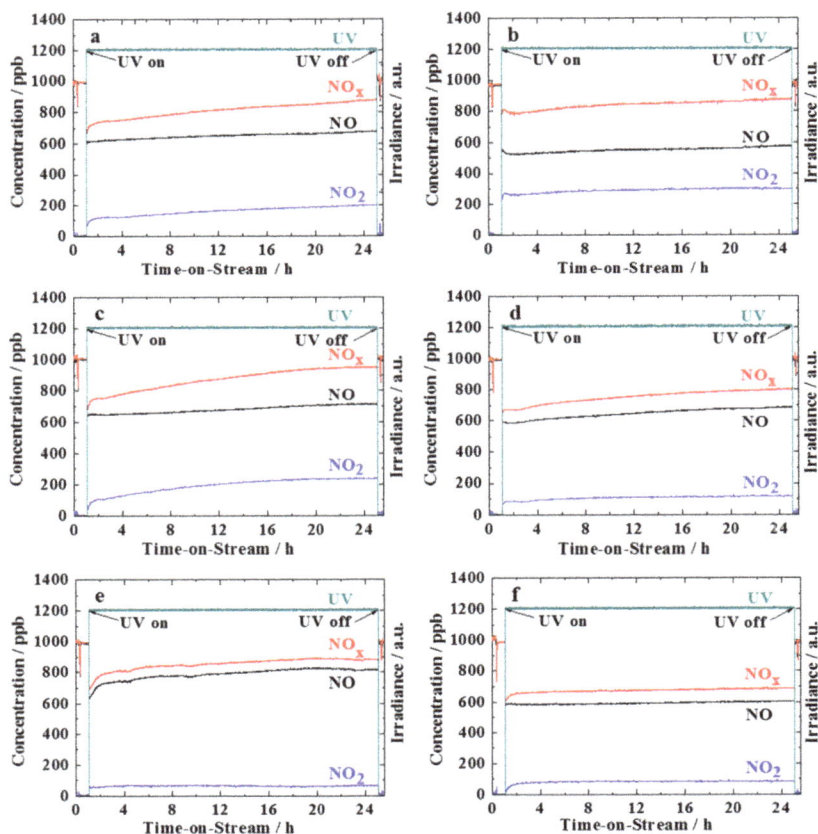

Figure 5. Concentrations of NO, NO₂, NOₓ and UV light irradiance in the photocatalytic oxidation of NO over (**a**) Hombikat; (**b**) P25; (**c**) TiO₂ from sol-gel synthesis; (**d**) composite HZSSD (50/50); (**e**) composite HZMM (50/50); (**f**) composite TZSG (50/50) (5 g TiO₂ or composite, 1000 ppb NO, 50% relative humidity, 3 dm³·min⁻¹ total volumetric flow rate, 10 W·m⁻² UV irradiance, 0.005 m² irradiated area).

Over the composites from mechanical mixing and from solid-state dispersion, the concentration of NO increases gradually leading to lower NO conversion at essentially constant NO₂ concentration (Figure 5d,e). Although no apparent differences in the textural and structural properties of the two composite classes (HZMM and HZSSD) were observed (see Section 2.1), the NO conversion of the composites from the solid-state dispersion is slightly higher than for the mechanical mixtures. This higher activity can be attributed to a closer proximity of the TiO₂ particles to the zeolite crystals after solid-state dispersion. Nevertheless, the Hombikat particles were not evenly dispersed over the ZSM-5 as seen from the SEM images in Figure 3d,e.

The highest NO conversion of 41% at an NO_2 selectivity as low as 19% is achieved over the composite from sol-gel synthesis (Table 1). This activity is even higher than that of the pure TiO_2 from sol-gel synthesis or Hombikat. In addition, a steady state is reached on these catalysts within less than 2 h of irradiation on-stream (Figure 5f). This steady state of high NO_x removal rate remains constant over the whole 24 h of the experiment without any detectable deactivation. This result can be attributed to the high dispersion and intense interaction of the TiO_2 particles with the surface of the crystals of zeolite ZSM-5 as confirmed by N_2 sorption, XRD and SEM (see Section 2.1). Besides serving as an adsorbent and reservoir for the NO_2 formed in the first step of NO oxidation, the zeolite might as well act as an efficient adsorbent for the nitrate and nitrites in solution, *i.e.*, nitric and nitrous acid, from the terminal NO_x oxidation. Thus, the well-dispersed TiO_2 particles over the ZSM-5 crystals from sol-gel not only decrease NO_2 selectivity, but also lead to an overall increase of the NO_x removal.

In order to further verify the role of zeolite ZSM-5 in the composite, non-porous sea sand with negligible adsorption capacity was used in the sol-gel synthesis as a 50/50 composite (denoted as TSSSG, for data from characterization by XRD, N_2 sorption and SEM, Figures S2–S4). This composite shows, however, very high NO_2 selectivity and low NO_x removal similar to the pure zeolite-free TiO_2 catalysts. The improvement in photocatalytic activity and reduction in NO_2 selectivity of the zeolite-containing composites are, therefore, due to the high adsorption capacity in the micropores (and possibly the hydrophilicity) of the zeolite ZSM-5.

As seen also from the results in Table 1, the photocatalytic activity of HZMM, HZSSD and TZSG decreases with increasing mass fraction of zeolite ZSM-5 and decreasing TiO_2 content in the composites. This suggests that a higher loading of ZSM-5 may lower the amount of TiO_2 available on the ZSM-5 surface and, therefore, decreases the photooxidation activity. Furthermore, increasing the TiO_2 content with decreasing ZSM-5 content at a mass ratio of 75/25 does not lead to a proportional increase of NO_x removal. This implies that a higher loading of TiO_2 forms thicker layers on the ZSM-5 crystallite surface where the excess amount of TiO_2 is inaccessible to the UV irradiation. It can be concluded that for the TZSG material a mass ratio of 50/50 is the optimum ratio to achieve the highest overall NO_x removal, even higher than for pure P25. Supporting this conclusion, the highest amount of nitrate, more than twice as high as for P25, is found on this sol-gel derived catalyst (Table 2).

Note, however, that the photocatalytic results obtained here are rather different from those for the photooxidation of NO over TiO_2/zeolite composites reported by Guo *et al.* [31], Hashimoto *et al.* [32] and Ichiura *et al.* [34]. NO conversions between 25% and 96% were achieved in these studies, however at largely different conditions of the photocatalytic experiments with respect to, e.g., irradiation time and intensity, relative humidity and feed concentration and low rates. For instance, the initial concentration of NO used in the above mentioned studies were 40 ppm [31], 10 ppm [32] and 0.25 ppm [34], respectively, which is different than the value of 1 ppm applied in the present study. In addition, the volumetric flow rates used in the above mentioned studies were 0.2 [31], 0.11 [32] and 1.5 $dm^3 \cdot min^{-1}$ [34], respectively, *i.e.*, significantly lower than the value of 3 $dm^3 \cdot min^{-1}$ used in the present work. These largely different conditions render a direct comparison of the photocatalytic results difficult and point at the importance of carrying out PCO of NO at standard conditions as given, e.g., in ISO 22197-1 [37] and as applied here.

2.3. Effect of Calcination Temperature

In an attempt to increase the photocatalytic activity of the composites according to [19], the calcination temperature of TiO_2 and the most active and selective TiO_2/ZSM-5 composite with equal mass fractions of the two components (TZSG (50/50)) from sol-gel synthesis was varied in the range of 423 to 823 K. With increasing calcination temperature, the reflexes in the XRD patterns of the TiO_2 component become sharper and more intense, indicating the presence of larger crystallites (Figure 6). For the highest calcination temperature of 823 K, the reflex at $2\theta = 27.5°$ points at the presence of the rutile phase of TiO_2. The average crystallite sizes of the anatase phase determined by the Scherrer equation from the (101) reflex at $2\theta = 25.4°$ increases slightly from 6 to 9 nm with the calcination temperature until 723 K, while much larger crystals of 23 nm are found after calcination at 823 K (Table 3). As expected for the increasing crystal size with increasing calcination temperature, the specific surface area of the composite also decreases gradually from 290 to 193 $m^2 \cdot g^{-1}$, respectively, in accordance with the increasing pore width (Table 3). This increase of the crystallite size with calcination temperature is in agreement with the literature [19] and can be attributed to the thermally promoted crystallite growth [48]. The formation of larger crystals at higher calcination temperatures is also observed for the TZSG-composite (Table 3). This may lead to a less intense interaction between TiO_2 particles and the crystal surface of zeolite ZSM-5.

Table 3. Crystallite sizes of TiO_2 as well as specific surface area A_{BET} and average pore width d_{BJH} of the composite TZSG (50/50) from sol-gel synthesis after calcination at different temperatures.

$T_{calcination}$/K	Crystallite Size/nm	$A_{BET}/(m^2 \cdot g^{-1})$	d_{BJH}/nm
423	6	290	4.16
523	8	278	4.49
623	8	265	4.57
723	9	219	5.80
823	23	193	7.62

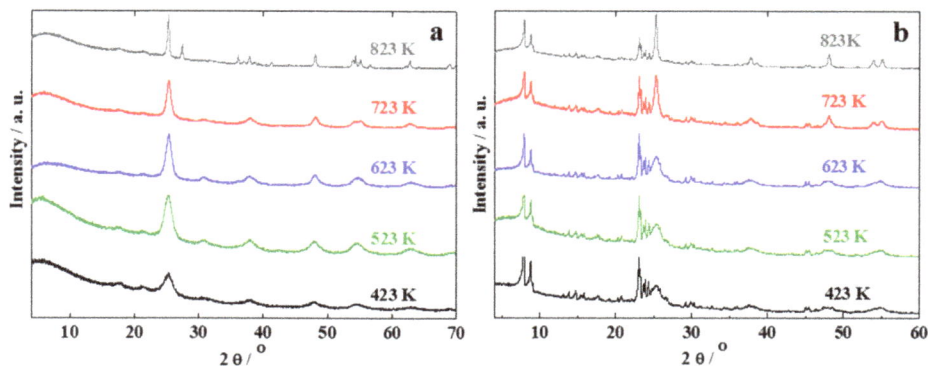

Figure 6. XRD patterns of (**a**) TiO$_2$ and (**b**) composite TZSG (50/50) from sol-gel synthesis after calcination at different temperatures.

The highest photocatalytic activity with respect to NO conversion is achieved for the composite calcined at 523 K (Table 4). At the lower calcination temperature of 423 K, the NO conversion is lower and more NO$_2$ is formed similar to the pure TiO$_2$ catalysts (*cf.* Table 1). It cannot be ruled out, however, that this is due to incomplete crystallization of the TiO$_2$ at this low calcination temperature. When the composite is calcined at 623 K or higher, the photocatalytic activity decreases with the calcination temperature as evident from the lower NO conversion and the lower amount of NO removed, however at lower selectivity for NO$_2$ than for the material calcined at 423 K.

Table 4. Conversion of NO (X_{NO}) and of NO$_x$ (X_{NO_x}), selectivity for NO$_2$ (S_{NO_2}), amount of NO removed (Z_{NO}), amount of NO$_2$ formed (Z_{NO_2}), and net amount NO$_x$ removed(Z_{NO_x}) in the photocatalytic oxidation of NO over different TiO$_2$ and TiO$_2$/ZSM-5 composite catalysts after 24 h UV irradiation (5 g TiO$_2$ or composite, 1000 ppb NO, 50% relative humidity, 3 dm$^3 \cdot$ min^{-1} total volumetric flow rate, 10 W$\cdot$ m^{-2} UV irradiance, 0.005 m^2 irradiated area).

$T_{calcinaton}$/K	X_{NO}/%	X_{NO_x}/%	Z_{NO}/ µmol$\cdot$ m$^{-2}\cdot$ h^{-1}	Z_{NO_2}/ µmol$\cdot$ m$^{-2}\cdot$ h^{-1}	Z_{NO_x}/ µmol$\cdot$ m$^{-2}\cdot$ h^{-1}	S_{NO_2}/%
423	30	19	494	287	280	48
523	41	33	655	131	524	19
623	38	30	608	119	447	19
723	35	28	548	109	400	20
823	27	21	429	150	292	25

This can be well explained by a lower surface area of the larger TiO$_2$ crystals and is also consistent with a less intense interaction of the larger TiO$_2$ particles with the outer surface of the zeolite crystals. For the material calcined at 823 K, the presence of rutile is expected to further deteriorate the photocatalytic activity [41–43]. Therefore, the calcination temperature used in this study was 523 K.

3. Experimental Section

3.1. Chemicals

Ethanol ($\geqslant$99.5%) and titanium (IV) isopropoxide (TIP) (97%) were purchased from Sigma-Aldrich (St. Louis, MO, USA). The ammonium form of zeolite ZSM-5 (NH$_4$-ZSM-5; n_{Si}/n_{Al} = 55) was supplied by Süd Chemie AG (Munich, Germany). TiO$_2$ was obtained as Hombikat from Sachtleben Chemie (NRW, Duisburg, Germany) and as P25 from Evonik Degussa (Frankfurt, Germany). All chemicals were used as received without further purification.

3.2. Preparation of the TiO$_2$/ZSM-5 Composites

3.2.1. Solid-State Dispersion

First, 1.25, 2.50 or 3.75 g of zeolite ZSM-5 was suspended in 20 cm^3 of deionized water with continuous stirring for 2 h. Subsequently, 3.75, 2.50 or 1.25 g of Hombikat was added to the ZSM-5 slurry with vigorous stirring for another 2 h. Then, the slurry was dried for 12 h in an air oven at 363 K, grounded and calcined at 523 K in air for 6 h. The composites prepared by this method are referred to as HZSSD. The label HZSSD represents the Hombikat/ZSM-5 composites prepared using solid-state dispersion. The mass fractions of Hombikat to ZSM-5 are 25/75, 50/50 and 75/25, respectively.

3.2.2. Sol-Gel Synthesis

A TiO$_2$ sol was prepared by acid-catalyzed hydrolysis of TIP according to [49]. 4.6, 9.2 or 14.4 cm^3 of TIP equivalent to 1.25, 2.50 or 3.75 g of TiO$_2$ were dissolved in 25 cm^3 of ethanol ($\geqslant$99.5%, Sigma-Aldrich, St. Louis, MO, USA) and vigorously stirred at room temperature for 2 h. Subsequently, 1 cm^3 of 1M HNO$_3$ (65%, VWR International, Radnor, PA, USA) was added to the solution under continuous stirring for 5 min until gelation takes place. Then, 25 cm^3 of deionized water was added to the gel. The resulting gel was stirred for another 3 h. To obtain the TiO$_2$/ZSM-5 composites with mass fractions of 25/75, 50/50 or 75/25, the TiO$_2$ sol was added to a suspension of 3.75, 2.50 or 1.25 g of ZSM-5 in 50 cm^3 of deionized water over 1 h at room temperature under stirring. The resulting gel was stirred for another 6 h and dried in an air oven at 353 K for 12 h. Finally, the obtained solid was ground to a powder and calcined at 523 K in air for 6 h. These composites were named TZSG indicating the composition of TiO$_2$/ZSM-5 prepared by sol-gel synthesis. For comparison, pure TiO$_2$ and a composite of TiO$_2$/sea sand (Merck) (TSSSG) were also prepared using the same sol-gel route. A TiO$_2$/ZSM-5 composite (50/50) for the purpose of studying the influence of calcination temperature was also prepared by sol-gel method. Before calcination, the dried powder was divided equally into five parts. Each of them was calcined at different temperatures from 423 to 823 K in air for 6 h.

3.2.3. Mechanical Mixture

For comparison, zeolite ZSM-5 (1.25, 2.50 or 3.75 g) and Hombikat (3.75, 2.50 or 1.25 g) were mixed in an agate pestle and mortar and ground to a fine powder followed by calcination at 523 K in air for 6 h. The obtained mixtures are denoted as HZMM indicating the preparation by mechanical mixing of Hombikat and zeolite ZSM-5. The mass fractions of Hombikat to ZSM-5 were maintained as in the solid-state dispersion and the sol-gel synthesis, *i.e.*, 25/75, 50/50 and 75/25, respectively.

3.3. Characterization of Photocatalysts

A micromeritics (Norcross, GA, USA) model ASAP2010 physisorption analyzer was used to record N_2-sorption isotherms at 77 K for the calcined materials. The samples were pre-treated at 523 K under the vacuum (3×10^{-11} MPa) for 6 h. The specific surface area A_{BET} was determined by the Brunauer-Emmett-Teller (BET) model. The specific pore volume was estimated at the point $P/P_0 = 0.995$, and the average pore width was calculated by the BJH model from the desorption branch of the isotherm. For characterization of the micropore volume according to the *t*-plot method, the pressure range of $P/P_0 = 0.35$–0.5 was considered.

Powder XRD patterns were taken at room temperature using a Siemens D5000 diffractometer (Karlsruhe, Germany) using Cu-Kα radiation ($\lambda = 0.154$ nm) in the 2θ-range of 4° and 90° with a step size of 0.02° and a counting time of 0.2 s for phase identification.

Selected samples were analyzed by scanning electron microscopy (SEM) using E-O-GmbH CamScan CS44 (Dortmund, Germany) operated at an accelerating voltage of 10 kV. Prior to analysis, the samples were grinded into a powder, mounted on SEM stubs using a carbon tape and then sputter-coated with a thin layer of gold to avoid charging during the microscopic measurement.

Diffuse reflectance UV-Vis absorption measurements were performed on a Lambda 650 instrument (LabSphere, North Sutton, NH, USA) of Perkin-Elmer. After grinding of the samples into fine powders, a thin layer was placed in a sample holder. All samples were measured within the range 200–800 nm. The reflected light was collected by a universal reflectance sphere for all wavelengths.

3.4. Photocatalytic Experiments

The photocatalytic experiments were carried out following the standard procedure based on ISO 22197-1 [37]. The experimental setup consists of a flat-bed reactor, a NO gas supply, a humidifier, a UV lamp (Ultra-Vitalux 300 W (OSRAM, München, Germany) UVA (315–400 nm) and UVB (280–315 nm)), a test gas generator model Sycos KT-GPTM (Ansyco, Karlsruhe, Germany) and a chemiluminescence NO_x analyzer AC 32 M (environment S.A, Poissy, France), as shown in Figure 7.

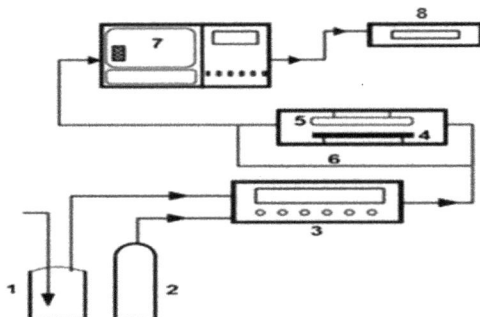

Figure 7. Schematic flow diagram of photocatalytic setup. (1) Humidifier; (2) NO gas supply; (3) test gas generator; (4) reactor; (5) UV lamp; (6) By pass; (7) NO_x analyzer; (8) Data recorder.

The reactor is made from non-adsorbing polymethyl methacrylate (PMMA) with 0.48 m length, 0.10 m width, and 0.06 m height. It is covered with a quartz glass plate permitting access of UV light to the reactor. The catalyst is filled as a powder into a sample holder made of plastic (PMMA) of the following dimensions (0.1 m length, 0.05 m width and 0.01 m height) and then placed in the middle of the reactor. The UV lamp is located above the reactor at a distance between the lamp and the reactor of 76 cm to achieve the UV irradiance of 10 W·m^{-2} recommended by the ISO standard [37]. The amount of NO, NO_2 and NO_x were continuously (5 s sampling frequency) monitored at the gas outlet using the chemiluminescence NO_x analyzer.

The photocatalytic NO oxidation experiments were carried out at ambient temperature and pressure. In a typical experiment, 5 g of the sample was placed in the sample holder covering an area of 0.005 m^2. Then, a stream of NO (1000 ppb) in air at 50% relative humidity with a total volumetric flow rate of 3 dm^3·min^{-1}; was led through the system using a bypass. Typically, after 15 min, the desired NO and water concentrations reached a steady state at the desired values. Subsequently, the humid NO-containing feed was supplied to the reactor with the UV light turned off. After achieving adsorption equilibrium as indicated by a constant NO concentration in the gas leaving the reactor, typically after 5 min, the UV light was switched on. After 24 h on-stream, the UV light was turned off again and the outlet concentration of NO was monitored under flow of the reactant feed through the reactor until it reached the initial concentration of 1000 ppb, typically after 15 min. It was proven

170

that no NO photolysis occurs in the absence of any catalyst, *i.e.*, in the empty reactor with UV light irradiation for 3 h.

In order to determine the amount of NO_2^- and NO_3^- anions on the sample surface after the photocatalytic experiments, selected samples were immersed in 25 cm^3 deionized water and the obtained aqueous solution was analyzed in an ion chromatograph (IC); (dionex, Sunnyvale, CA, USA). The conversion of NO (X_{NO}), Conversion of NO_x (X_{NO_x}), the selectivity for NO_2 (S_{NO_2}), the amount of NO removed ($Z_{NO}/\mu mol \cdot m^{-2} \cdot h^{-1}$), the amount of NO_2 formed ($Z_{NO_2}/\mu mol \cdot m^{-2} \cdot h^{-1}$), and the net amount NO_x removed ($Z_{NO_x}/\mu mol \cdot m^{-2} \cdot h^{-1}$) were calculated according to the following equations [37,50]:

$$X_{NO} = ((C_{NO,in} - C_{NO,out})/C_{NO,in})\ 100\%, \tag{1}$$

$$X_{NO_x} = ((C_{NO_x,in} - C_{NO_x,out})/C_{NO_x,in})\ 100\%, \tag{2}$$

$$S_{NO_2} = (C_{NO_2,out}/(C_{NO,in} - C_{NO,out}))\ 100\%, \tag{3}$$

$$Z_{NO} = ((f/22.4\ dm^3 \cdot mol^{-1}) \int_0^{24} (C_{NO,in} - C_{NO,out})\ dt)/A\ t, \tag{4}$$

$$Z_{NO_2} = ((f/22.4\ dm^3 \cdot mol^{-1}) \int_0^{24} (C_{NO_2,out} - C_{NO_2,in})\ dt)/A\ t, \tag{5}$$

$$Z_{NO_x} = (Z_{NO} - Z_{NO_2}), \tag{6}$$

where C_{NO}^{in} and $C_{NO_2}^{in}$ are the inlet concentrations of NO and NO_2, C_{NO}^{out} and $C_{NO_2}^{out}$ are the outlet concentration and calculated as the average outlet concentration over 24 h of irradiation time; t is the irradiation time in min; f is the overall volumetric flow rate in dm$^3 \cdot$ min^{-1}; A is the irradiated surface area in m^2; 22.4 dm$^3 \cdot$ mol^{-1} is the molar volume of an ideal gas at the standard state. The mass balance for NO_x was closed within $\pm 10\%$. The reported conversion, selectivity and Z-values are accurate within an experimental error of $\pm 4\%$.

4. Conclusions

Composites of TiO_2 from different sources such as Hombikat, P25, or from sol-gel synthesis, with crystalline zeolites such as ZSM-5 can be readily prepared over the complete range of mass fractions by mechanical mixing or solid-state dispersion. No apparent difference in the textural, morphological or optical properties in the composites from the two preparation techniques were observed for TiO_2/ZSM-5 composites. A significantly more homogeneous distribution of small TiO_2 particles covering the outer surface of the zeolite ZSM-5 crystals was achieved by sol-gel synthesis of TiO_2 in the presence of the zeolite crystals. This results in an intimate contact of both components in the composites. At higher mass fractions of the TiO_2, however, increasingly large mesopores within the TiO_2 phase are formed, reducing the contact area to the zeolite crystals.

The activity of the composites in the photocatalytic oxidation (PCO) of NO in a flatbed reactor under continuous flow according to ISO 22197-1 is distinctly different from that of the pure TiO_2 materials. While the NO conversion is lower, the selectivity for the undesired NO_2 is lower and a higher fraction of nitrate and nitrite may be formed. The role of the zeolite can, thus, be understood to be an adsorptive component for NO_2, which is continuously released to the photocatalytically active TiO_2 particles and further converted to the desired nitrate. Another function of the zeolite may be to adsorb the nitric acid formed by photooxidation of NO_x, which otherwise deposits on the TiO_2 surface and leads to deactivation by blocking of active sites. This function, would, however, be limited by the overall adsorption capacity of the zeolite which, in the experiments conducted here, was not reached.

The highest NO conversion of 41% at a NO_2 selectivity as low as 19% stable for 24 h on-stream was reached over the TiO_2/ZSM-5 composite with equal amounts of the two components after calcination at 523 K from sol-gel synthesis. This activity and stability for complete NO oxidation was even higher than for pure TiO_2 from sol-gel synthesis, Hombikat, or P25. These favorable photocatalytic properties could be clearly attributed to the adsorptive properties of the zeolite ZSM-5 in the composite photocatalyst. They have to be balanced, however, with the photocatalytic function provided by the TiO_2 and its amount in the composite.

The approach of utilizing zeolites as adsorptive components in photocatalytic composites for oxidative NO_xremoval may be applied to other zeolites and photoactive catalyst components. Further studies towards increasing the effectiveness of these composites should, besides long-term stability studies under the application conditions, target the balance of surface hydrophobicity/hydrophilicity with respect to adsorption capacity, *i.e.*, specific pore volume and surface area. As the contact between the photocatalytically active and the sorptive function is of prime importance, the effect of the diameters and the shape of the zeolite crystals may also be adjusted for improving the activity and stability of composites catalysts for photooxidative NO_x removal.

Acknowledgments: The authors are grateful to Sebastian Zimmermann, Institut of Chemical Technology, Universität Leipzig, for help with the photocatalytic apparatus and to Gerald Wagner, Institute for Mineralogy, Crystallography and Material Science; Universität Leipzig, for the SEM pictures. The authors acknowledge the Ministry of Higher Education and Scientific Research-Libya for financial support.

Author Contributions: Roger Gläser provided the idea and design for the study. Akram Tawari performed the experiments. Akram Tawari, Roger Gläser and Wolf-Dietrich Einicke analyzed the data. Akram Tawari drafted the manuscript. Wolf-Dietrich Einicke and Roger Gläser revised it.

Conflicts of Interest: The authors declare no conflict of interest.

References

1. Farrell, A. Multi-lateral emission trading: Lessons from inter-state NO_x control in the United States. *Energy Policy* **2001**, *29*, 1061–1072.

2. Castro, T.; Madronich, S.; Rivale, S.; Muhlia, A.; Mar, B. The influence of aerosols on photochemical smog in Mexico City. *Atmos. Environ.* **2001**, *35*, 1765–1772.

3. Devahasdin, S.; Fan, C.; Li, K.Y.; Chen, D.H. TiO_2 photocatalytic oxidation of nitric oxide: Transient behavior and reaction kinetics. *J. Photochem. Photobiol. A* **2003**, *156*, 161–170.

4. Cooper, C.D.; Alley, F.C. *Air Pollution Control: A Design Approach*, 4th ed.; Waveland Press: Long Grove, IL, USA, 2011; pp. 535–543.

5. Tang, X.L.; Hao, J.M.; Xu, W.G.; Li, J.H. Low temperature selective catalytic reduction of NO_x with NH_3 over amorphous MnO_x catalysts prepared by three methods. *Catal. Commun.* **2007**, *8*, 329–334.

6. Lim, T.H.; Jeong, S.M.; Kim, S.D.; Gyenis, J. Photocatalytic decomposition of NO by TiO_2 particles. *J. Photochem. Photobiol. A* **2000**, *134*, 209–217.

7. Pleskov, Y.V.; Fujishima, A.; Hashimoto, K.; Watanabe, T. TiO_2 photocatalysis: Fundamentals and applications. *Russ. J. Electrochem.* **1999**, *35*, 1137–1138.

8. Fujishima, A.; Honda, K. Electrochemical photolysis of water at a semiconductor electrode. *Nature* **1972**, *238*, 37–38.

9. Ibusuki, T.; Takeuchi, K. Removal of Low concentration nitrogen-oxides through photoassisted heterogeneous catalysis. *J. Mol. Catal.* **1994**, *88*, 93–102.

10. Fujishima, A.; Rao, T.N.; Tryk, D.A. TiO_2 photocatalysts and diamond electrodes. *Electrochim. Acta* **2000**, *45*, 4683–4690.

11. Fujishima, A.; Zhang, X. Titanium dioxide photocatalysis: Present situation and future approaches. *C. R. Chim.* **2006**, *9*, 750–760.

12. Wang, H.; Wu, Z.; Zhao, W.; Guan, B. Photocatalytic oxidation of nitrogen oxides using TiO_2 loading on woven glass fabric. *Chemosphere* **2007**, *66*, 185–190.

13. Poon, C.S.; Cheung, E. NO removal efficiency of photocatalytic paving blocks prepared with recycled materials. *Constr. Build. Mater.* **2007**, *21*, 1746–1753.

14. Maggos, T.; Bartzis, J.G.; Liakou, M.; Gobin, C. Photocatalytic degradation of NO_x gases using TiO_2-containing paint: A real scale study. *J. Hazard. Mater.* **2007**, *146*, 668–673.

15. Hüsken, G.; Hunger, M.; Brouwers, H.J.H. Experimental study of photocatalytic concrete products for air purification. *Build. Environ.* **2009**, *44*, 2436–2474.

16. Hassan, M.M.; Dylla, H.; Mohammad, L.N.; Rupnow, T. Evaluation of the durability of titanium dioxide photocatalyst coating for concrete pavement. *Constr. Build. Mater.* **2010**, *24*, 1456–1461.

17. Martinez, T.; Bertron, A.; Ringot, E.; Escadeillas, G. Degradation of NO using photocatalytic coatings to different substrates. *Build. Environ.* **2011**, *46*, 1808–1816.

18. Guo, S.; Wu, Z.; Zhao, W. TiO_2-based building materials above and beyond traditional applications. *Chin. Sci. Bull.* **2009**, *54*, 1137–1142.

19. Hashimoto, K.; Wasada, K.; Toukai, N.; Kominami, H.; Kera, Y. Photocatalytic oxidation of nitrogen monoxide over titanium (IV) oxide nanocrystals large size areas. *J. Photochem. Photobiol. A* **2000**, *136*, 103–109.

20. Nakamura, I.; Sugihara, S.; Takeuchi, K. Mechanism for NO photooxidation over the oxygen-deficient TiO$_2$ powder under visible light irradiation. *Chem. Lett.* **2000**, *11*, 1276–1277.

21. Dalton, J.S.; Janes, P.A.; Jones, N.G.; Nicholson, J.A.; Hallam, K.R.; Allen, G.C. Photocatalytic oxidation of NO$_x$ gases using TiO$_2$: A surface spectroscopic approach. *Environ. Pollut.* **2002**, *120*, 415–422.

22. Ichiura, H.; Kitaoka, T.; Tanaka, H. Photocatalytic oxidation of NO$_x$ using composite sheets containing TiO$_2$ and a metal compound. *Chemosphere* **2003**, *51*, 855–860.

23. Takeda, N.; Torimoto, T.; Sampath, S.; Kuwabata, S.; Yoneyama, H. Effect of inert supports for titanium-dioxide loading on enhancement of photodecomposition rate of gaseous propionaldehyde. *J. Phys. Chem.* **1995**, *99*, 9986–9991.

24. Anderson, C.; Bard, A.J. Improved photocatalytic activity and characterization of mixed TiO$_2$/SiO$_2$ and TiO$_2$/Al$_2$O$_3$ materials. *J. Phys. Chem. B* **1997**, *101*, 2611–2616.

25. Matos, J.; Laine, J.; Herrmann, J.M. Synergy effect in the photocatalytic degradation of phenol on a suspended mixture of titania and activated carbon. *Appl. Catal. B* **1998**, *18*, 281–291.

26. Fichtner, S.; Hofmann, J.; Möller, A.; Schrage, C.; Giebelhausen, J.M.; Böhringer, B.; Gläser, R. Decomposition of 2-chloroethylethylsulfide on copper oxides to detoxify polymer-based spherical activated carbons from chemical warfare agents. *J. Hazard. Mater.* **2013**, *262*, 789–795.

27. Portela, R.; Tessinari, R.F.; Suarez, S.; Rasmussen, S.B.; Alonso, M.D.H.; Canela, M.C.; Avila, P.; Sanchez, B. Photocatalysis for continuous air purification in wastewater treatment plants: From lab to reality. *Environ. Sci. Technol.* **2012**, *46*, 5040–5048.

28. Mahalakshmi, M.; Priya, S.V.; Arabindoo, B.; Palanicharnly, M.; Murugesan, V. Photocatalytic degradation of aqueous propoxur solution using TiO$_2$ and H beta zeolite-supported TiO$_2$. *J. Hazard. Mater.* **2009**, *161*, 336–343.

29. Corma, A. From microporous to mesoporous molecular sieve materials and their use in catalysis. *Chem. Rev.* **1997**, *97*, 2373–2419.

30. Zhang, J.L.; Minagawa, M.; Matsuoka, M.; Yamashita, H.; Anpo, M. Photocatalytic decomposition of NO on Ti-HMS mesoporous zeolite catalysts. *Catal. Lett.* **2000**, *66*, 241–243.

31. Guo, G.F.; Hu, Y.; Jiang, S.M.; Wei, C.H. Photocatalytic oxidation of NO$_x$ over TiO$_2$/HZSM-5 catalysts in the presence of water vapor: Effect of hydrophobicity of zeolites. *J. Hazard. Mater.* **2012**, *223*, 39–45.

32. Hashimoto, K.; Wasada, K.; Osaki, M.; Shono, E.; Adachi, K.; Toukai, N.; Kominami, H.; Kera, Y. Photocatalytic oxidation of nitrogen oxide over titania-zeolite composite catalyst to remove nitrogen oxides in the atmosphere. *Appl. Catal. B* **2001**, *30*, 429–436.

33. Ichiura, H.; Kitaoka, T.; Tanaka, H. Preparation of composite TiO$_2$-zeolite sheets using a papermaking technique and their application to environmental improvement—Part I—Removal of acetaldehyde with and without UV irradiation. *J. Mater. Sci.* **2002**, *37*, 2937–2941.

174

34. Ichiura, H.; Kitaoka, T.; Tanaka, H. Removal of indoor pollutants under UV irradiation by a composite TiO_2-zeolite sheet prepared using a papermaking technique. *Chemosphere* **2003**, *50*, 79–83.

35. Takeuchi, M.; Kimura, T.; Hidaka, M.; Rakhmawaty, D.; Anpo, M. Photocatalytic oxidation of acetaldehyde with oxygen on TiO_2/ZSM-5 photocatalysts: Effect of hydrophobicity of zeolites. *J. Catal.* **2007**, *246*, 235–240.

36. Jansson, I.; Suárez, S.; Javier Garcia-Garcia, F.; Sánchez, B. Zeolite-TiO_2 hybrid composites for pollutant degradation in gas phase. *Appl. Catal. B* **2015**, *178*, 100–107.

37. ISO. *Fine Ceramics (Advanced Ceramics, Advanced Technical Ceramics)—Test Method for Air-Purification Performance of Semi Conducting Photocatalytic Materials—Part 1: Removal of Nitric oxide*, 1st ed.; ISO: Geneva, Switzerland, 2007.

38. Hu, Y.; Tsai, H.L.; Huang, C.L. Phase transformation of precipitated TiO_2 nanoparticles. *Mater. Sci. Eng. A* **2003**, *344*, 209–214.

39. Wongkalasin, P.; Chavadej, S.; Sreethawong, T. Photocatalytic degradation of mixed azo dyes in aqueous wastewater using mesoporous-assembled TiO_2 nanocrystal synthesized by a modified sol-gel process. *Colloids Surf. A* **2011**, *384*, 519–528.

40. Paola, A.D.; Bellardita, M.; Palmisano, L. Brookite, the least known TiO_2 photocatalyst. *Catalysts* **2013**, *3*, 36–73.

41. Zhang, J.; Zhou, P.; Liu, J.; Yu, J. New understanding of the difference of photocatalytic activity among anatase, rutile and brookite TiO_2. *Phys. Chem. Chem. Phys.* **2014**, *16*, 20382–20386.

42. Rodella, C.B.; Nascente, P.A.P.; Franco, R.W.A.; Magon, C.J.; Mastelaro, V.R.; Florentino, A.O. Surface characterization of V_2O_5/TiO_2 catalytic system. *Phys. Status Solidi. A* **2001**, *187*, 161–169.

43. Tanaka, K.; Capule, M.F.V.; Hisanaga, T. Effect of crystallinity of TiO_2 on its photocatalytic action. *Chem. Phys. Lett.* **1991**, *187*, 73–76.

44. Sang, S.Y.; Chang, F.X.; Liu, Z.M.; He, C.Q.; He, Y.L.; Xu, L. Difference of ZSM-5 zeolites synthesized with various templates. *Catal. Today* **2004**, *93*, 729–734.

45. Xu, Y.M.; Langford, C.H. Photoactivity of titanium dioxide supported on MCM41, zeolite X, and zeolite Y. *J. Phys. Chem. B* **1997**, *101*, 3115–3121.

46. Sing, K.S.W.; Everett, D.H.; Haul, R.A.W.; Moscou, L.; Pierotti, R.A.; Rouquerol, J.; Siemieniewska, T. Reporting physisorption data for gas solid systems with special reference to the determination of surface-area and porosity. *Pure Appl. Chem.* **1985**, *57*, 603–619.

47. Kamegawa, T.; Ishiguro, Y.; Kido, R.; Yamashita, H. Design of composite photocatalyst of TiO_2 and Y-zeolite for degradation of 2-propanol in the gas phase under UV and visible light irradiation. *Molecules* **2014**, *19*, 16477–16488.

48. Mahshid, S.; Askari, M.; Ghamsari, M.S. Synthesis of TiO_2 nanoparticles by hydrolysis and peptization of titanium isopropoxide solution. *Mater. J. Process. Technol.* **2007**, *189*, 296–300.

49. Xu, Y.M.; Langford, C.H. Enhanced photoactivity of a titanium (IV) oxide supported on ZSM5 and zeolite A at low coverage. *J. Phys. Chem.* **1995**, *99*, 11501–11507.

50. Ma, J.; Wu, H.; Liu, Y.; He, H. Photocatalytic removal of NO_x over visible light responsive oxygen-deficient TiO_2. *J. Phys. Chem. C* **2014**, *118*, 7434–7441.

The Fabrication of Ga₂O₃/ZSM-5 Hollow Fibers for Efficient Catalytic Conversion of *n*-Butane into Light Olefins and Aromatics

Jing Han, Guiyuan Jiang, Shanlei Han, Jia Liu, Yaoyuan Zhang, Yeming Liu, Ruipu Wang, Zhen Zhao, Chunming Xu, Yajun Wang, Aijun Duan, Jian Liu and Yuechang Wei

Abstract: In this study, the dehydrogenation component of Ga_2O_3 was introduced into ZSM-5 nanocrystals to prepare Ga_2O_3/ZSM-5 hollow fiber-based bifunctional catalysts. The physicochemical features of as-prepared catalysts were characterized by means of XRD, BET, SEM, STEM, NH_3-TPD, *etc.*, and their performances for the catalytic conversion of *n*-butane to produce light olefins and aromatics were investigated. The results indicated that a very small amount of gallium can cause a marked enhancement in the catalytic activity of ZSM-5 because of the synergistic effect of the dehydrogenation and aromatization properties of Ga_2O_3 and the cracking function of ZSM-5. Compared with Ga_2O_3/ZSM-5 nanoparticles, the unique hierarchical macro-meso-microporosity of the as-prepared hollow fibers can effectively enlarge the bifunctionality by enhancing the accessibility of active sites and the diffusion. Consequently, Ga_2O_3/ZSM-5 hollow fibers show excellent catalytic conversion of *n*-butane, with the highest yield of light olefins plus aromatics at 600 °C by 87.6%, which is 56.3%, 24.6%, and 13.3% higher than that of ZSM-5, ZSM-5 zeolite fibers, and Ga_2O_3/ZSM-5, respectively.

Reprinted from *Catalysts*. Cite as: Han, J.; Jiang, G.; Han, S.; Liu, J.; Zhang, Y.; Liu, Y.; Wang, R.; Zhao, Z.; Xu, C.; Wang, Y.; Duan, A.; Liu, J.; Wei, Y. The Fabrication of Ga_2O_3/ZSM-5 Hollow Fibers for Efficient Catalytic Conversion of *n*-Butane into Light Olefins and Aromatics. *Catalysts* **2016**, *6*, 13.

1. Introduction

Light olefins, including ethene, propene, butene, *etc.* are important basic chemical raw materials and the demand for them stays at a high level. Aromatic hydrocarbons, such as benzene, toluene, and xylene, are also significant materials for organic chemical production, which are widely used in the production of synthetic fiber, resin, and rubber. For the past few years, crude processing capacity of refinery continues to improve, resulting in the by-production of a large amount of C4 hydrocarbons. Among C4 hydrocarbons, many processes on the conversion of C4 alkenes have been reported, including catalytic cracking, disproportionation, *etc.* However, C4 alkanes, as important components of C4 hydrocarbons, due to their high stability, the chemical utilization efficiency is still low. Currently, they are mostly used as low-added value fuel. So, using less valuable but industrially abundant C4 alkanes as feedstock to produce light olefins and aromatics has been attracting increasing attention [1–3].

Compared with the current main process of steam cracking for the production of light olefins, catalytic cracking, due to the introduction of catalyst, can reduce the reaction temperature and energy consumption, and it also can improve the selectivity to light olefins, especially to that of propylene [4]. Up to now, three kinds of catalysts have been proposed for catalytic cracking of hydrocarbons, including zeolites [5–7], metal oxides [8,9], and composite catalysts [10,11]. Among various catalysts, ZSM-5 zeolite is a typical and superior candidate because of its excellent stability, adjustable acidity, and special pore structure [3,12–14]. To further improve the catalytic cracking performances of ZSM-5 zeolite, many modifications have been reported, including alkaline earth metal [15], transition metal [16], rare earth elements [17], phosphorus modification, *etc.* [18]. The above modifications can modulate the amount of acidic sites and the acid strength of ZSM-5 zeolites, thus enhancing the selectivity to light olefins and promoting the catalytic performances of ZSM-5.

In addition to the regulation of acidity, the optimization of the pore structure is another effective strategy to enhance the catalytic performances of zeolite catalysts. In this context, nano ZSM-5 zeolites [19], mesoporous ZSM-5 zeolites [20], nanosheets of zeolite [21], and other hierarchical ZSM-5 zeolites [12] have been reported. Due to the introduction of pores with different levels, the accessibility of active sites of hierarchical ZSM-5 zeolites can be improved greatly. Meanwhile, the transportation capability for feedstock of large sizes could be also enhanced. Among various hierarchical ZSM-5 zeolites, hierarchical ZSM-5 fibers have received much concern because the hierarchical ZSM-5 fibers not only possess the high catalytic activity of zeolite, but also have high mass transfer performance and low pressure drop. Previously, we reported a versatile and facile method for the fabrication of hierarchical ZSM-5 zeolite fibers with macro-meso-microporosity by coaxial electrospinning, and it was found that suitable acidity and the hierarchical porosity contribute to the excellent catalytic performances in the catalytic cracking of *iso*-butane [12].

Although many catalysts have been proposed for catalytic cracking of C4 alkanes, there are few reports on the catalysts for efficient conversion of *n*-butane, the most stable component in C4 hydrocarbons. To promote the catalytic conversion of *n*-butane, the introduction of dehydrogenation component in the current acid-based zeolite to construct the bifunctional catalyst may provide a good solution [22,23]. Many metal oxide-based catalysts have been reported for the dehydrogenation of alkanes, such as vanadia-based [24,25], chromium-based [26–28], gallium-based [29,30], *etc.* So in the present study, using *n*-butane as feedstock, we chose Ga_2O_3 as the dehydrogenation component and introduced it into the ZSM-5 nanocrystals to fabricate Ga_2O_3/ZSM-5 hollow fiber based bifunctional catalysts. The physicochemical features of as-prepared catalysts were characterized by means of XRD, BET, SEM, STEM, NH_3-TPD, *etc.*, and their performance for the catalytic conversion of *n*-butane to produce light olefins and aromatics were investigated.

2. Results and Discussion

2.1. Catalyst Characterization

Figure 1 shows the XRD patterns of the as-synthesized catalysts. From Figure 1 the characteristic diffraction peaks of ZSM-5 could be seen, showing that the catalysts are typical of MFI topology and the introduction of gallium on ZSM-5 does not change the crystalline structure of ZSM-5. No obvious diffraction peaks corresponding to gallium species are observed in the XRD patterns of Ga/ZSM-5, indicating that gallium species may be well dispersed on the surface of ZSM-5 zeolite [31] or the amount of them is too low to show obvious diffraction peak. According to the Ga $2p_{3/2}$ XPS spectra (Figure 2), the Ga $2p_{3/2}$ binding energy value (BEs) for the as-prepared Ga/ZSM-5 (1118.3 eV) is consistent with that for Ga_2O_3, indicating that the gallium exists on the surface of ZSM-5 in the form of Ga_2O_3. In comparison with that of the single bulk Ga_2O_3, the broadening of the half peak width of Ga $2p_{3/2}$ of the as-prepared Ga_2O_3/ZSM-5 occurs, which is connected with the increase of the Ga dispersion [32].

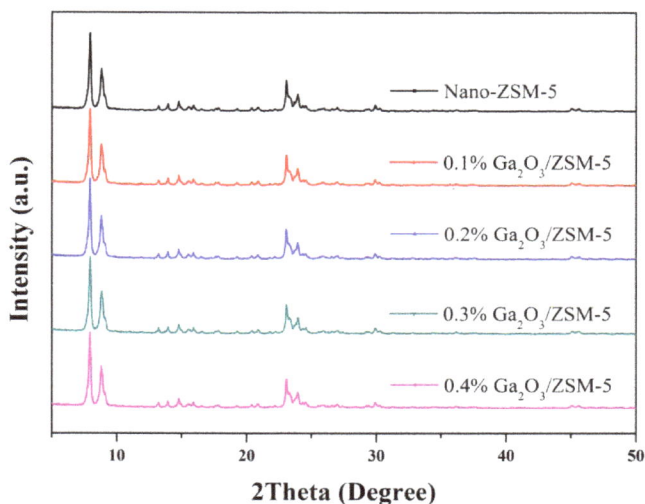

Figure 1. XRD patterns of the as-synthesized Ga_2O_3/ZSM-5 and Nano-ZSM-5.

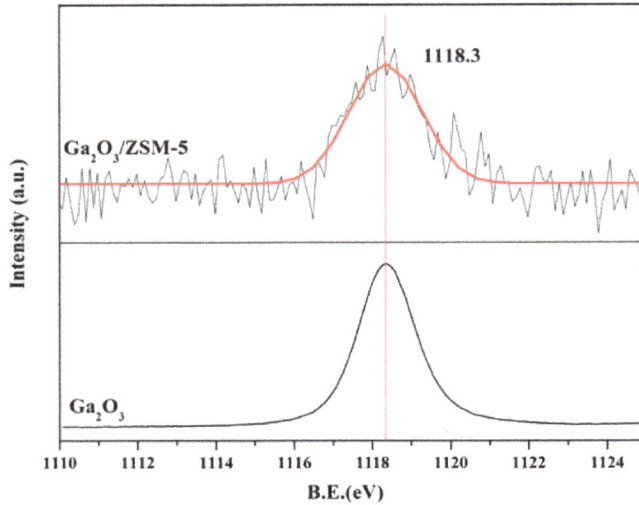

Figure 2. Ga $2p_{3/2}$ XPS spectra of Ga_2O_3 and $Ga_2O_3/ZSM-5$.

Figure 3A shows the SEM image of the synthesized Nano-ZSM-5 with the average particle size of 137 nm. After the introduction of Ga_2O_3 on Nano-ZSM-5 by impregnation (Figure 3B), the morphology and particle size of $Ga_2O_3/ZSM-5$ do not show obvious change. Then, $Ga_2O_3/ZSM-5$ was used as building blocks for the preparation of $Ga_2O_3/ZSM-5$ hollow fibers via coaxial electrospinning. Specifically, a suspension of $Ga_2O_3/ZSM-5$ nanocrystals in polyvinylpyrrolidon (PVP)/ethanol solution used as the outer fluid, and paraffin oil acted as the inner liquid. The addition of PVP in the suspension was to provide appropriate viscosity for the fluent electrospinning, and paraffin oil, the inner liquid, was to present hollow structure via calcination. The $Ga_2O_3/ZSM-5/PVP$ composite fibers were successfully prepared (Figure 3C), and after temperature-programmed calcination, the PVP and paraffin oil were removed, and hierarchical $Ga_2O_3/ZSM-5$ hollow fibers were obtained (Figure 3D). From Figure 3D, hollow structure and comparatively uniform diameter of the as-prepared fibers could be well seen. Figure 3E presents the high magnification SEM image of the red square in Figure 3D, and it clearly shows that the wall of the $Ga_2O_3/ZSM-5$ zeolite hollow fibers is composed of $Ga_2O_3/ZSM-5$ nanoparticles, with uniform macropores on the hollow fiber level. The TEM image (Figure 3F) of the fibers further proves the continuous hollow structure character of the fibers. The EDAX elemental maps present the element distribution of Si, O, Al, and Ga (Figure 4), and bright green dots indicates that Ga element is well dispersed on the as-prepared $Ga_2O_3/ZSM-5$ hollow fibers.

180

(A)   (B)   (C)

(D)   (E)   (F)

Figure 3. SEM images of the as-prepared Nano-ZSM-5 (**A**); 0.3% Ga_2O_3/ZSM-5 (**B**); 0.3% Ga_2O_3/ZSM-5/PVP fibers before calcination (**C**); the low (**D**) and magnified (**E**) SEM images of 0.3% Ga_2O_3/ZSM-5 hollow fibers after calcination at 550 °C, respectively; TEM image (**F**) of one hollow fiber after calcination at 550 °C.

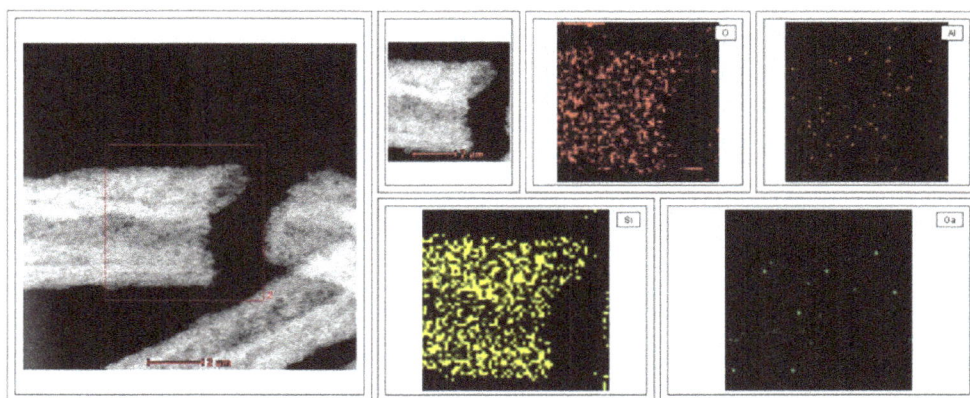

Figure 4. EDAX elemental maps of Si, O, Al, and Ga of 0.3% Ga_2O_3/ZSM-5 zeolite hollow fibers.

Figure 5 shows the nitrogen adsorption-desorption isotherms and BJH pore size distribution (inset) of the Ga_2O_3/ZSM-5 hollow fibers. According to the nitrogen adsorption-desorption measurement, the Brunauer-Emmett-Teller (BET) surface area of 377 m^2/g and the pore volume of 0.275 cm^3/g were determined. From Figure 5, it can be seen that the present isotherms exhibit a typical hysteresis loop at $p/p_0 > 0.1$, indicating mesopores are formed, arising from the interparticle voids of $Ga_2O_3/ZSM-5$. Hence, the as-prepared $Ga_2O_3/ZSM-5$ hollow fibers exhibit a good hierarchical macro-meso-microporosity, with micropores in ZSM-5 nanoparticles, mesopores formed by the stacking of the $Ga_2O_3/ZSM-5$ nanoparticles, and continuous macropores on the hollow fibers.

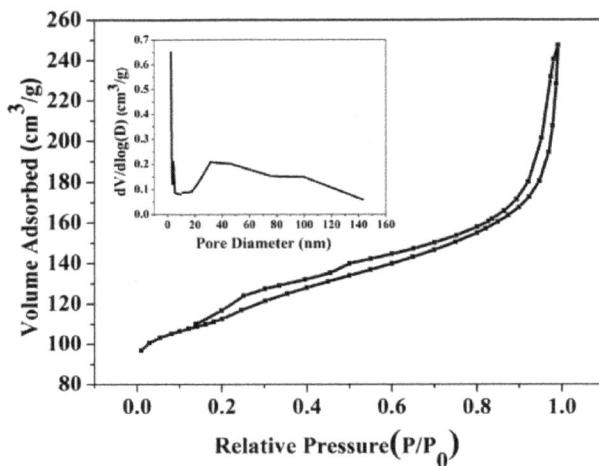

Figure 5. Nitrogen adsorption-desorption isotherms and BJH pore size distribution (inset) of the 0.3% $Ga_2O_3/ZSM-5$ zeolite hollow fibers.

The temperature-programmed desorption of ammonia (NH_3-TPD) experiments were conducted to obtain the acidic properties of the as-prepared $Ga_2O_3/ZSM-5$. Figure 6 presents NH_3-TPD profiles of $Ga_2O_3/ZSM-5$. From Figure 6, it can be seen that there are two desorption group peaks for ZSM-5, one is in the range of 150–250 °C, corresponding to the weak acid sites, and the other, *i.e.*, the strong acid sites, reside in the range of 300–450 °C. After the introduction of Ga_2O_3, there is no shift of the diffraction peaks, indicating that the strength of acidic sites does not change. Among the samples, except 0.2% $Ga_2O_3/ZSM-5$, the amount of weak acid does not show apparent change, and only the amount of strong acid shows a slight decrease (Table S1).

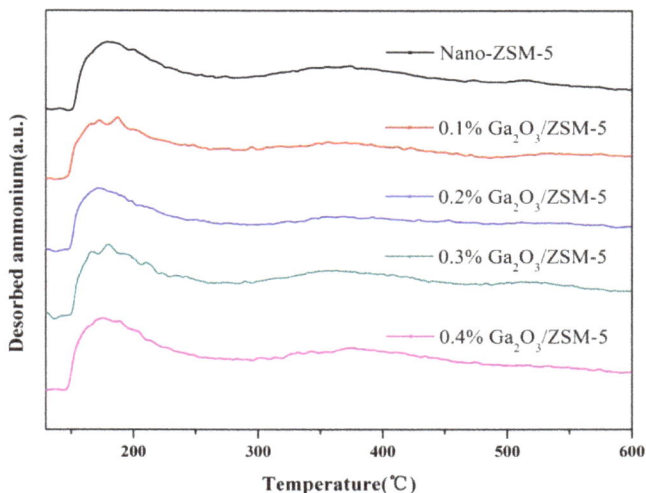

Figure 6. NH_3-TPD profiles of $x\%$ Ga_2O_3/ZSM-5($x = 0, 0.1, 0.2, 0.3, 0.4$).

2.2. Catalytic Preformances

Before investigating the catalytic performance of Ga_2O_3/ZSM-5 hollow fibers, to check whether an optimal introducing of Ga_2O_3 on ZSM-5 exists, Ga_2O_3/ZSM-5 catalysts with different Ga_2O_3 loading were evaluated. Figure 7 shows the conversion of n-butane (A), yield of ethene and propene (B), and yield of $C_2^=$, $C_3^=$, $C_4^=$ plus BTX ($C_2^=$, $C_3^=$, $C_4^=$ plus BTX refer to ethene, propene, butenes, butadiene and benzene, toluene, xylene) (C) as a function of reaction temperature. From Figure 7, it is apparent that with increase of the reaction temperature, the conversion of n-butane, the yield of ethene and propene and the yield of $C_2^=$, $C_3^=$, $C_4^=$ plus BTX on all catalysts markedly increase. And at the same temperatrue, increasing the Ga_2O_3 loading results in the increase of the conversion and the yield of $C_2^=$, $C_3^=$, $C_4^=$ plus BTX. As for the yield of ethene and propene, when the reaction temperature is lower than 600 °C, it increases with the increasing loading of Ga_2O_3, while when the temperature is above 625 °C, the yield of ethene and propene decreases with increasing the loading of Ga_2O_3. Such a result could be attributed to the synergistic effect of the dehydrogenation and aromatization properties of Ga_2O_3 [33] and the cracking function of ZSM-5. At lower temperatures, due to the dehydrogenation property, the presence of Ga_2O_3 contributes to the activation and dehydrogenation of n-butane. Hence n-butane may undergo dehydrogenation to produce butene, which is much easier to be converted than n-butane, and then, butene will undergo catalytic cracking reaction to produce ethene and propene as primary products. When temperature is above 625 °C, the produced ethene and propene will undergo oligomerization to produce higher olefins (C2–C9), and then,

183

C7–C9 alkylcyclohexenes would be formed by cyclization and dehydrogenation. Finally, C7–C9 alkylcyclohexenes are converted to corresponding aromatics [34]. Thus, although the increase of the yield of ethene and propene flattens out with the increase of the temperature, the generation of the benzene, toluene, and xylene results in the yield of $C_2^=$, $C_3^=$, $C_4^=$ plus BTX continuing to increase. From Figure 7, it is can be seen that when the Ga_2O_3 loading amount is above 0.3%, the enhancing capacity of the conversion of n-butane and the yield of $C_2^=$, $C_3^=$, $C_4^=$ plus BTX is not obvious. According to the literature [35], one possible explanation is that a very small amount of gallium is sufficient to cause a marked enhancement in the activity of ZSM-5, and too much gallium loading will lead to excessive dehydrogenation and secondary reaction, resulting in no beneficial improvement in catalytic conversion of n-butane. In addition, according to the NH_3-TPD results (Figure 6), the introduction of Ga_2O_3 slightly decreased the amount of strong acid of ZSM-5, so with the increase of the Ga_2O_3 loading, the decreased acidity property results in the decrease of cracking performance. The present result indicates that excellent catalytic activity could be obtained by regulating of the Ga_2O_3, leading to a good balance between the dehydrogenation of Ga_2O_3 and the cracking function of ZSM-5. Given that 0.3% Ga_2O_3/ZSM-5 was chosen as building blocks for the preparation of Ga_2O_3/ZSM-5 hollow fibers via coaxial electrospinning, and their catalytic conversion of n-butane into light olefins and aromatics was further investigated.

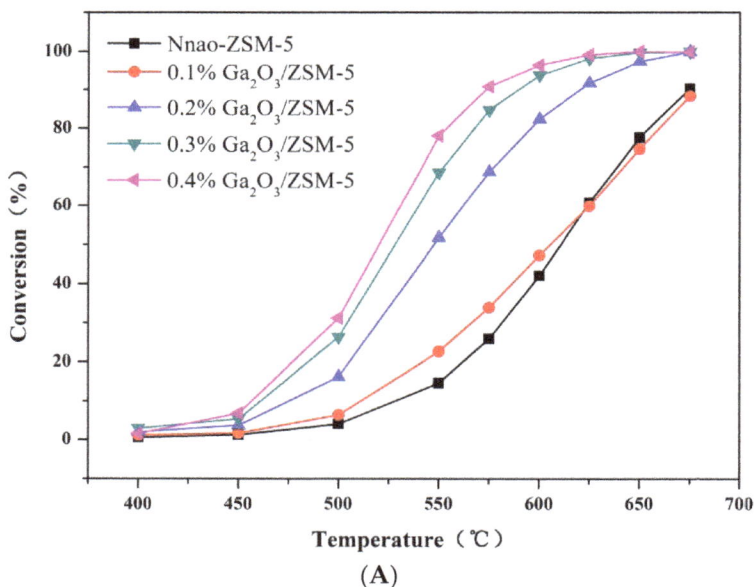

Figure 7. *Cont.*

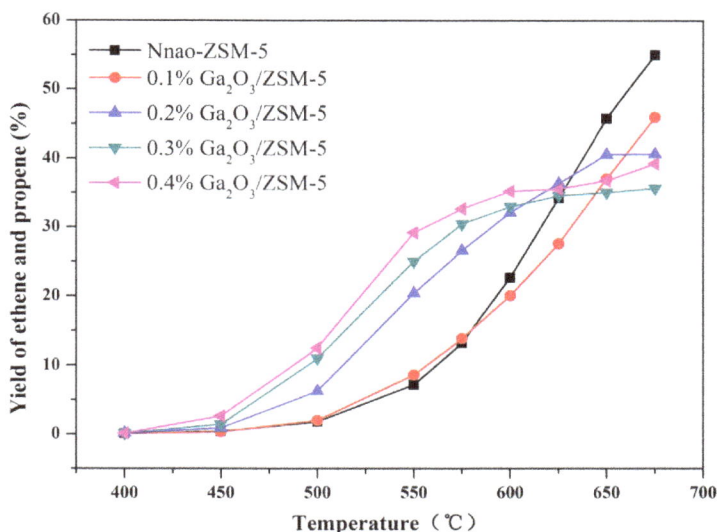

(B)

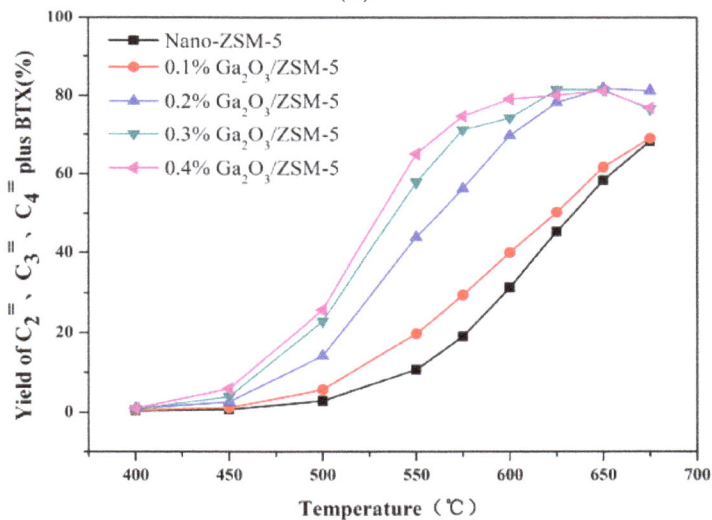

(C)

Figure 7. The conversion of *n*-butane (**A**); the yield of ethane and propene (**B**) and $C_2^=$, $C_3^=$, $C_4^=$ plus BTX (**C**) of the catalysts with different Ga_2O_3 loadings as a function of reaction temperatures.

Figure 8 shows the conversion of *n*-butane (A) and the yield of $C_2^=$, $C_3^=$, $C_4^=$ plus BTX (B) on the as-prepared 0.3% Ga_2O_3/ZSM-5 hollow fibers. For comparison, the catalytic performance of Nano-ZSM-5, 0.3% Ga_2O_3/ZSM-5 nanoparticles and the hierarchical ZSM-5 hollow fibers electrospun by the as-prepared Nano-ZSM-5 were also investigated. From Figure 8, it can be seen that with increasing the reaction temperature, the conversion of n-butane increases on all catalysts. Among the four catalysts, Ga_2O_3/ZSM-5 hollow fibers exihits the best catalytic reactivity in the whole temperature range. At the temperature of 575 °C the conversion of n-butane is almost 100%. Similar phenomenon is also reflected in the yield of $C_2^=$, $C_3^=$, $C_4^=$ plus BTX. When the reaction temperatures are lower than 650 °C, the yield of $C_2^=$, $C_3^=$, $C_4^=$ plus BTX on the four catalysts follows the order of Ga_2O_3/ZSM-5 hollow fibers > Ga_2O_3/ZSM-5 > ZSM-5 hollow fibers > Nano-ZSM-5. According to the catalytic behavior of four catalysts, specifically, compared with Nano-ZSM-5 and ZSM-5 hollow fibers, the catalytic performances of Ga_2O_3/ZSM-5 and Ga_2O_3/ZSM-5 hollow fibers are enhanced, respectively, indicating that the introduction of Ga_2O_3 is beneficial to the dehydrogenation of n-butane and thus promotes the catalytic activity of the catalysts efficiently. Meanwhile, in comparison with the present Nano-ZSM-5 and Ga_2O_3/ZSM-5 as well as magnesium-containing HZSM-5 reported in the literature [15], the catalytic reactivities of hollow fibers of both ZSM-5 and Ga_2O_3/ZSM-5 are improved, indicating that the hierarchical pore structure of hollow fibers contributes to enhancing the whole catalytic performance. There exists a synergistic effect between Ga_2O_3/ZSM-5 (or ZSM-5) and the hierarchical porosity of the hollow fibers. The unique hierarchical macro-meso-microporosity structure of the as-prepared hollow fibers can effectively enhance the accessibility of the feedstock to catalytic active sites and facilitates the mass transfer of targeted products, including ethene, propylene, aromatics, *etc.* Thus, the secondary reactions of ethene and propene as well as the carbon deposition could be hindered [12]. Ga_2O_3/ZSM-5 hollow fibers, which effectively combine the cracking function of ZSM-5, the hierarchical macro-meso-microporosity of hollow fibers, and the dehydrogenation of Ga_2O_3, show the best catalytic behavior among the four catalysts, with the highest yield of $C_2^=$, $C_3^=$, $C_4^=$ plus BTX at 600 °C by 87.6%, which is 56.3%, 24.6%, and 13.3% higher than ZSM-5, ZSM-5 zeolite fibers, and 0.3% Ga_2O_3/ZSM-5, respectively. The stability results of Ga_2O_3/ZSM-5 hollow fibers show that the catalytic activity of the catalyst decreases with time on stream, and carbon deposition is the cause of catalyst deactivation since the activity of the Ga_2O_3/ZSM-5 hollow fibers was fully recovered after regeneration (Figures S1–S3).

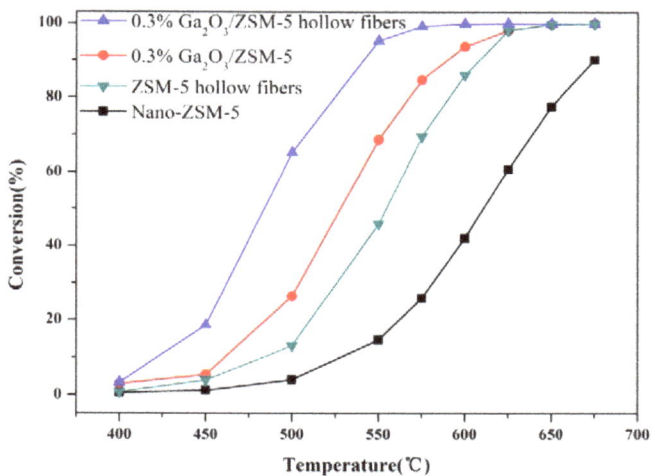

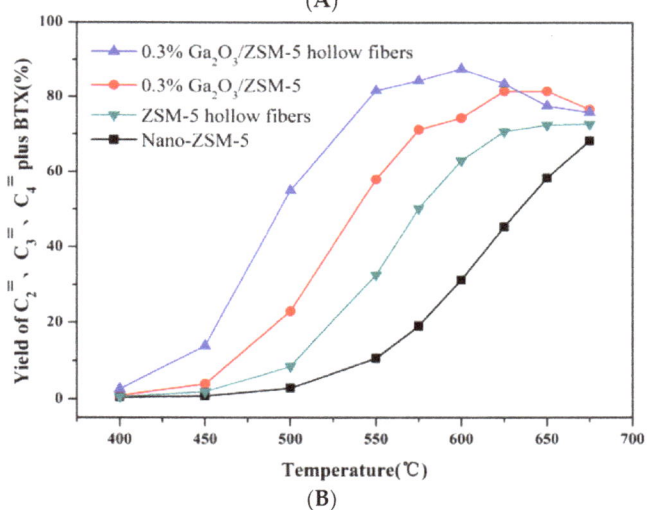

Figure 8. The conversion of *n*-butane (**A**) and the yield of $C_2^=$, $C_3^=$, $C_4^=$ plus BTX (**B**) as a function of reaction temperatures: 0.3% Ga_2O_3/ZSM-5 hollow fibers, 0.3% Ga_2O_3/ZSM-5, ZSM-5 hollow fibers, and Nano-ZSM-5.

3. Experimental Section

3.1. Catalyst Preparation

3.1.1. ZSM-5 Nanocrystals Preparation

ZSM-5 nanocrystals were prepared by hydrothermal crystallization from a clear solution with the composition of $9TPAOH:0.125Al_2O_3:25SiO_2:599H_2O:1.3NaOH$

187

according to previous literature [12,36] except that the addition of NaOH and the synthesis was carried out at 100 °C for 72 h. When the synthesis was completed, the nanocrystals products were washed, dried, and calcined at 550 °C for 6 h in muffle furnace followed by ion-exchanged, dried, and calcined process. The zeolite prepared according to the above methods is denoted Nano-ZSM-5 (with an actual Si/Al of 55 determined by XRF).

3.1.2. Ga_2O_3/ZSM-5 Preparation

The as-prepared Nano-ZSM-5 was subsequently impregnated by the incipient wetness technique using aqueous solution of $Ga(NO)_3 \cdot xH_2O$, dried for 10 h at 100 °C, and calcined at 600 °C for 6 h. Catalysts prepared in this way with x wt. % of Ga_2O_3 are denoted x% Ga_2O_3/ZSM-5.

3.1.3. Hierarchical Ga_2O_3/ZSM-5 Hollow Fibers Preparation

The outer fluid was prepared as follows: 1.5 g dry Ga_2O_3/ZSM-5 nanocrystals were added to 12.16 g absolute ethanol in a beaker (sealed by plastic wrap) and the mixture was subjected to ultrasonic treatment at 100 W for 8 h in order to the nanocrystals be fully dispersed in absolute ethanol. Then, 2.5 g PVP powder was added into the suspension followed by stirring to dissolve PVP powder completely. At last, in order to exclude bubbles in the suspension, another 0.5 h sonication treatment was conducted.

The electrospinning experimental device and the process of electrospinning are similar to that described in the literature [12] except that the flow rate of the paraffin oil is 0.8 mL·h^{-1}. The as-prepared fibers are denoted x% Ga_2O_3/ZSM-5 hollow fibers.

3.2. Catalyst Characterization

X-ray powder diffraction (XRD) patterns in the range of 5°–50° were recorded on a powder X-ray diffractometer(Shimadzu XRD 6000) (Shimadzu, Tokyo, Japan) using CuKα radiation (λ = 0.15406 nm) with a scanning rate of 2°/min, voltage 40 kV, and current 30 mA. Quanta 200F (FEI, Hillsboro, OR, USA) scanning electron microscpy (SEM) was used to observe morphology of the catalysts, and it was also employed for EDX line scan. TEM images were obtained by a JEOL JEM 2100 electron microscope (JEOL, Tokyo, Japan) equipped with a field emission source at an accelerating voltage of 200 kV. The BET specific surface area and pore volume of the samples were determined by adsorption-desorption of nitrogen at liquid nitrogen temperature, using a Micromeritics TriStar II 3020 porosimetry analyzer (Micromeritics, Norcross, GA, USA). X-ray photoelectron spectroscopy (XPS) was applied to analyze the change of surface composition performed on a PerkinElmer PHI-1600 ESCA (PerkinElmer, Waltham, MA, USA) spectrometer using

Mg·Ka (hv = 1253.6 eV, 1 eV = 1.603×10^{-19} J) X-ray source. The binding energy values were corrected for charging effect by referring to the adventitious C1s line at 284.6 eV.

Acidic properties of the catalysts were characterized by the temperature-programmed desorption of ammonia (NH_3-TPD) method. 0.1 g sample was pretreated in nitrogen at 600 °C for 1 h, cooled to room temperature, and adsorbed NH_3 for 30 min. After flushing with pure nitrogen gas for 45 min, TPD started at a rate of 10 °C/min from 100 °C to 600 °C and the signal was monitored with a thermal conductivity detector (TCD). The TG-DSC test was performed to analyze the amount of carbon deposition using METTLER TOLEDO TGA/DSC 1 (Mettler Toledo, Zurich, Switzerland), at a heating rate of 10 °C/min from 30 °C to 800 °C in an oxygen atmosphere.

3.3. Reaction Testing

Catalytic tests were performed in a fixed-bed flow reactor by passing a gaseous of *n*-butane (2 mL·min^{-1}, 99.9%) in nitrogen at a flow rate of 38 mL/min, and the catalyst load was 200 mg. The products were analyzed on-line using a gas chromatograph (SP-2100) (Beifen-Ruili, Beijing, China) equipped with a 30 m GS-ALUMINA capillary column and a FID detector (Beifen-Ruili, Beijing, China), and the contents of them are calculated on hydrocarbon basis (Tables S2–S7).

4. Conclusions

The dehydrogenation component of Ga_2O_3 was introduced to acid-based ZSM-5 nanocrystals, using Ga_2O_3/ZSM-5 nanoparticles as building blocks, Ga_2O_3/ZSM-5 hollow fibers with hierarchical macro-meso-microporosity were successfuly prepared by coaxial electrospinning. High conversion activity of *n*-butane and good yield of $C_2^=$, $C_3^=$, $C_4^=$ plus BTX were demonstrated. Superior catalytic performances are attributed to the good banlance of the cracking function of ZSM-5 and the dehydrogenation of Ga_2O_3, and the synergetic effect of bifunctionality and hierarchical porosity. The present results help to cast new light on the design of bifunctional fiber-based catalysts for efficient catalytic conversion of light alkanes.

Acknowledgments: The authors thank the support of this work by the National Basic Research Program of China (973 Program, No. 2012CB215001), National Science Foundation of China (Grant No. U1162117), Beijing Higher Education Young Elite Teacher Project (YETP0696), and Prospect Oriented Foundation of China University of Petroleum, Beijing (Grant No. ZX20140257).

Author Contributions: J.H., S.H., J.L., Y.Z., Y.L., and R.W. performed the experiments and conducted the catalytic activity tests. J.H., G.J., and Z.Z. conceived and designed the experiments, analyzed the experimental data, and wrote the paper. C.X., Y.W., A.D., J.L. and Y.W. interpreted the results, and gave advice about the data analysis as well as the preparation of the manuscript.

Conflicts of Interest: The authors declare no conflict of interest.

References

1. Maia, A.J.; Oliveira, B.G.; Esteves, P.M.; Louis, B.; Lam, Y.L.; Pereira, M.M. Isobutane and *n*-butane cracking on Ni-ZSM-5 catalyst: Effect on light olefin formation. *Appl. Catal. A* **2011**, *403*, 58–64.

2. Meng, X.; Wang, Z.; Zhang, R.; Xu, C.; Liu, Z.; Wang, Y.; Guo, Q. Catalytic conversion of C4 fraction for the production of light olefins and aromatics. *Fuel Process. Technol.* **2013**, *116*, 217–221.

3. Rahimi, N.; Karimzadeh, R. Catalytic cracking of hydrocarbons over modified ZSM-5 zeolites to produce light olefins: A review. *Appl. Catal. A* **2011**, *398*, 1–17.

4. Pollesel, P.; Bellussi, G. Industrial applications of zeolite catalysis: Production and uses of light olefins. *Stud. Surf. Sci. Catal.* **2005**, *158*, 1201–1212.

5. Janda, A.; Bell, A.T. Effects of Si/Al ratio on the distribution of framework Al and on the rates of alkane monomolecular cracking and dehydrogenation in H-MFI. *J. Am. Chem. Soc.* **2013**, *135*, 19193–19207.

6. Lin, L.; Qiu, C.; Zhuo, Z.; Zhang, D.; Zhao, S.; Wu, H.; Liu, Y.; He, M. Acid strength controlled reaction pathways for the catalytic cracking of 1-butene to propene over ZSM-5. *J. Catal.* **2014**, *309*, 136–145.

7. Wang, Y.; Yokoi, T.; Namba, S.; Kondo, J.N.; Tatsumi, T. Catalytic cracking of *n*-hexane for producing propylene on MCM-22 zeolites. *Appl. Catal. A* **2015**, *504*, 192–202.

8. Pant, K.K.; Kumar, V.A.; Kunzru, D. Potassium-containing calcium aluminate catalysts for pyrolysis of *n*-heptane. *Appl. Catal. A* **1997**, *162*, 193–200.

9. Jeong, S.M.; Chae, J.H.; Lee, W.-H. Study on the Catalytic Pyrolysis of Naphtha over a $KVO_3/\alpha-Al_2O_3$ Catalyst for Production of Light Olefins. *Ind. Eng. Chem. Res.* **2001**, *40*, 6081–6086.

10. Al-Yassir, N.; Le van Mao, R.; Heng, F. Cerium promoted and silica-alumina supported molybdenum oxide in the zeolite-containing hybrid catalyst for the selective deep catalytic cracking of petroleum naphthas. *Catal. Lett.* **2005**, *100*, 1–6.

11. Na, J.; Liu, G.; Zhou, T.; Ding, G.; Hu, S.; Wang, L. Synthesis and Catalytic Performance of ZSM-5/MCM-41 Zeolites With Varying Mesopore Size by Surfactant-Directed Recrystallization. *Catal. Lett.* **2013**, *143*, 267–275.

12. Liu, J.; Jiang, G.; Liu, Y.; Di, J.; Wang, Y.; Zhao, Z.; Sun, Q.; Xu, C.; Gao, J.; Duan, A.; *et al.* Hierarchical macro-meso-microporous ZSM-5 zeolite hollow fibers with highly efficient catalytic cracking capability. *Sci. Rep.* **2014**.

13. Wan, J.; Wei, Y.; Liu, Z.; Li, B.; Qi, Y.; Li, M.; Xie, P.; Meng, S.; He, Y.; Chang, F. A ZSM-5-based Catalyst for Efficient Production of Light Olefins and Aromatics from Fluidized-bed Naphtha Catalytic Cracking. *Catal. Lett.* **2008**, *124*, 150–156.

14. Zhao, G.; Teng, J.; Xie, Z.; Jin, W.; Yang, W.; Chen, Q.; Tang, Y. Effect of phosphorus on HZSM-5 catalyst for C4-olefin cracking reactions to produce propylene. *J. Catal.* **2007**, *248*, 29–37.

15. Wakui, K.; Satoh, K.; Sawada, G.; Shiozawa, K.; Matano, K.; Suzuki, K.; Hayakawa, T.; Yoshimura, Y.; Murata, K.; Mizukami, F. Dehydrogenation cracking of *n*-butane over modified HZSM-5 catalysts. *Catal. Lett.* **2002**, *81*, 83–88.

16. Lu, J.; Zhao, Z.; Xu, C.; Zhang, P.; Duan, A. FeHZSM-5 molecular sieves-Highly active catalysts for catalytic cracking of isobutane to produce ethylene and propylene. *Catal. Commun.* **2006**, *7*, 199–203.

17. Wang, X.; Zhao, Z.; Xu, C.; Duan, A.; Zhang, L.; Jiang, G. Effects of Light Rare Earth on Acidity and Catalytic Performance of HZSM-5 Zeolite for Catalytic Cracking of Butane to Light Olefins. *J. Rare Earths* **2007**, *25*, 321–328.

18. Jiang, G.; Zhang, L.; Zhao, Z.; Zhou, X.; Duan, A.; Xu, C.; Gao, J. Highly effective P-modified HZSM-5 catalyst for the cracking of C4 alkanes to produce light olefins. *Appl. Catal. A* **2008**, *340*, 176–182.

19. Jia, C.-J.; Liu, Y.; Schmidt, W.; Lu, A.-H.; Schüth, F. Small-sized HZSM-5 zeolite as highly active catalyst for gas phase dehydration of glycerol to acrolein. *J. Catal.* **2010**, *269*, 71–79.

20. Zhao, L.; Shen, B.; Gao, J.; Xu, C. Investigation on the mechanism of diffusion in mesopore structured ZSM-5 and improved heavy oil conversion. *J. Catal.* **2008**, *258*, 228–234.

21. Choi, M.; Na, K.; Kim, J.; Sakamoto, Y.; Terasaki, O.; Ryoo, R. Stable single-unit-cell nanosheets of zeolite MFI as active and long-lived catalysts. *Nature* **2009**, *461*, 246–249.

22. Martens, J.A.; Blomsma, E.; Jacobs, P.A. Isomerization and Hydrocracking of Heptane over Bimetallic Bifunctional PtPd/H-Beta and PtPd/USY Zeolite Catalysts. *J. Catal.* **1997**, *165*, 241–248.

23. Kim, J.; Kim, W.; Seo, Y.; Kim, J.-C.; Ryoo, R. *n*-Heptane hydroisomerization over Pt/MFI zeolite nanosheets: Effects of zeolite crystal thickness and platinum location. *J. Catal.* **2013**, *301*, 187–197.

24. Jackson, S.; Rugmini, S. Dehydrogenation of *n*-butane over vanadia catalysts supported on θ-alumina. *J. Catal.* **2007**, *251*, 59–68.

25. Vanlingen, J.; Gijzeman, O.; Weckhuysen, B.; Vanlenthe, J. On the umbrella model for supported vanadium oxide catalysts. *J. Catal.* **2006**, *239*, 34–41.

26. Cabrera, F.; Ardissone, D.; Gorriz, O.F. Dehydrogenation of propane on chromia/alumina catalysts promoted by tin. *Catal. Today* **2008**, *133–135*, 800–804.

27. Santhoshkumar, M.; Hammer, N.; Ronning, M.; Holmen, A.; Chen, D.; Walmsley, J.; Oye, G. The nature of active chromium species in Cr-catalysts for dehydrogenation of propane: New insights by a comprehensive spectroscopic study. *J. Catal.* **2009**, *261*, 116–128.

28. Karamullaoglu, G.; Onen, S.; Dogu, T. Oxidative dehydrogenation of ethane and isobutane with chromium-vanadium-niobium mixed oxide catalys. *Chem. Eng. Process.* **2002**, *41*, 337–347.

29. Nakagawa, K.; Okamura, M.; Ikenaga, N.; Suzuki, N.; Kobayashi, T. Dehydrogenation of ethane over gallium oxide in the presence of carbon dioxide. *Chem. Commun.* **1998**, *9*, 1025–1026.

30. Sattler, J.J.; Gonzalez-Jimenez, I.D.; Luo, L.; Stears, B.A.; Malek, A.; Barton, D.G.; Kilos, B.A.; Kaminsky, M.P.; Verhoeven, T.W.; Koers, E.J.; *et al*. Platinum-promoted Ga/Al$_2$O$_3$ as highly active, selective, and stable catalyst for the dehydrogenation of propane. *Angew. Chem. Int. Ed. Engl.* **2014**, *53*, 9251–9256.

31. Joly, J.F.; Ajot, H.; Merlen, E.; Raatz, F.; Alario, F. Parameters affecting the dispersion of the gallium phase of gallium H-MFI aromatization catalysts. *Appl. Catal. A* **1991**, *79*, 249–263.

32. Drogone, L.; Moggi, P.; Predieri, G.; Zanoni, R. Niobia and silica-niobia catalysts from sol-gel synthesis: An X-ray photoelectron spectroscopic characterization. *Appl. Surf. Sci.* **2002**, *187*, 82–88.

33. Sun, Y.; Brown, T.C. Catalytic cracking, dehydrogenation, and aromatization of isobutane over Ga/HZSM-5 and Zn/HZSM-5 at low pressures. *Int. J. Chem. Kinet.* **2002**, *34*, 467–480.

34. Shin, F.; Tatsuski, Y.; Takayuki, K. Reaction Scheme of Aromatization of Butane over Ga Loaded HZSM-5 Catalyst. *J. Jpn. Petrol. Inst.* **1998**, *41*, 37–44.

35. Qiu, P.; Jack, H.L.; Michael, P.R. Characterization of Ga/ZSM-5 for the catalytic aromatization of dilute ethylene streams. *Catal. Lett.* **1998**, *52*, 37–42.

36. Mintova, S.; Hölzl, M.; Valtchev, V.; Mihailova, B.; Bouizi, Y.; Bein, T. Closely packed zeolite nanocrystals obtained via transformation of porous amorphous silica. *Chem. Mater.* **2004**, *16*, 5452–5459.

Catalytic Cracking of Triglyceride-Rich Biomass toward Lower Olefins over a Nano-ZSM-5/SBA-15 Analog Composite

Xuan Hoan Vu, Sura Nguyen, Tung Thanh Dang, Binh Minh Quoc Phan, Duc Anh Nguyen, Udo Armbruster and Andreas Martin

Abstract: The catalytic cracking of triglyceride-rich biomass toward C_2–C_4 olefins was evaluated over a hierarchically textured nano-ZSM-5/SBA-15 analog composite (ZSC-24) under fluid catalytic cracking (FCC) conditions. The experiments were performed on a fully automated Single-Receiver Short-Contact-Time Microactivity Test unit (SR-SCT-MAT, Grace Davison) at 550 °C and different catalyst-to-oil mass ratios (0–1.2 $g \cdot g^{-1}$). The ZSC-24 catalyst is very effective for transformation of triglycerides to valuable hydrocarbons, particularly lower olefins. The selectivity to C_2–C_4 olefins is remarkably high (>90%) throughout the investigated catalyst-to-oil ratio range. The superior catalytic performance of the ZSC-24 catalyst can be attributed to the combination of its medium acid site amount and improved molecular transport provided by the bimodal pore system, which effectively suppresses the secondary reactions of primarily formed lower olefins.

Reprinted from *Catalysts*. Cite as: Vu, X.H.; Nguyen, S.; Dang, T.T.; Phan, B.M.Q.; Nguyen, D.A.; Armbruster, U.; Martin, A. Catalytic Cracking of Triglyceride-Rich Biomass toward Lower Olefins over a Nano-ZSM-5/SBA-15 Analog Composite. *Catalysts* **2015**, *5*, 1692–1703.

1. Introduction

Lower olefins, also known as light olefins (ethene, propene, and butenes), are important feedstock for the production of valuable polymers and chemicals such as polyethylene (PE), polypropylene (PP), methyl tert-butyl ether (MTBE), and ethyl *tert*-butyl ether (ETBE). The major fraction of produced lower olefins currently stems from petroleum feedstock [1,2], and the increasing demand along with the oil depletion has led to an increased interest in production of those olefins from renewable feedstock such as biomass [3]. In this respect, the processing of triglyceride-rich biomass by fluid catalytic cracking (FCC) units in petroleum refineries represents a promising option for the future to produce renewable liquid fuels and lower olefins [4,5]. The utilization of the existing refining infrastructure and configuration for the conversion of triglyceride-rich biomass would require little additional capital investment. However, the yields of gasoline and lower olefins decrease sharply when cracking highly unsaturated triglyceride feedstock

193

over conventional FCC zeolite catalysts (typically ultra-stable zeolite Y (USY)). This is because the large micropore size of zeolite Y (pore opening size of 0.74 nm; pore intersection size (supercage) of 1.3 nm) tends to promote the fast aromatization of unsaturated fatty acids, resulting in a large fraction of heavy aromatic species that are less crackable [6,7].

To avoid such a phenomenon, medium pore size offering zeolite ZSM-5 has been investigated in the cracking of triglycerides [8–10]. With the pore opening size of 0.52–0.56 nm and pore intersection size of 0.8 nm, ZSM-5 selectively directs the cracking process toward the formation of C_5–C_{10} hydrocarbons corresponding to the gasoline-boiling range while avoiding the formation of poly-aromatic species. Unfortunately, a high gas yield with low concentrations of light olefins is generally observed [9,10], implying the occurrence of undesired secondary reactions because of the diffusion limitations caused by the relatively small pore size of ZSM-5. Accordingly, the modification of ZSM-5 zeolites to enhance the selectivity to C_2–C_4 olefins has been extensively studied in recent years. It has been reported that the shortened diffusion path length and reduced amount of acid sites were favorable for the formation of lower olefins in the cracking of triglyceride-rich biomass [11–13].

In our previous work [14], we have successfully developed a novel nano-ZSM-5-based composite by dispersing nano-ZSM-5 in well-ordered, highly condensed mesoporous SBA-15 analogs via a two-step process. The resulting nano-ZSM-5/SBA-15 analog composites (ZSC) exhibited high hydrothermal stability and improved acidic properties while their accessibility was enhanced by the bimodal pore system. This was clearly proved by gas-phase cracking of model compounds 1,3,5-triisopropylbenzene and cumene in a conventional tube reactor and comparison to H-ZSM-5 and Al-SBA-15 as benchmarks [14]. In the present contribution, we like to report on the application of such ZSC catalysts for the efficient conversion of triglyceride-rich biomass toward lower olefins under FCC conditions. The standard test procedure using the SR-SCT-MAT unit was employed for the catalyst evaluation as it allows simulating an industrial FCC unit more accurately, thereby improving the practical relevance of the results [15].

2. Results and Discussion

2.1. Physico-Chemical Properties of the Catalysts

The catalysts ZSC-24, Al-SBA-15, and commercial H-ZSM-5 were thoroughly characterized by various techniques and model reactions as presented in the earlier work [14]. However, the main catalyst characteristics are summarized in Table 1. In general, Al-SBA-15 possesses the largest external surface (hexagonal mesopores) and the smallest amount of acid sites. Contrarily, commercial H-ZSM-5 shows the least external surface and the highest amount of acid sites, mostly located in micropores. These named properties of ZSC-24, which was prepared from the ZSM-5 precursor solution with 24 h precrystalization time, are in between, as the previous studies evidenced that this material indeed comprises hexagonal mesoporous domains wherein ZSM-5 nanoparticles are highly dispersed.

Table 1. Physico-chemical properties of ZSC-24, Al-SBA-15, and commercial H-ZSM-5.

Catalyst	SiO_2/Al_2O_3 [a] (mol/mol)	S_{BET} $(m^2 \cdot g^{-1})$	$S_{ext/meso}$ [b] $(m^2 \cdot g^{-1})$	V_{micro} [b] $(m^3 \cdot g^{-1})$	V_t $(m^3 \cdot g^{-1})$	Acid site amount [c] $(mmol\ NH_3 \cdot g^{-1})$
ZSC-24	60	361	233	0.058	0.84	0.34
Al-SBA-15	90	446	446	0	0.93	0.18
H-ZSM-5	22	373	110	0.113	0.22	1.24

[a] AAS and ICP; [b] t-plot method, [c] NH$_3$-TPD.

2.2. Catalyst Evaluation

Catalytic cracking of triglyceride-rich biomass is generally initiated by thermal decomposition of triglyceride molecules into free fatty acids by means of free radical mechanism. Subsequently, the acid zeolite-based catalyst controls the process and converts free fatty acids into oxygen-containing products (mainly CO, CO_2, and water) and a mixture of hydrocarbons lumped into gaseous hydrocarbon, gasoline, light cycle oil (LCO), and heavy cycle oil (HCO) fractions [6]. It has been reported that the conversion and product selectivity heavily depend on the unsaturation degree of triglycerides under FCC conditions [6,7]. Therefore, the catalytic cracking of triolein comprising predominantly unsaturated triglycerides (technical grade triolein, Santa Cruz Biotechnology) as model feedstock was first carried out over the ZSC-24 catalyst under different cracking severities to study the catalytic behavior and to optimize the cracking conditions. Then the catalytic performance of ZSC-24 relative to that of reference catalysts Al-SBA-15 and H-ZSM-5 was evaluated in the cracking of waste cooking oil (WCO; obtained from a local restaurant) consisting of less unsaturated triglycerides. The effect of the unsaturation level can be assessed by comparing the catalytic results obtained from the cracking of triolein and those obtained from the cracking of WCO. The fatty acid and elemental composition of triolein and WCO are given in Table 2.

2.2.1. Catalytic Cracking of Triolein

The cracking of triolein was carried out over the ZSC-24 catalyst under different severities by varying the catalyst-to-oil (CTO) mass ratio from 0.2 to 1.2 g·g^{-1} at 550 °C. In general, the effect of thermal cracking becomes significant at elevated temperatures (>460 °C); thus, the thermal cracking of triolein over an inert material (glass beads) was conducted in a blank test at 550 °C to reveal the effect of thermal degradation under the investigated conditions. The catalytic results are given in Table 3 and Figure 1.

195

Table 2. Fatty acid and elemental compositions of technical grade triolein and waste cooking oil (WCO).

Feedstock	Triolein	WCO
Fatty acid composition as wt. % methyl esters		
Dodecanoic acid [C12:0]	-	1.2
Palmitic acid [C16:0]	1.1	37.2
Stearic acid [C18:0]	1.9	4.9
Oleic acid [C18:1]	75.6	48.8
Linoleic acid [C18:2]	21.4	7.9
Elemental composition, wt. %		
Carbon	79.4	79.4
Hydrogen	12.0	12.6
Oxygen	8.6	7.8
Nitrogen	0	0.2

[Cx:y] where x is the number of carbon atoms and y is the number of double bonds.

Table 3. The catalytic performance of ZSC-24 catalyst in the cracking of triolein under different severities by varying the CTO ratio from 0 (glass beads) to 1.2 $g \cdot g^{-1}$ at 550 °C.

CTO ratio ($g \cdot g^{-1}$)		0	0.2	0.4	0.8	1.2
Conversion (wt. %)		24.7	62.5	70.9	77.5	79.6
	Total gas	6.4	21.0	28.0	38.9	41.0
	Dry gas	1.3	2.1	2.3	3.0	4.0
	LPG	1.7	13.1	18.0	28.9	33.0
	C_2–C_4 olefins	1.9	13.5	18.2	29.4	33.5
Product yields	CO, CO_2	3.5	5.8	7.6	7.1	4.1
(wt. %)	C_{5+} Gasoline	17.0	34.9	35.2	29.7	28.8
	LCO	29.6	22.4	18.7	14.4	12.1
	HCO	45.7	15.1	10.4	8.1	8.3
	Coke	0.1	1.0	1.6	2.4	2.3
	Water	1.1	5.6	6.1	6.5	7.5
Selectivity to C_2–C_4 olefins (%) [a]		67.7	90.9	91.7	93.7	92.0

wt. % on a feed basis. [a] fraction of C_2–C_4 olefins per total C_2–C_4 hydrocarbons.

From Table 3, it can be seen that the thermal cracking (blank test) of triolein occurs noticeably at 550 °C, giving a conversion of 24.7 wt. %. However, the low fraction of short-chain products suggests that most of the heavy oxygenated compounds formed by the decomposition of triglycerides were not converted in the absence of catalyst, which is in line with the previous reports [6,8]. In the presence of the ZSC-24 catalyst, the yields of light products such as gas and gasoline rise dramatically at the expense of heavy fractions such as HCO and LCO. This indicates that the heavy oxygenated compounds have been effectively converted over ZSC-24 into valuable hydrocarbons, particularly gasoline and lower olefins.

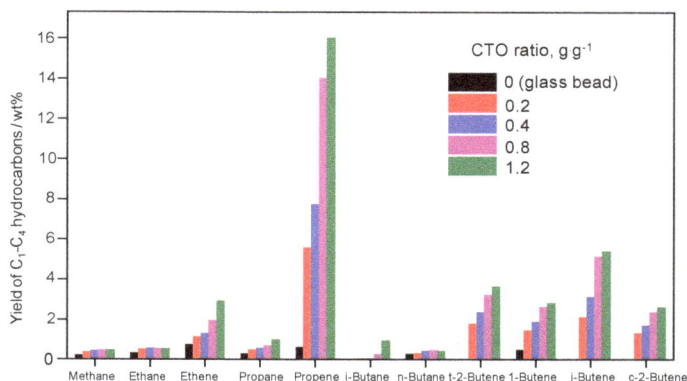

Figure 1. Yields of C_1–C_4 hydrocarbons in the cracking of triolein under different severities by varying the CTO ratio from 0 (glass beads) to 1.2 g·g^{-1} at 550 °C.

As shown in Table 3, the CTO ratio increase from 0.2 to 1.2 g·g^{-1} steadily enhances the conversion and the C_2–C_4 olefin yield from 62.5 and 13.5 wt. % to 79.6 and 33.5 wt. %, respectively. This corresponds to increases by factors of approximately 1.27 and 2.48 and can be explained by secondary cracking reactions of intermediate fatty acids being mainly activated by the catalysts' acid sites [8,10,16] and leading to the production of desirable hydrocarbons. Hence, increasing the CTO ratio provided more active acid sites in the reaction system; therefore, more gasoline and lower olefins were produced.

Unlike the conversion and C_2–C_4 olefin yield, the gasoline yield apparently goes through a maximum when varying the CTO ratio at 550 °C (Table 3). In fact, the gasoline yield shows a little gain from 34.9 to 35.2 wt. % with the CTO ratio increase from 0.2 to 0.4 g·g^{-1} and then drops sharply to 29.7 and 28.8 wt. % at the CTO ratios of 0.8 and 1.2 g·g^{-1}, respectively. This implies that the over-cracking of gasoline hydrocarbons has been promoted as the CTO ratio exceeded 0.4 g·g^{-1}, giving rise to a substantial reduction in the gasoline yield [17]. Thus, the CTO ratio should not be higher than 0.4 g·g^{-1} under the investigated conditions as the production of gasoline and light olefins are targeted.

Since the main target of this work is to enhance the selectivity to lower olefins over the ZSC-24 catalyst, the yield of gaseous hydrocarbons was analyzed in detail and the C_2–C_4 olefin selectivity was estimated (Figure 1 and Table 3). Interestingly, the selectivity to C_2–C_4 olefins is remarkably high, independent of the cracking severity, signifying that the ZSC-24 catalyst is a highly selective catalyst for the conversion of triglyceride-rich biomass to lower olefins. Botas *et al.* [12] studied the cracking of rapeseed oil over nano-ZSM-5 using a fixed bed reactor. It was found that the shortened diffusion path lengths along with the reduced acid site amount prevented the further transformation of primary C_2–C_4 olefins, thereby increasing the yield and selectivity to lower olefins. Considering this, it is reasonable that the superior selectivity to C_2–C_4 olefins in the cracking of triolein over ZSC-24 originates from the combination of the catalyst's medium acid site amount and improved diffusion characteristics provided by the bimodal pore system.

2.2.2. Catalytic Cracking of WCO

The performance of ZSC-24 compared to the reference catalysts Al-SBA-15 and H-ZSM-5 was evaluated in the catalytic cracking of real feedstock WCO under optimized cracking conditions ($T = 550\ °C$ and CTO ratio = $0.4\ g \cdot g^{-1}$). First, the effect of thermal cracking of WCO was checked in a blank test at $550\ °C$ using glass beads as inert material.

Table 4 gives the overall data on the catalytic performance of the different catalysts. One can see that there is a significant difference in the conversion of WCO. Commercial H-ZSM-5 looks highly active, showing an almost complete conversion of 91.5 wt. %, followed by ZSC-24 (70.6 wt. %) and Al-SBA-15 (58.3 wt. %). Without the presence of the catalyst, the thermal degradation of WCO yields the least conversion of 26.4 wt. %. A good correlation of the conversion and the acid site amount can be observed (Tables 1 and 4). This further supports the important role of acid sites in the conversion of intermediate fatty acids into lighter and more valuable hydrocarbons [8,10].

Table 4. Performance of ZSC-24 and the reference catalysts Al-SBA-15 and H-ZSM-5 in the cracking of WCO at $550\ °C$ and a CTO ratio of $0.4\ g \cdot g^{-1}$.

Catalyst		Glass beads	ZSC-24	H-ZSM-5	Al-SBA-15
Conversion (wt. %)		26.4	70.6	91.5	58.3
Product yields (wt. %)	Total gas	7.9	30.7	48.8	17.0
	Dry gas	1.6	2.6	8.2	2.2
	LPG	2.1	22.1	33.7	8.6
	C_2–C_4 olefins	2.3	22.7	30.4	8.5
	CO/CO_2	4.2	5.9	6.9	6.3
	C_{5+} Gasoline	17.3	32.2	35.0	34.3
	LCO	29.0	17.7	4.0	30.4
	HCO	44.6	11.7	4.5	11.2
	Coke	0.1	1.4	1.2	1.7
	Water	1.1	6.3	6.5	5.3
Selectivity to C_2–C_4 olefin (%) [a]		67.6	93.2	73.5	83.3

wt. % on a feed basis; [a] fraction of C_2–C_4 olefins per total C_2–C_4 hydrocarbons.

Regarding the product yields, it is obvious that the thermal cracking of WCO gives the lowest yields of desirable products, *i.e.*, gasoline (17.3 wt. %) and lower olefins (2.3 wt. %). This is because of lacking acidity; thus, most of the intermediate oxygenated compounds were not converted [6]. The presence of the catalysts generally increases the fractions of gaseous and gasoline hydrocarbons by the cracking of LCO and HCO compounds. However, there are pronounced differences in the product distribution because of the different acid site amount and porosity. With the lowest acid site amount and largest mesopores, Al-SBA-15 produces the most LCO (30.4 wt. %), but the least C_2–C_4 olefins (8.5 wt. %). In contrast, H-ZSM-5 having the highest acid site amount exhibits the largest fractions of gasoline and C_2–C_4 olefins (35.0 wt. % and 30.4 wt. % respectively). The superior selectivity toward gasoline over H-ZSM-5 is well documented [8,16]. With respect to Al-SBA-15 and H-ZSM-5, the fractions of gasoline and lower olefins obtained with ZSC-24 are ranking in between. However, the outstanding yield for C_2–C_4 olefins over H-ZSM-5 mainly results from the high conversion rather than from selectivity since the latter is lowest (73.5 wt. %) in this series. Remarkably, the highest selectivity to C_2–C_4 olefins has been achieved over ZSC-24 (93.2%), confirming the advantage of ZSC-24 over the reference catalysts. The lower selectivity to C_2–C_4 olefins over Al-SBA-15 in comparison with ZSC-24 can be attributed to the shape selectivity of microporous ZSM-5 domains in the latter. However, the low selectivity to C_2–C_4 olefins over H-ZSM-5 is probably due to its very high amount of acid sites (mostly located in micropores) and the diffusion constraints imposed by the small pore size of H-ZSM-5 [13]. This might lead to elongated residence times and undesired consecutive reactions of the olefins. Tago *et al.* [18,19] studied the effect of crystal sizes and acidic properties on the catalytic performance of H-ZSM-5 in the cracking of n-hexane at high temperature (550–650 °C). Reducing the crystal size to nano-scale led to rapid diffusion of light olefins out of the intercrystalline micropores of nanosized zeolites as result of the shortened diffusion path lengths, which helped avoid the further transformation of these olefins. This clearly indicates the detrimental effect of internal mass transfer limitation in H-ZSM-5. On the other hand, increasing the number of acid sites enhanced the conversion of n-hexane but decreased the selectivity to light olefins. They postulated that the higher amount of acid sites promoted hydrogen transfer reactions which consumed light olefins to produce gasoline aromatics and light paraffins. A similar behavior can be seen from Figure 2 wherein H-ZSM-5 displays higher concentrations of light paraffins and aromatic hydrocarbons than ZSC-24, confirming the acceleration of hydrogen transfer reactions over H-ZSM-5 due to its higher acid site amount. Thus, the remarkable selectivity to C_2–C_4 olefins over the ZSC-24 catalyst can be attributed to the improved molecular transport being well balanced with the medium acid site amount which enables ZSC-24 not only to promote the formation of light olefins, but also effectively to protect them from further transformation.

To discuss the effect of feedstock composition, the results obtained from the cracking of triolein and WCO under the same conditions (ZSC-24, 550 °C and CTO ratio = 0.4 g·g^{-1}) are used (Tables 3 and 4). It appears that the cracking of WCO produces more light olefins but less gasoline than the cracking of triolein. However, these observed differences in the yields of gasoline and lower olefins are not significant compared to that reported by Dupain *et al.* [6], who converted triglyceride-based feedstock with FCC catalysts under realistic FCC conditions (525 °C and 4 s). They found that the cracking of saturated stearic acid produced considerably higher yields of gasoline and light olefins (approximately 57 and 14 wt. %, respectively) than the cracking of unsaturated rapeseed oil (approximately 34 and 3 wt. % for gasoline and light olefins, respectively). It was postulated that the fast aromatization of unsaturated fatty acid over the large pore zeolite Y-based catalyst led to the formation of polyaromatics that are recalcitrant against further cracking. Taking this into account, the minor shift in the yields of gasoline and lower olefins over ZSC-24 confirms that the formation of polyaromatics has been largely suppressed.

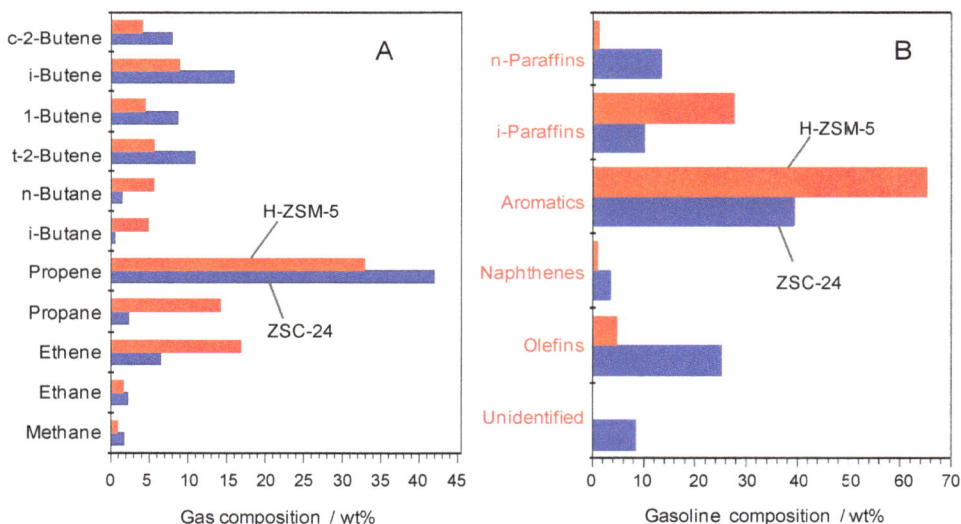

Figure 2. Gaseous hydrocarbon (**A**) and gasoline (**B**) compositions in the catalytic cracking of WCO over ZSC-24 and H-ZSM-5 at 550 °C and a CTO ratio of 0.4 (g·g^{-1}) (wt. % on a product basis).

3. Experimental Section

3.1. Catalyst Preparation

The representative nano-ZSM-5/SBA-15 analog composite, pre-crystallized for 24 h (denoted as ZSC-24), was prepared according to the synthesis procedure as reported in the previous work [14]. Briefly, the zeolite seed solution was pre-crystallized for 24 h to partially form nano-ZSM-5 crystals and then added to the surfactant solution to convert unreacted precursors into an ordered mesoporous SBA-15 analog. Finally, the resulting nano-ZSM-5/SBA-15 analog composite (ZSC-24) was activated by protonation. For comparison, Al-SBA-15 was prepared under the same conditions as for synthesis of ZSC-24, but using conventional silica-based precursors instead of pre-formed seeds. Additionally, commercial H-ZSM-5 (Zeocat PZ-2/25, ZeoChem AG, Uetikon am See, Switzerland) was used to provide a benchmark. All catalyst samples used in this study were fresh since they provide a better understanding of structure, property, and function relationships.

3.2. Catalyst and Feedstock Characterization

The nitrogen physisorption studies were carried out at $-196\,^{\circ}C$ with an ASAP 2010 apparatus (Micromeritics GmbH, Aachen, Germany). The temperature-programmed desorption of ammonia (NH3-TPD) measurements using TCD detector (TCD, GOW-MAC Instrument Co., Bethlehem, PA, USA) were performed in a homemade set-up using a quartz tube reactor in the range of 100–550 °C. The Al and Si contents were determined by inductively coupled plasma atomic emission spectroscopy (ICP-AES; 715-ES, Varian, Inc., Palo Alto, PA, USA) and atomic absorption spectroscopy (AAS; Analyst 300, PerkinElmer, Inc., Waltham, MA, USA), respectively. More details of these characterization methods and experiment parameters are described elsewhere [14,20].

The fatty acid compositions of technical grade triolein (obtained from Santa Cruz Biotechnology, Inc, Heidelberg, Germany) and waste cooking oil (collected from a local restaurant) were analyzed on a gas chromatography-mass spectrometry combination (GC-MS; QP2010S, Shimadzu Europa GmbH, Duisburg, Germany). For this purpose, a derivatization was carried out with trimethylsulfonium hydroxide solution (0.25 M in methanol, Fluka, Sigma-Aldrich Chemie GmbH, Taufkirchen, Germany) to transform triglycerides completely into methyl esters. The C, H, N elemental analysis of triolein and WCO was performed on a TruSpec CHNS Micro analyser (Leco Instrumente GmbH, Mönchengladbach, Germany).

3.3. Catalyst Tests

The fixed bed microactivity test equipment (MAT) is widely used for studying catalytic cracking at the laboratory scale. However, the traditional MAT unit provides cracking results of limited precision and accuracy due to the problems related to the fluid dynamics differences between a fixed bed and a fluidized bed, the large contact time (longer than industrial FCC operations, typically about 2 to 10s), low temperature in the preheating section, or low pressure in the catalyst bed. Most of these shortcomings encountered with the traditional MAT unit can be avoided by using a novel Single Receiver Short-Contact-Time Microactivity Test unit (SR-SCT-MAT, Grace GmbH & Co.KG, Frankfurt, Germany) [15]. Details of the experimental setup of the SR-SCT-MAT unit, the testing conditions, and product analyses used in the present work are described elsewhere [13].

In a typical run, 1.75 g of feedstock (either triolein or WCO) was fed into the reactor which contained a desired amount of catalyst diluted with glass beads to maintain a constant-volume reaction independent of the catalyst-to-oil (CTO) mass ratio used. The CTO mass ratio was varied by keeping the weight of feedstock constant (1.75 g) and changing the catalyst weight. The cracking reaction was carried out at ambient pressure, 550 °C, CTO mass ratios of 0–1.2 $(g \cdot g^{-1})$ and a reaction time of 12 s. After the reaction, stripping of the catalyst was done by using a nitrogen purge. The gaseous and liquid products were collected in the single receiver cooled to 18 °C via an external cooling system. All catalytic experiments were repeated at least two times to check the reproducibility and mass balances in all runs were between 95% and 100% of the injected feed.

The products comprised mainly hydrocarbons along with oxygenated compounds (water, CO, and CO_2) and coke. The gaseous hydrocarbon fraction was divided into dry gas (hydrogen, methane, ethane, and ethene) and liquefied petroleum gas (LPG; propane, propene, butenes, and butanes). The liquid hydrocarbons were lumped in terms of boiling ranges: C_{5+} gasoline (<221 °C), light cycle oil (LCO; 221–360 °C), and heavy cycle oil (HCO; >360 °C).

The gaseous products were analyzed according to the ASTM D1945-3 method using a Refinery Gas Analyzer (Agilent 7890A, Santa Clara, CA, USA). The liquid organic products were classified according to the boiling ranges: C_{5+} gasoline, LCO, and HCO as mentioned above by means of simulated distillation (ASTM D2887) on a Simulated Distillation gas chromatograph (Agilent 7890A, Santa Clara, CA, USA). For several liquid samples, PIONA (Paraffin, i-Paraffin, Olefin, Naphthene, and Aromatic) analyses were performed to determine the composition of gasoline. Water content was measured by Karl Fischer titration (MKS-520, Kem, Kyoto, Japan) and coke amount on the spent catalyst was determined by an elemental analyzer (CS600, Leco Instrumente GmbH, Mönchengladbach, Germany).

The yield toward different products (Y_i, wt. %) is defined as gram of product i per gram of the feed. The standard MAT conversion is defined as $100\% - (Y_{HCO} + Y_{LCO})$. The selectivity to C_2–C_4 olefins is defined as the fraction of C_2–C_4 olefins per total fraction of C_2–C_4 hydrocarbons.

4. Conclusions

We have shown that ZSC-24 is an efficient catalyst for the cracking of triglyceride-rich biomass toward lower olefins under FCC conditions. Compared to Al-SBA-15 and commercial H-ZSM-5, the ZSC-24 catalyst exhibits the highest selectivity to C_2–C_4 olefins (>90%) irrespective of the cracking severity and feedstock composition. The effect of the catalyst's porosity and acid site amount on the conversion and product distribution was found to be significant. Medium acid site amount, shape selectivity, and improved molecular transport obtained by the high dispersion of nano-scaled ZSM-5 domains in a mesoporous SBA-15 analog matrix might be responsible for the enhanced formation of desired lower olefins. These findings might open the way for future research in the rational design of new FCC catalysts for efficient conversion of triglyceride-rich biomass toward lower olefins. However, it should be noted that the fresh catalysts used in this study might not provide the same results as if the catalysts had been aged.

Acknowledgments: Financial support from VIED and LIKAT is gratefully acknowledged.

Author Contributions: X.H.V., T.T.D. and U. A. designed the experiments. T.T.D., X.H.V. and S.N. conducted the experiments. B.M.Q.P., D.A.N., U.A. and A. M. analyzed the data. X.H.V. wrote the first draft of the manuscript which was then revised by all other authors.

Conflicts of Interest: The authors declare no conflict of interest.

References

1. Ren, T.; Patel, M.; Blok, K. Olefins from conventional and heavy feedstocks: Energy use in steam cracking and alternative processes. *Energy* **2006**, *31*, 425–451.
2. Rahimi, N.; Karimzadeh, R. Catalytic cracking of hydrocarbons over modified ZSM-5 zeolites to produce light olefins: A review. *Appl. Catal. A* **2011**, *398*, 1–17.
3. Maher, K.D.; Bressler, D.C. Pyrolysis of triglyceride materials for the production of renewable fuels and chemicals. *Biores. Technol.* **2007**, *98*, 2351–2368.
4. Melero, J.A.; Iglesias, J.; Garcia, A. Biomass as renewable feedstock in standard refinery units. Feasibility, opportunities and challenges. *Energy Environ. Sci.* **2012**, *5*, 7393–7420.
5. Huber, G.W.; Corma, A. Synergies between bio- and oil refineries for the production of fuels from biomass. *Angew. Chem. Int. Ed.* **2007**, *46*, 7184–7201.
6. Dupain, X.; Costa, D.J.; Schaverien, C.J.; Makkee, M.; Moulijn, J.A. Cracking of a rapeseed vegetable oil under realistic FCC conditions. *Appl. Catal. B* **2007**, *72*, 44–61.
7. Rao, T.V.M.; Dupain, X.; Makkee, M. Fluid catalytic cracking: Processing opportunities for Fischer-Tropsch waxes and vegetable oils to produce transportation fuels and light olefins. *Micropor. Mesopor. Mater.* **2012**, *164*, 148–163.
8. Chen, D.; Tracy, N.I.; Crunkleton, D.W.; Price, G.L. Comparison of canola oil conversion over MFI, BEA, and FAU. *Appl. Catal. A* **2010**, *384*, 206–212.

9. Idem, R.O.; Katikaneni, S.P.R.; Bakhshi, N.N. Catalytic conversion of canola oil to fuels and chemicals: Roles of catalyst acidity, basicity and shape selectivity on product distribution. *Fuel Process. Technol.* **1997**, *51*, 101–125.

10. Katikaneni, S.P.R.; Adjaye, J.D.; Bakhshi, N.N. Studies on the catalytic conversion of canola oil to hydrocarbons: influence of hybrid catalysts and steam. *Energy Fuels* **1995**, *9*, 599–609.

11. Katikaneni, S.P.R.; Adjaye, J.D.; Idem, R.O.; Bakhshi, N.N. Catalytic conversion of canola oil over potassium-impregnated HZSM-5 catalysts: C_2–C_4 olefin production and model reaction studies. *Ind. Eng. Chem. Res.* **1996**, *35*, 3332–3346.

12. Botas, J.A.; Serrano, D.P.; García, A.; Ramos, R. Catalytic conversion of rapeseed oil for the production of raw chemicals, fuels and carbon nanotubes over Ni-modified nanocrystalline and hierarchical ZSM-5. *Appl. Catal. B* **2014**, *145*, 205–215.

13. Vu, X.H.; Schneider, M.; Bentrup, U.; Dang, T.T.; Phan, B.M.Q.; Nguyen, D.A.; Armbruster, U.; Martin, A. Hierarchical ZSM-5 materials for an enhanced formation of gasoline-range hydrocarbons and light olefins in catalytic cracking of triglyceride-rich biomass. *Ind. Eng. Chem. Res.* **2015**, *54*, 1773–1782.

14. Vu, X.H.; Bentrup, U.; Hunger, M.; Kraehnert, R.; Armbruster, U.; Martin, A. Direct synthesis of nanosized-ZSM-5/SBA-15 analog composites from preformed ZSM-5 precursors for improved catalytic performance as cracking catalyst. *J. Mater. Sci.* **2014**, *49*, 5676–5689.

15. Wallenstein, D.; Seese, M.; Zhao, X. A novel selectivity test for the evaluation of FCC catalysts. *Appl. Catal. A* **2002**, *231*, 227–242.

16. Twaiq, F.A.; Zabidi, N.A.M.; Bhatia, S. Catalytic conversion of palm oil to hydrocarbons: performance of various zeolite catalysts. *Ind. Eng. Chem. Res.* **1999**, *38*, 3230–3237.

17. Melero, J.A.; Clavero, M.M.; Calleja, G.; Garcia, A.; Miravalles, R.; Galindo, T. Production of biofuels via the catalytic cracking of mixtures of crude vegetable oils and nonedible animal fats with vacuum gas oil. *Energy Fuels* **2010**, *24*, 707–717.

18. Tago, T.; Konno, H.; Nakasaka, Y.; Masuda, T. Size-controlled synthesis of nano-zeolites and their application to light olefin synthesis. *Catal. Surv. Asia* **2012**, *16*, 148–163.

19. Konno, H.; Tago, T.; Nakasaka, Y.; Ohnaka, R.; Nishimura, J.; Masuda, T. Effectiveness of nano-scale ZSM-5 zeolite and its deactivation mechanism on catalytic cracking of representative hydrocarbons of naphtha. *Micropor. Mesopor. Mater.* **2013**, *175*, 25–33.

20. Vu, X.H.; Eckelt, R.; Armbruster, U.; Martin, A. High-temperature synthesis of ordered mesoporous aluminosilicates from ZSM-5 nanoseeds with improved acidic properties. *Nanomaterials* **2014**, *4*, 712–725.

High Selectively Catalytic Conversion of Lignin-Based Phenols into *para-/m*-Xylene over Pt/HZSM-5

Guozhu Liu, Yunxia Zhao and Jinhua Guo

Abstract: High selectively catalytic conversion of lignin-based phenols (*m*-cresol, *p*-cresol, and guaiacol) into *para-/m*-xylene was performed over Pt/HZSM-5 through hydrodeoxygenation and *in situ* methylation with methanol. It is found that the *p*-/*m*-xylene selectivity is uniformly higher than 21%, and even increase up to 33.5% for *m*-cresol (with phenols/methanol molar ratio of 1/8). The improved *p*-/*m*-xylene selectivity in presence of methanol is attributed to the combined reaction pathways: methylation of *m*-cresol into xylenols followed by HDO into *p*-/*m*-xylene, and HDO of *m*-cresol into toluene followed by methylation into *p*-/*m*-xylene. Comparison of the product distribution over a series of catalysts indicates that both metals and supporters have distinct effect on the *p*-/*m*-xylene selectivity.

Reprinted from *Catalysts*. Cite as: Liu, G.; Zhao, Y.; Guo, J. High Selectively Catalytic Conversion of Lignin-Based Phenols into *para-/m*-Xylene over Pt/HZSM-5. *Catalysts* **2016**, *6*, 19.

1. Introduction

p-/*m*-Xylene, an important intermediate for making polyethylene terephthalate (PET), is industrially derived from the reforming of petroleum naphtha. With depleting stocks of fossil fuels and growing concern on the excessive emission of greenhouse gases, great efforts have been made to develop alternative method of producing sustainable *p*-/*m*-xylene from biomass-based compounds [1–4]. Shiramize and Toste [1] proposed a method to produce *p*-xylene with 2,5-dimethyfuran (derived from lignocellulosic biomass) and acrolein (from glycerol, a side product of biodiesel production) through four step-reactions with an overall yield of 34%. Lyons *et al.* [2] used ethylene (derived from biomass or shale gas) as an alternative feedstock, and synthesized *p*-xylene through a very complex process involving trimerizaiton, dehydrogenation, Diels-Alder reaction, and further dehydrogenation. Williams *et al.* [3] suggested a renewable route to *p*-xylene from cycloaddition/dehydration of biomass-based dimethylfuran and ethylene with a high selectivity of 75% using Y zeolite as catalyst. Very recently, Cheng *et al.* [4] reported the production of *p*-xylene from 2-methylfuran (a product of catalytic fast pyrolysis of lignocellulosic biomass) and propylene using ZSM-5 catalysts. For all of those methods lignocellulosic biomass or its derivatives were used as feedstocks.

Lignin, a residue of pulp/paper industries and lignocellulosics-to-ethanol bioprocess [5], is a highly cross-linked, oxygenated aromatic polymer consisted of methoxylated phenylpropane units [6]. In this view, lignin is an abundant, inexpensive and renewable resource of aromatic compounds, such as BTX (benzene, toluene, and xylenes), and may be an attractive alternative resource for producing p-/m-xylene. Generally, there are two typical pathways to convert lignin into BTX. Firstly, depolymerization of lignin has been performed to produce amount of phenols (guaiacol, anisole, phenol, cresols, *etc.*) [7–9], and then catalytic deoxygenation of lignin-derived phenols is used to produce benzene and toluene. For example, phenol was deoxygenated with high selectivity (above 60%) to benzene at 513 K over Ni/HZSM-5 [10] and at 723 K with CoMoP/MgO catalyst [11]. Similarly, cresol (p-cresol or m-cresol) was also hydrotreated by many researchers at atmospheric pressure over different catalysts to obtain toluene with high selectivity of toluene [12–20]. Another pathway to convert lignin into aromatics is catalytic fast pyrolysis producing xylenes with selectivity less than 23%, as reported by Huber *et al.* [21]. Vichaphund *et al.* [22] used metal/HZSM-5 prepared by ion-exchange and impregnation methods producing aromatic compounds from catalytic fast pyrolysis of Jatropha residues. The highest xylenes selectivity is below 30%. Kim *et al.* [23] pyrolyzed native lignin with different HZSM-5 producing xylenes with selectivity less than 10%. Therefore, it is impossible for both the deoxygenation of lignin-derived phenols and fast catalytic pyrolysis of lignin to directly produce p-/m-xylene with high selectivity (larger than 10%) [24].

Side-chain alkylation of toluene with methanol to form styrene and ethylbenzene was studied and the formation of styrene and ethylbenzene was significant on Na(Cs)Y zeolites of exchange degree higher than about 40% [25]. Gas-phase acylation of phenol with acetic acid to synthesize o-hydroxyacetophenone was studied on solid acids and zeolites [26,27]. The gas-phase alkylation of phenols with methanol was studied on different catalysts (SiO_2-Al_2O_3, HBEA, HZSM5, and HMCM22) [28,29]. The main products are cresols and xylenols which are greatly depended on the zeolite pore structure and surface acid properties. The methylation of toluene with methanol is a well-documented reaction to increase the yield of p-/m-xylene. This reaction could be catalyzed by various parent zeolites, such as HZSM-5, MOR, MCM-22, *etc.*, with a high selectivity of p-xylene up to 90%. However, the maximum conversion of toluene in these processes was generally below 50.0% [30]. Meanwhile, the alkylation of m-cresol with methanol was observed to produce xylenols [31], which also could be converted into xylenes via HDO [32]. Therefore, if the methanol was co-feeding with hydrogen, the methylation of cresols may also occur, and then p-/m-xylene could be produced through the *in situ* HDO of xylenols, which is favorable for the high-selective conversion of lignin-based phenols into p-/m-xylene.

The objective of this work is to explore and test a new method to improve the p-/m-xylene selectivity through catalytic hydrodeoxygenation and methylation of lignin-based phenols. The hydrodeoxygenation (HDO) of m-cresol and co-catalytic conversion of m-cresol with methanol at different molar ratio over Pt/HZSM-5 were studied to investigate the effect of methanol on co-feeding process. In order to explain the reason why p-/m-xylene selectivity increases, the co-catalytic conversion of m-cresol with methanol under N_2, and conversion of the intermediate were carried out. Additionally, we applied this procedure (co-conversion with methanol) to different catalysts and other two phenols (guaiacol and p-cresol).

2. Results and Discussion

In this paper, the HZSM-5 (SiO_2/Al_2O_3 of 25, 57, and 107), desilicated HZSM-5, and HBeta (SiO_2/Al_2O_3 of 25) zeolites were denoted as Z-25, Z-57, Z-107, Z-57D, and B-25, respectively.

In this section, the hydrodeoxygenation of m-cresol was firstly carried out over Pt/Z-57. Then catalytic hydrodeoxygenation and methylation of m-cresol at different molar ratio were studied to investigate the methanol effect on the p-/m-xylene selectivity. Immediately following, co-catalytic conversion under N_2 and the intermediate conversion were researched to illustrate the methanol effect on p-/m-xylene selectivity. Based on the discussion above, possible reaction routes for catalytic hydrodeoxygenation and methylation of m-cresol over Pt/Z-57 in the presence of H_2 were proposed. Thirdly, in order to validate the reaction routes, catalytic hydrodeoxygenation, and methylation of m-cresol were also conducted over a series of catalysts. Fourthly, the co-conversion with methanol was applied to other kinds of lignin-based phenols to verify its effectiveness.

2.1. Catalytic Hydrodeoxygenation of m-Cresol over Pt/Z-57

Figure 1 presents the conversion of m-cresol and product distribution with the reaction temperature over Pt/Z-57 catalyst. With the increasing reaction temperature, all the m-cresol conversion is greater than 95% and slightly changes. However, the product distribution significantly changes. For instance, the toluene selectivity increases from 9.8% at 573 K to 39.8% at 673 K and alters a little at higher temperature, while the selectivity of alkanes (including methylcyclohexane, 1,3-dimethylcyclopentane, and C_{3-6}) sharply declines from 88.2% at 573 K to 4.4% at 723 K. The p-/m-xylene and o-xylene selectivities have similar variation trend as the toluene selectivity, with separately reaching 16.3% and 4.7% at 673 K and changing a little at higher temperature. Selectivities of benzene and polyalkylated benzenes also raise and reach up to 13.8% and 3.3% at 723 K, respectively. The different selectivity tendencies between aromatic compounds and alkanes could be explained by the fact that at low temperature hydrogenation of toluene and benzene dominates, and H_2 availability on the catalyst surface decreases with the increasing temperature leading to the rapid decrease of alkanes selectivity [33]. As a result, the aromatic compounds selectivity increases. At the same time, the increase of isomerization, transmethylation, and demethylation activities of *m*-cresol over acid sites causes more *o*-cresol, xylenols (2,5-xylenol, 2,3-xylenol, and 3,4-xylenol) and phenol (0.9%, 0.0%, and 0.1% at 573 K to 3.3%, 2.1%, and 4.0% at 723 K). The amount of polyalkylated phenols is tiny in the temperature range of 573 K–723 K. Except alkanes, aromatics, and phenolics, other products such as indane, naphthalene, and their derivatives, also exist in small amounts.

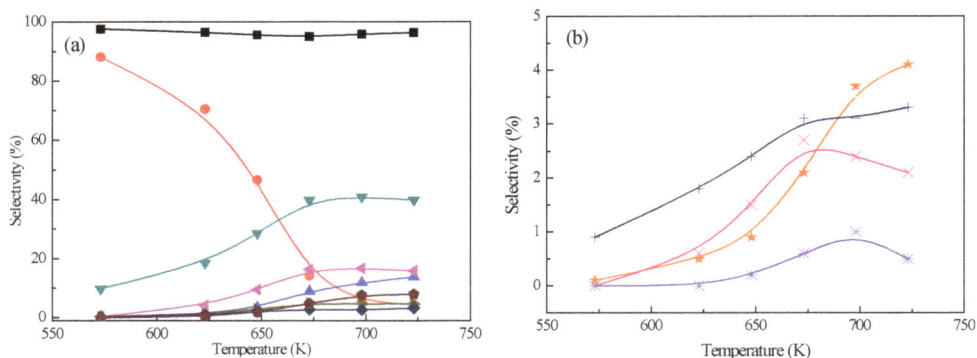

Figure 1. Effect of reaction temperature on (**a**) the conversion and deoxygenated product distribution and (**b**) oxygenated product distribution from *m*-cresol HDO. ■: *m*-cresol conversion, ●: alkanes, ▲: benzene, ▼: toluene, ◀: *p*-/*m*-xylene ▶: *o*-xylene, ◆: polyalkylated aromatics, ⬠: others, ★: phenol, +: *o*-cresol; ✕: xylenols, ✳: polyalkylated phenols. (Reaction conditions: p = 2 MPa, W/F = 75 $g_{cat} \cdot h/mol_{cresol}$, H_2/cresol = 5, catalyst = 1.0 wt. % Pt/Z-57).

Generally, the catalytic HDO of oxygen-containing compounds under different conditions using different catalysts is known to proceed through two reaction pathways [34,35]. The one path is hydrogenolysis or direct deoxygenation (DDO) of phenols to produce aromatic hydrocarbons. The other one is the coupled ring saturation/rapid dehydration/hydrogenation (HYD) to produce cycloparaffins. The second path involves the intermediate product cyclohexanol or substituted cyclohexanol. In this study, there is undetectable amount of methylcyclohexanol in the products, implying that the DDO reaction route mainly occurs during the *m*-cresol HDO process under the tested conditions (p = 2 MPa, W/F = 75 $g_{cat} \cdot h/mol_{cresol}$, H_2/cresol = 5). Therefore, alkanes in the products may be generated from the hydrogenation of toluene.

2.2. Catalytic Hydrodeoxygenation and Methylation of m-Cresol over Pt/Z-57

2.2.1. Catalytic hydrodeoxygenation and Methylation at *m*-Cresol/Methanol Molar Ratio of 1/1

By co-feeding methanol, various methylation reactions of reactants or intermediates may occur as follows: (1) methylation of cresol into xylenols and methylanisole [31]; (2) methylation of benzene with methanol to give toluene [36]; (3) methylation of toluene producing xylenes (*o*-, *m*- and *p*-xylene) [30,37–39]; and (4) polyalkylated benzenes and polyalkylated phenols from the methylation of xylenes and xylenols with methanol, respectively.

Catalytic hydrodeoxygenation and methylation of *m*-cresol were carried out to examine the possible effect of methanol on the selective conversion into *p*-/*m*-xylene. The product distribution (at *m*-cresol/methanol molar ratio of 1/1) is presented as a function of temperature (see Figure 2).

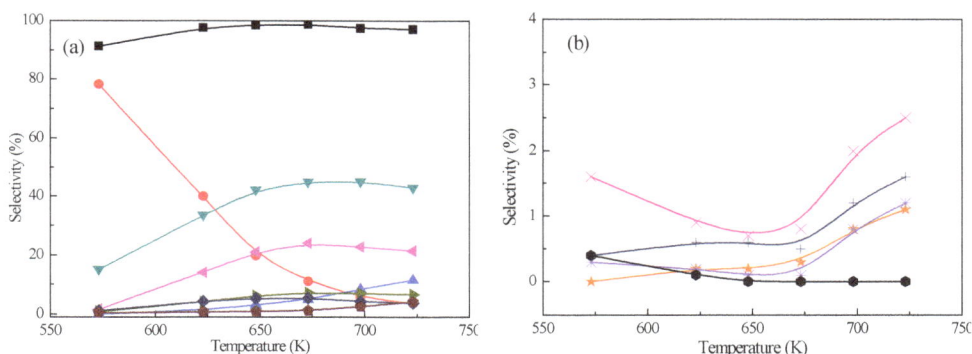

Figure 2. (**a**) The *m*-cresol conversion and deoxygenated product distribution and (**b**) oxygenated product distribution from converting *m*-cresol with methanol. ■: *m*-cresol conversion, ●: alkanes, ▲: benzene, ▼: toluene, ◄: *p*-/*m*-xylene ►: *o*-xylene, ◆: polyalkylated aromatics, ⬟: others, ★: phenol, +: *o*-cresol; ✕: xylenols, ✳: polyalkylated phenols, ●: methylanisole. (Reaction conditions: *m*-cresol/methanol molar ratio = 1/1, p = 2 MPa, W/F = 75 $g_{cat} \cdot h/mol_{cresol}$, H_2/reagent = 5, catalyst = 1.0 wt. % Pt/Z-57).

At 573 K, the alkanes selectivity of 78.2% drops 11.3% compared with that without methanol (88.2% at 573 K), while further decreases to 3.6% at 723 K. This result implies that the activity of toluene hydrogenation may be suppressed by methanol. Compared with co-feeding methanol, the benzene selectivity decreases (from 0.07% to 0.0% at 573 K and from 13.8% to 11.5% at 723 K), while the toluene selectivity enhances (from 9.8% to 15.1% at 573 K and from 39.9% to 42.9% at 723 K). This can be partly attributed to the methylation of benzene to toluene. The methylation of toluene could be confirmed by the change in the xylenes selectivity, *i.e.*, the *p-/m*-xylene selectivity (a maximum of 24.0% at 673 K) is about 1.4 times as that without methanol (a maximum of 16.7% at 698 K). Moreover, the deoxygenation of xylenols is also another reaction to produce *p-/m*-xylene. Similarly, the further methylation of xylenes also takes place which can be demonstrated by the increase of polyalkylated benzenes selectivity after adding methanol (*i.e.*, from 2.8% without methanol to 5.3% with methanol at 673 K).

The decreased phenol and *o*-cresol selectivities are observed from 0.1% and 0.9% to 0.0% and 0.4% at 573 K, and from 4.1% and 3.3% to 1.1% and 1.6% at 723 K. There are several reasons for this result: (1) phenol and *o*-cresol converts into cresols and xylenols via an alkylation reaction, respectively; and (2) phenol and *o*-cresol undergo HDO in the presence of methanol. A small amount of polyalkylated phenols (1.2% at 723 K) are observed compared with negligible amount in the absence of methanol, which may result from the secondary methylation of xylenols.

Above all, the introduction of methanol in the HDO of *m*-cresol bring several new reactions, including methylation of cresol and toluene, as well as xylenes and other compounds, which obviously improve the selectivity of *p-/m*-xylene by 44%, possibly through direct methylation of toluene and HDO of xylenols.

2.2.2. Effect of *m*-Cresol/Methanol Ratio

Catalytic hydrodeoxygenation and methylation of *m*-cresol at different *m*-cresol/methanol (C/M) ratio were also experimentally studied towards a further understanding of reaction routes (shown in Figure 3). At 698 K, the alkanes selectivity is between 5.6% and 9.6% in the investigated C/M ratio range. With the C/M ratio increasing from 1/1 to 1/8, the benzene and toluene selectivities decrease from 8.3% and 45.1% to 1.5% and 17.8%, while the *p-/m*-xylene, *o*-xylene, and polyalkylated benzene selectivities increase from 22.8%, 6.8%, and 4.2% to 33.5%, 10.2%, and 17.3%, which also confirm that the further methylation of benzene, toluene, and xylenes produces toluene, xylenes (*p-/m*-xylene and *o*-xylene), and polyalkylated benzenes. Especially, when the C/M ratio is greater than or equal to 1/4, the most abundant compound in the product are xylenes. With an increase of methanol at the C/M ratio of 1/8, the *p-/m*-xylene selectivity reaches 33.5%, which is as high as 1.5 times of that at the C/M ratio equal to 1/1, and *ca.* 2.0 times as high as that without methanol. The remarkable improvement in the *p-/m*-xylene selectivity may be a result of enhanced methylation reaction rate, as well as the other reactions brought by the methanol, such as methanol to aromatics (MTA) process. Of course, an enhanced methylation reaction rate should be taken as one of the most important pathways as discussed below.

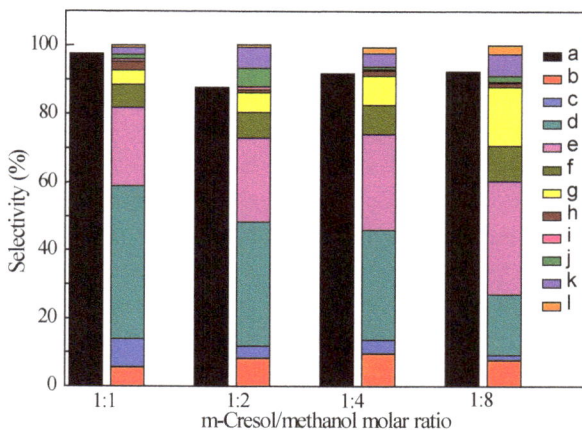

Figure 3. Effect of the *m*-cresol/methanol molar ratio on the product distribution. a: *m*-cresol conversion, b: alkanes, c: benzene, d: toluene, e: *p*-/*m*-xylene, f: *o*-xylene, g: polyalkylated aromatics, h: others, i: phenol, j: *o*-cresol; k: xylenols, l: polyalkylated phenols. (Reaction conditions: $T = 698$ K, $p = 2$ MPa, $W/F = 75$ $g_{cat} \cdot h/mol_{cresol}$, H_2/ reagent = 5, catalyst = 1.0 wt. % Pt/Z-57).

At 698 K, the phenol selectivity is small (less than 1%) in the investigated C/M ratio range. With the increase of the methanol amount in the reactant, the *o*-cresol and xylenols firstly increase from 1.2% and 2.0% (the C/M ratio of 1/1) to 5.4% and 6.1% (the C/M ratio of 1/2), and then decrease to 0.8% and 3.9% (the C/M ratio of 1/4), and lastly increase to 1.8% and 6.4% (the C/M ratio of 1/8), indicating that there is a complex interaction between the methanol and *o*-cresol and xylenols needing further investigation.

Therefore, increasing the methanol/*m*-cresol molar ratio would further improve the selectivity of *p*-/*m*-xylene, but also introduce additional reactions, such as the conversion of methanol to hydrocarbons or aromatics [40]. Obviously, those reactions, to a certain degree, contribute to the considerable improvement of *p*-/*m*-xylene selectivity, which needs further investigation and clarification.

2.2.3. Co-Catalytic Conversion under N_2 and the Intermediate Conversion

To identify the other reactions (MTA, MTH, *etc.*) of methanol other than methylation, co-catalytic conversion of *m*-cresol with methanol at the C/M ratio of 1/4 was carried out under N_2. Under these conditions, HDO reactions of *m*-cresol and xylenols to form toluene and xylenes do not take place [41]. Thus, the observed alkanes, and aromatics (benzene, toluene, *p*-/*m*-xylene, *o*-xylene, and polyalkylated benzenes) under N_2 (see Table 1), may be products from MTA and MTH processes. It should be noted in the temperature range of 573 K to 723 K, the *m*-cresol conversion under N_2 is less than 51%, while the *m*-cresol conversion under H_2 is more than 91%, and the p-/m-xylene selectivity (2.6% at 698 K) under N_2 is *ca.* 1/11 as much as that with H_2 (28.6% at 723 K), implying that the MTA process, as a minor pathway, only gives a small amount of p-/m-xylene in presence of H_2. That is to say, the increased selectivity of p-/m-xylene as the increasing C/M ratio is mainly from the toluene methylation and deoxygenation of xylenols. As far as the MTA is concerned, toluene also plays an important role in remarkably enhancing the selectivity of the xylenes (*ca.* 8.4 times) as reported by Ilias and Bhan [42], which may also attribute to high selectivity of p-/m-xylene. Similarly, the alkane's selectivity decreases from 26.8% (with H_2) to 1.0% at 648 K (without H_2), confirming that the MTH process indeed produces partial alkanes in the co-catalytic conversion of m-cresol in spite of further hydrogenation of toluene in the presence of H_2.

Table 1. Effect of carrier gas on the product distribution from converting m-cresol with methanol.

Reaction Conditions	H_2						N_2					
	573 K	623 K	648 K	673 K	698 K	723 K	573 K	623 K	648 K	673 K	698 K	723 K
m-Cresol conversion, %	97.1	91.7	93.1	92.8	91.8	92.2	6.2	9.3	19.4	39.5	44.3	51.0
Selectivity, %												
Alkanes	85.4	44.8	26.8	18.1	9.6	7.1	0.0	0.0	1.0	1.9	2.0	1.6
Benzene	0.0	0.6	1.2	2.1	4.0	5.2	0.0	0.0	0.2	0.1	0.2	0.2
Toluene	8.5	21.7	27.9	29.8	32.4	32.1	0.0	0.8	1.0	0.8	0.7	0.7
p-/m-Xylene	2.1	17.0	24.2	27.3	28.0	28.6	1.4	0.9	2.1	2.4	2.6	2.3
o-Xylene	1.0	5.1	7.2	8.3	8.5	8.9	0.0	0.0	0.4	0.5	0.5	0.5
[a] PAB	1.0	7.5	9.0	9.2	8.6	9.2	0.0	2.8	8.2	10.5	9.9	8.3
Phenol	0.0	0.1	0.1	0.1	0.3	0.3	0.0	0.0	0.0	0.0	0.0	0.0
o-Cresol	0.2	0.1	0.3	0.4	0.8	0.7	5.6	0.9	0.5	0.7	1.0	1.2
Xylenols	0.9	1.9	1.9	2.5	3.9	3.5	64.2	58.0	52.3	52.6	52.8	54.3
[b] PAP	0.1	0.2	0.4	0.8	1.7	1.8	6.3	8.8	11.8	13.5	15.7	16.9
[c] MA	0.8	0.7	0.7	0.8	0.5	0.5	22.5	27.1	22.0	16.6	14.0	13.5

[a] PAB is polyalkylated benzenes; [b] PAP is polyalkylated phenols; [c] MA is methylanisole; Reaction conditions: m-cresol/methanol molar ratio = 1/4, catalyst = 1.0 wt. % Pt/Z-57, W/F = 75 $g_{cat} \cdot h/g_{m\text{-cresol}}$, reaction temperature = 573–723 K, carrier gas = H_2 or N_2, pressure = 2 MPa.

On the other side, the xylenol's selectivity under N_2 (54.3% at 723 K) is around 15.5 times compared to that with H_2 (3.5% at 723 K), indicating that the increased p-/m-xylene under H_2 may be partially formed from the deoxygenation of xylenols [32]. Without H_2, the polyalkylated phenols and methylanisole selectivities (16.9% at 723 K and 27.1% at 623 K) raise sharply compared with those with H_2 (1.8% at 723 K and 0.7% at 623 K), illustrating that the reaction rate of xylenols methylation and cresol methylation is much faster in the absence of H_2.

In order to confirm the deoxygenation of xylenols to p-/m-xylene, the conversion of 2,4-xylenol was performed under H_2. Table 2 reports the conversion of 2,4-xylenol and product distribution with the reaction temperature over Pt/Z-57 catalyst. In the investigated temperature range, the 2,4-xylenol conversion is above 97% and the

212

alkanes (including dimethycyclohexane, methylcyclohexane, and pentane) selectivity sharply drops from 95.3% at 573 K to 5.7% at 723 K. With the increasing temperature from 573 K to 723 K, the benzene, toluene, p-/m-xylene, o-xylene, and polyalkylated benzene's selectivities increase from 0.0%, 0.7%, 2.0%, 0.5%, and 0.2% to 2.7%, 22.5%, 40.0%, 12.2%, and 13.6%, respectively. The total selectivity of the oxygenated product is less than 5.0%. Obviously, it can be concluded that the deoxygenation of 2,4-xylenol could effectively generate p-/m-xylene under the experimental conditions.

Table 2. Product selectivity and conversion for 2,4-xylenol hydrodeoxygenation.

Temperature, K	2,4-Xylenol Conversion, %	Selectivity, %								
		Alkanes	Benzene	Toluene	p-/m-Xylene	o-Xylene	[a] PAB	Cresol	Xylenols	[b] PAP
573	99.4	95.3	0.0	0.7	2.0	0.5	0.2	0.7	0.5	0.0
623	99.7	93.0	0.0	0.5	3.2	0.9	0.9	0.9	0.5	0.0
648	99.8	84.7	0.2	1.7	8.4	2.2	1.6	0.8	0.3	0.0
673	97.3	42.3	0.5	5.9	29.8	8.6	9.1	0.7	3.1	0.2
698	97.4	16.4	1.1	13.1	39.2	11.8	13.6	0.7	3.3	0.7
723	99.0	5.7	2.7	22.5	40.0	12.2	13.6	0.5	1.7	0.8

[a] PAB is polyalkylated benzenes; [b] PAP is polyalkylated phenols; Reaction conditions: catalyst = 1.0 wt. % Pt/Z-57, $W/F = 75$ $g_{cat} \cdot h/g_{m-cresol}$, reaction temperature = 573 K–723 K, carrier gas = H_2, pressure = 2 MPa.

To sum up, the MTA and MTH of methanol occur in the co-catalytic conversion m-cresol with methanol, but only a small part of p-/m-xylene is produced from the MTA process. Nevertheless, the methylation of toluene and deoxygenation of xylenols are indeed the important pathways of p-/m-xylene formation.

2.2.4. Catalyst Stability

An experiment of co-conversion of m-cresol with methanol (C/M ratio of 1/4) as a function of time on stream was carried out to test the catalyst stability. The variation of the m-cresol conversion and product selectivities over Pt/Z-57 at 673 K are plotted in Figure 4. The catalyst shows deactivation during *ca.* 8 h of on-stream reaction on account of the decrease in the conversion from 91.8% to 76.6%. The main products are always p-/m-xylene and toluene, and their selectivities decrease from 28.0% and 32.4% to 21.0% and 18.5%. Nevertheless, the selectivities of xylenols, o-cresol, and polyalkylated phenols increase from 3.9%, 0.8%, and 1.7% to 16.5%, 7.8%, and 6.6%. Thus, it can be seen that the deactivation is attributed to the formation of oxygenated compounds which may lead to the formation of coke on catalysts. The variation of the alkanes, o-xylene, and polyalkylated benzene's selectivities is insignificant. Benzene, phenol, methylanisole, and other products are in minor amounts and their selectivities change slightly with increasing the time on stream.

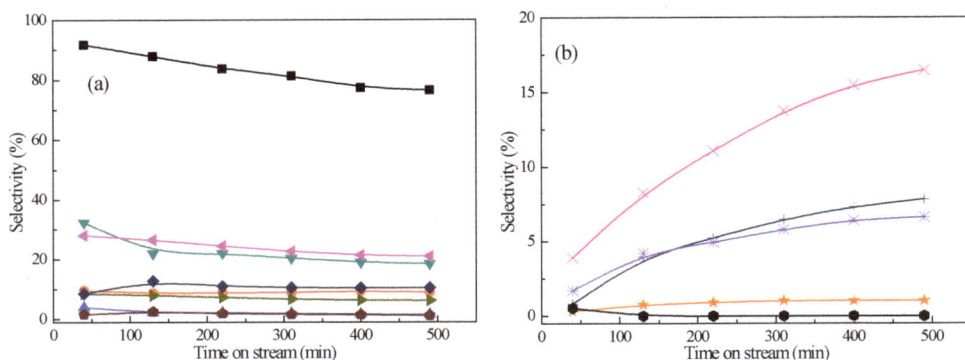

Figure 4. Effect of time on stream on (**a**) the *m*-cresol conversion and deoxygenated product distribution and (**b**) oxygenated product distribution from converting *m*-cresol with methanol. ━■━: *m*-cresol conversion, ━●━: alkanes, ━▲━: benzene, ━▼━: toluene, ━◀━: *p*-/*m*-xylene ━▶━: *o*-xylene, ━◆━: polyalkylated aromatics, ━⬟━: others, ━★━: phenol, ━+━: *o*-cresol; ━✕━: xylenols, ━✱━: polyalkylated phenols, ━●━: methylanisole. (Reaction conditions: *m*-cresol/methanol molar ratio = 1/4, catalyst = 1.0 wt. % Pt/Z-57, W/F = 75 g$_{cat}$·h/mol$_{m\text{-cresol}}$, *T* = 673 K, P$_{H2}$ = 2 MPa).

The TGA measurement was carried out to investigate the deposition of the spent catalyst. The weight change during oxidation with DTG profiles are shown in Figure 5. The weight loss below 200 °C may be attributed to the removal of water and residual organics absorbed on the catalyst. The weight loss above 200 °C may be attributed to the combustion of coke deposited on catalyst surface [43]. Obviously, the weight loss (0.8 wt. %) below 200 °C is just 14.5% of the total weight loss (5.5 wt. %), indicating that coke deposition might be one reason for the catalyst deactivation under the experimental conditions.

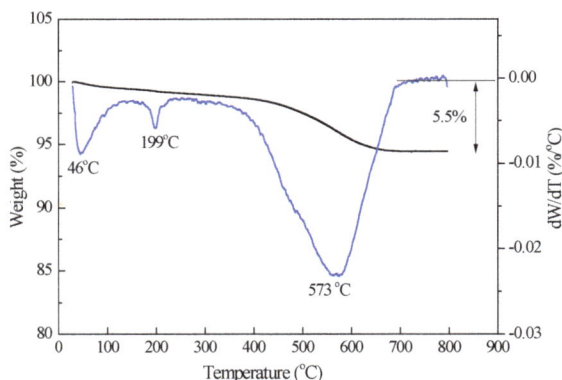

Figure 5. TGA profiles of the spent 1.0 wt. % Pt/Z-57 catalysts.

214

2.2.5. A Possible Reaction Network for Catalytic Hydrodeoxygenation and Methylation of *m*-Cresol

Based on the discussion above, a possible reaction network for catalytic hydrodeoxygenation and methylation of *m*-cresol over Pt/HZSM-5 is shown in Figure 6. First of all, *m*-cresol could be converted into toluene via HDO reaction. Then toluene could be further converted into alkanes and xylenes (*p*-/*m*-xylene and *o*-xylene) through methylation. Of course, the further methylation of xylenes generates polyalkylated benzenes. Two other parallel alkylation reaction pathways of *m*-cresol with methanol are described below: *O*-alkylation of *m*-cresol to methylanisole, and *C*-alkylation into xylenols which could be also formed by disproportionation of *m*-cresol. The HDO of xylenols generates xylenes accompanied with the formation of polyalkylated phenols as by-products. Phenol could be deoxygenated to benzene which could be converted into toluene through methylation. *m*-Cresol could be isomerized to *o*-cresol via an isomerization reaction which could be converted into methyanisole and xylenols by methylation, as well as toluene via HDO.

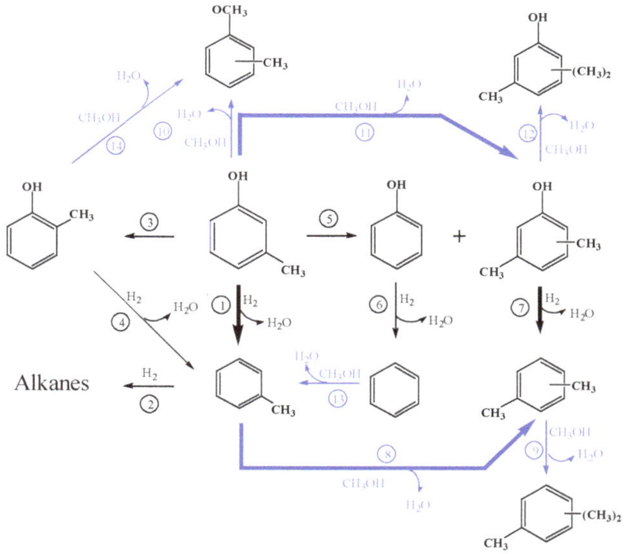

Figure 6. A possible reaction routes for catalytic hydrodeoxygenation and methylation of *m*-cresol over Pt/HZSM-5. The blue line represents the reaction involved in methanol.

2.3. Catalyst Screening

In this section, catalytic hydrodeoxygenation and methylation of *m*-cresol (C/M ratio of 1/4) over different catalysts were studied to investigate the effect of the nature of the supporter and metal on the product distribution. Table 3 shows the conversion of *m*-cresol and product distribution at 698 K over seven kinds of catalysts. At 698 K, the abundant compounds are toluene and *p*-/*m*-xylene on the catalysts except 1.0 wt. % Pt/B-25. For the Pt/HZSM-5 catalysts with parent supporter loading same metal content, the toluene and *p*-/*m*-xylene selectivities are, respectively, in the range of 32.4% to 36.9% and 27.8% to 28.7%, and the conversion of *m*-cresol and product distribution does not exhibit a distinct relation to the SiO_2/Al_2O_3 ratio (acid amount). Nevertheless, for Pt/Z-57 with lower Pt content, the conversion of *m*-cresol declines from 91.8% to 83.4%, the alkanes selectivity decreases (from 9.6% to 8.1%), and the total oxygenated compounds (phenol, *o*-cresol, xylenols, and polyalkylated phenols) selectivity increases (from 6.7% to 10.0%) with lower Pt loading. This is probably associated with the reducing hydrogenation capacity of the catalyst with low metal loading. The lower selectivities of benzene and toluene, and the higher selectivities of *p*-/*m*-xylene and polyalkylated benzenes, are also observed over 0.6 wt. % Pt/Z-57 compared to 1.0 wt. % Pt/Z-57. This could be explained that the deoxygenation activity of oxygenated compounds decreases because of the deactivation, and the aromatics' methylation rate enhances over 0.6 wt. % Pt/Z-57. Over 10.0 wt. % Ni/Z-107, the decrease in the *m*-cresol conversion (from 93.7% over 1.0 wt. % Pt/Z-107 to 85.1%), and the selectivities of *p*-/*m*-xylene and polyalkylated benzenes (from 28.7% and 9.3% over 1.0 wt. % Pt/Z-107 to 22.7% and 6.2%) may be attributed to the coverage of the active sites by high metal loading, while the increase of the alkane's selectivity (from 8.0% over 1.0 wt. % Pt/Z-107 to 11.4%) is possible relative to the high hydrogenation selectivity over Ni/Z-107. Otherwise, the total oxygenated compounds selectivity on 10.0 wt. % Ni/Z-107 is *ca.* two times as much as that on 1.0 wt. % Pt/Z-107, which may result from the catalyst deactivation.

After introducing mesopores by desilication, the *m*-cresol conversion and the alkane's selectivity increases from 91.8% and 9.6% to 96.7% and 10.4%, while the total oxygenated compounds selectivity is just 56.7% of that over 1.0 wt. % Pt/Z-57, demonstrating that the generation of mesopores and the improved Pt dispersion by desilication are conducive to the *m*-cresol transformation, aromatics' hydrogenation, and deoxygenation of oxygenated compounds. It should be noted that the *p*-/*m*-xylene selectivity (34.3%) over Pt/Z-57D is the highest among the seven catalysts, and the benzene and toluene selectivities clearly reduce (from 4.0% and 32.4% over Pt/Z-57 to 1.3% and 23.3% over Pt/Z-57D). The reason for this may be correlated with the increase of the aromatics methylation rate in the generated mesopores thanks to appropriate aperture size and the decrease in the effective diffusion length by desilication [44]. For Pt/B-25, the *m*-cresol conversion (97.2%) is the largest among the seven catalysts. Meanwhile, the selectivities of larger molecules (such as polyalkylated benzenes, xylenols, and polyalkylated phenols) are also the maximal. This is related to the 12-membered ring channels' convenience to further alkylation of p-/m-xylene, cresols, and xylenols, which causes the consumption of p-/m-xylene and the increase of larger molecules.

Table 3. Effect of catalysts on the product distribution from converting m-cresol with methanol.

Type of Catalyst	1.0 wt. % Pt/Z-25	0.6 wt. % Pt/Z-57	1.0 wt. % Pt/Z-57	1.0 wt. % Pt/Z-57D	1.0 wt. % Pt/Z-107	10.0 wt. % Ni/Z-107	1.0 wt. % Pt/B-25
m-Cresol conversion %	94.5	83.4	91.8	96.7	93.7	85.1	97.2
				Selectivity %			
Alkanes	7.0	8.1	9.6	10.4	8.0	11.4	4.8
Benzene	4.2	2.6	4.0	1.3	3.3	5.5	0.4
Toluene	36.9	28.7	32.4	23.3	35.3	33.3	4.5
p-/m-Xylene	27.8	29.6	28.0	34.3	28.7	22.7	14.2
o-Xylene	8.3	8.9	8.5	10.4	8.6	6.9	4.2
[a] PAB	7.8	10.4	8.6	16.0	9.3	6.2	45.9
Phenol	0.3	0.4	0.3	0.2	0.3	1.3	0.3
o-Cresol	1.6	2.1	0.8	0.6	1.4	4.8	0.9
Xylenols	3.6	6.0	3.9	2.0	3.4	6.7	8.5
[b] PAP	1.2	1.5	1.7	1.0	1.0	0.8	12.5
Oxygenated compounds	6.5	10.0	6.7	3.8	6.1	13.6	22.1
others	1.4	0.9	1.6	0.4	0.7	0.4	4.0

[a] PAB is polyalkylated benzenes; [b] PAP is polyalkylated phenols; Reaction conditions: m-cresol/methanol molar ratio = 1/4, W/F = 75 $g_{cat} \cdot h/g_{m\text{-cresol}}$, reaction temperature = 698 K, carrier gas = H_2, pressure = 2 MPa.

As a consequence, the nature of the supporter and metal of a catalyst has significant influence on the catalytic hydrodeoxygenation and methylation of m-cresol. The optimal catalyst should have suitable pore structure, metal loading, and acid amount. Among the investigated catalysts, 1.0 wt. % Pt/Z-57D is the best with highest p-/m-xylene selectivity of 34.3% in the experimental condition.

2.4. Catalytic Hydrodeoxygenation and Methylation of Other Phenols

Catalytic hydrodeoxygenation and methylation of other lignin-derived model compounds (guaiacol and p-cresol) were carried out over the optimum catalyst (Pt/Z-57D) at the phenolic compound/methanol ratio of 1/4 at 673 K. As shown in Table 4, the reactivity of the tested phenolics decreases in the order guaiacol (100.0%) > m-cresol (95.3%) > p-cresol (93.0%). The selectivities of alkanes (33.8% > 27.3% > 17.5%) and benzene (6.6% > 1.3% > 1.1%) are in the same order as that of the reactivity, while the order of the selectivities of toluene (14.4% < 18.0% < 23.1%) and p-/m-xylene (21.3% < 22.2% < 30.0%) are opposite. The higher selectivities of alkylated compounds (toluene, p-/m-xylene, polyalkylated benzenes, and xylenols) from cresol conversion than those from guaiacol conversion may be attributed to the methyl group directly connecting with the aromatic ring in the cresol molecule. The different product distribution from p-cresol conversion with methanol and m-cresol conversion with methanol may relate to the electronic effect caused by the different position of the methyl group.

Table 4. Effect of phenols on the product distribution.

Reactant	Guaiacol	p-Cresol	m-Cresol
Conversion, %	100.0	93.0	95.3
Selectivity, %			
Alkanes	33.8	27.3	17.5
Benzene	6.6	1.3	1.1
Toluene	14.4	18.0	23.1
p-/m-Xylene	21.3	22.2	30.0
o-Xylene	6.2	6.7	9.0
[a] PAB	11.2	14.4	13.6
Phenol	0.6	0.6	0.2
o-Cresol	1.6	1.4	0.9
Xylenols	2.0	5.0	2.6
[b] PAP	1.4	2.4	1.4

[a] PAB is polyalkylated benzenes; [b] PAP is polyalkylated phenols; Reaction conditions: m-cresol (guaiacol and p-cresol)/methanol molar ratio = 1/4, catalyst = 1.0 wt. % Pt/Z-57D, W/F = 75 $g_{cat} \cdot h/mol_{phenols}$, reaction temperature = 673 K, carrier gas = H_2, pressure = 2 MPa.

As a result, although the conversion and product distribution have some difference from various kinds of phenolics with methanol, catalytic hydrodeoxygenation and methylation of lignin-derived model compounds can effectively generate p-/m-xylene. This result could be as a reference for direct conversion of lignin-derived model compounds (lignin phenols) with methanol.

3. Experimental Section

3.1. Materials

m-Cresol (purity 99.7%, Aladdin, Shanghai, China), p-cresol (purity 99.0%, Aladdin, Shanghai, China), guaiacol (purity 99.0%, J&K, Tianjin, China), 2,4-xylenol (purity 99.0%, J&K, Tianjin, China), and methanol (purity 99.0%, J&K, Tianjin, China) were used as received. The HZSM-5 (SiO_2/Al_2O_3 of 25, 57 and 107) and HBeta (SiO_2/Al_2O_3 of 25) zeolites were purchased from Shanghai Novel Chemical Technology Co. Ltd. (Shanghai, China) and $H_2PtCl_6 \cdot 6H_2O$ as the active component precursor was obtained from Tianjin Guangfu Fine Chemical Research Institute (Tianjin, China). Other compounds (cyclohexane, toluene, o-xylene, m-xylene, p-xylene, phenol, 2,5-xylenol, 2,6-xyelnol, and 3,4-xylenol, GCS specifications) for identifying and calibrating products were also purchased from Tianjin Guangfu Fine Chemical Research Institute (Tianjin, China). Hydrogen (purity 99.96%) was obtained from Tianjin Six-Party Industrial Gases Co., Ltd. (Tianjin, China).

3.2. Catalyst Preparation and Characterization

The desilicated HZSM-5 sample was prepared by heating the parent HZSM-5 (SiO_2/Al_2O_3 of 57) in a 0.5 M NaOH solution (10 mL per gram of zeolite) at 333 K under stirring for 60 min. Pt/HZSM-5 was prepared by impregnating HZSM-5 zeolite with aqueous solutions of $H_2PtCl_6 \cdot 6H_2O$. The prepared sample was dried at 393 K for 12 h, and then calcined at 773 K for 4 h in air. After calcination, catalysts were ground to a fine powder and pressed into pellets. These pellets were then crushed and sieved to give particles sized 250–400 μm (40–60 mesh). Ni/HZSM-5 and Pt/HBeta were prepared as the same method as Pt/HZSM-5. In this paper, the HZSM-5 (SiO_2/Al_2O_3 of 25, 57 and 107), desilicated HZSM-5 and HBeta (SiO_2/Al_2O_3 of 25) zeolites were denoted as Z-25, Z-57, Z-107, Z-57D, and B-25, respectively.

The BET (Brunauer-Emmett-Teller) surface area and pore volume of catalysts were determined by using N_2 isothermal (77 K) adsorption on Micromeritics Tristar 3000 physisorption analyzer (Atlanta, GA, USA). The SiO_2/Al_2O_3 ratio of the fresh catalysts was determined using an X-ray Fluorescence spectrometer (Philips Magix PW 2403 XRF, Amsterdam, Holland). NH_3-temperature programmed desorption (NH_3-TPD) and CO pulse chemisorptions were performed on an Altamira-300 temperature programming unit equipped with a thermal conductivity detector (TCD). The chemical composition, textural properties, calculated dispersion, and NH_3 uptake of the parent and desilicated materials are exhibited in Table 5.

The coke deposited on the spent catalysts was quantified by thermogravimetry (TGA-Q500, TA, New Castle, DE, USA). The TG experiments were carried out by raising the sample temperature from room temperature up to 1073 K at a rate of 10 K/min in a mixture flow of N_2 balance gas (40 mL/min) and air sample gas (40 mL/min).

Table 5. Chemical composition, textural properties, NH_3 uptake, and calculated dispersion of the catalysts.

Catalysts	[a] Metal Content, wt. %	[b] SiO_2/Al_2O_3	[c] S_{BET}, m^2/g	[d] V_o, cm^3/g	[e] V_m, cm^3/g	NH_3 Uptake (μmol/g)	[f] Calculated Dispersion, %
Pt/Z-25	1.0	29.2	334	0.11	0.10	959.2	14.1
Pt/Z-57	0.6	-	368	0.11	0.04	627.6	13.7
Pt/Z-57	1.0	50.6	400	0.12	0.08	717.6	10.4
Pt/Z-57D	1.0	38.6	342	0.10	0.12	513.0	17.2
Pt/Z-107	1.0	100.4	342	0.11	0.10	413.8	18.0
Ni/Z-107	10.0	-	264	0.10	0.05	472.3	0.6
Pt/B-25	1.0	21.6	503	0.18	0.18	938.8	7.0

[a] Metal content calculated based on the wet impregnating procedure; [b] The SiO_2/Al_2O_3 ratio was measured by XRF; [c] S_{BET} is the specific surface area analyzing according to Brunauer-Emmett-Teller method; [d] V_o is the micropore volume and calculated from the desorption branch of the N_2 isotherm; [e] V_m is the meso-and macropore volumes and calculated from the desorption branch of the N_2 isotherm; [f] Calculated dispersion was measured by CO pulse chemisorption.

3.3. Experiment Procedure

Experiments were carried out in a continuous-flow tubular fixed-bed reactor made of stainless steel (i.d. = 10 mm, l = 1000 mm), which was described in detail elsewhere [45]. In a typical run, 3 g of catalyst diluted with SiC was packed into the stainless steel reactor, and then reduced at 723 K for 4 h in a flow of H_2 (120 mL/min) controlled by a mass flow controller. Reactants (*m*-cresol or a mixture of *m*-cresol (*p*-cresol or guaiacol) with methanol, $W/F_{m-cresol}$ = 75 $g_{cat} \cdot h/mol_{m-cresol}$) were feed through a HPLC pump. The pressure of H_2 was 2 MPa and the molar ratio of H_2/reagent kept constant at five. The catalyst bed temperature was measured by a K-type thermocouple and maintained using a tube furnace equipped with a temperature controller. The reaction temperature arrived at a heating rate of 5 K/min from room temperature and in this study it changed from 573 K to 723 K. A water-cooled condenser was used to separate condensable from volatile products. The gas product stream was collected in a gas bag. The liquid product was sampled for analysis with an off-line GC. Each experiment was repeated three times, and the results were deviated *ca.* 4%.

A Bruker 456-GC (Karlsruhe, Germany) equipped with BR-5ms column (30 m × 0.25 mm × 0.25 μm) and a flame ionization detector (FID) was used to quantify the liquid products. The GC oven temperature was programmed as follow: maintained 318 K for 2 min, and ramped from 318 K to 523 K with a heating rate of 10 K/min, and finally kept 523 K for 5 min. Standards of known concentration containing observed reactants and products, *i.e.*, a mixture of *m*-cresol, cylclohexane, toluene, *p*-xylene, phenol, and 2,4-dimethyl phenol, were used to calibrate the conversion and yield from FID area. Identification of products was carried out by liquid product injections to an Agilent 7890A gas chromatograph (GC) equipped with HP-5ms column (30 m × 0.25 mm × 0.25 μm) and a 5975C mass selective detector (MSD), with temperature starting at 323 K for 5 min, and then 5 K/min up to 443 K and, lastly, dwelling at 443 K for 2 min. Each sample was analyzed at least twice, and the deviation error of analysis was about ±1%.

The *W/F* (catalyst weight/molar feed flow rate) values reported were found by dividing the total weight of catalyst in the reactor bed by the molar flow rate of *m*-cresol (or other kinds of lignin-based phenols) injected during a given experiment. Product selectivity is defined as the mass fraction of the material in the total product.

4. Conclusions

A novel method of high-selectively catalytic hydrodeoxygenation and methylation of lignin-based phenols into xylenes was explored over a series of Pt/HZSM-5 under H_2. A maximum *p*-/*m*-xylene selectivity of 33.5% was obtained at the *m*-cresol/methanol molar ratio of 1/8, which was two times as much as that without methanol. The improved *p*-/*m*-xylene selectivity with methanol was mainly attributed to two reaction pathways: methylation of *m*-cresol into xylenols followed by HDO into *p*-/*m*-xylene, and HDO of *m*-cresol into toluene followed by methylation into *p*-/*m*-xylene. Otherwise, the catalyst nature (metals and supporters) had obvious influence on the product distribution. Catalysts with mesoporous can influence the diffusion of the compound and the selectivity of xylenes. Catalysts loading Pt have higher hydrogenation activity than those loaded with Ni.

Acknowledgments: We gratefully acknowledge the National Program for New Century Excellent Talents (NCET-13-0408) for financial support on this work.

Author Contributions: Guozhu Liu designed the experiments. Jinhua Guo conducted the experiments and analyzed the data. Jinhua Guo and Yunxia Zhao wrote the draft of the manuscript which was then revised by all other authors.

Conflicts of Interest: The authors declare no conflict of interest.

References

1. Shiramizu, M.; Toste, F.D. On the Diels-Alder approach to solely biomass-derived polyethylene terephthalate (PET): Conversion of 2,5-dimethylfuran and acrolein into *p*-xylene. *Chemistry* **2011**, *17*, 12452–12457.

2. Lyons, T.W.; Guironnet, D.; Findlater, M.; Brookhart, M. Synthesis of *p*-xylene from ethylene. *J. Am. Chem. Soc.* **2012**, *134*, 15708–15711.

3. Williams, C.L.; Chang, C.C.; Do, P.; Nikbin, N.; Caratzoulas, S.; Vlachos, D.G.; Lobo, R.F.; Fan, W.; Dauenhauer, P.J. Cycloaddition of Biomass-Derived Furans for Catalytic Production of Renewablep-Xylene. *ACS Catal.* **2012**, *2*, 935–939.

4. Cheng, Y.T.; Wang, Z.; Gilbert, C.J.; Fan, W.; Huber, G.W. Production of *p*-xylene from biomass by catalytic fast pyrolysis using ZSM-5 catalysts with reduced pore openings. *Angew. Chem.* **2012**, *51*, 11097–11100.

5. Huber, G.W.; Iborra, S.; Corma, A. Synthesis of transportation fuels from biomass: Chemistry, catalysts, and engineering. *Chem. Rev.* **2006**, *106*, 4044–4098.

6. Effendi, A.; Gerhauser, H.; Bridgwater, A.V. Production of renewable phenolic resins by thermochemical conversion of biomass: A review. *Renew. Sustain. Energy Rev.* **2008**, *12*, 2092–2116.

7. Mullen, C.A.; Boateng, A.A. Catalytic pyrolysis-GC/MS of lignin from several sources. *Fuel Process. Technol.* **2010**, *91*, 1446–1458.

8. Karagöz, S.; Bhaskar, T.; Muto, A.; Sakata, Y. Hydrothermal upgrading of biomass: Effect of K_2CO_3 concentration and biomass/water ratio on products distribution. *Bioresour. Technol.* **2006**, *97*, 90–98.

9. Bai, X.; Kim, K.H.; Brown, R.C.; Dalluge, E.; Hutchinson, C.; Lee, Y.J.; Dalluge, D. Formation of phenolic oligomers during fast pyrolysis of lignin. *Fuel* **2014**, *128*, 170–179.

10. Zhang, X.; Wang, T.; Ma, L.; Zhang, Q.; Jiang, T. Hydrotreatment of bio-oil over Ni-based catalyst. *Bioresour. Technol.* **2012**, *127*, 306–311.

11. Yang, Y.; Gilbert, A.; Xu, C.C. Hydrodeoxygenation of bio-crude in supercritical hexane with sulfided CoMo and CoMoP catalysts supported on MgO: A model compound study using phenol. *Appl. Catal. A* **2009**, *360*, 242–249.

12. Whiffen, V.M.; Smith, K.J. A Comparative Study of 4-Methylphenol Hydrodeoxygenation Over High Surface Area MoP and Ni_2P. *Top. Catal.* **2012**, *55*, 981–990.

13. Foster, A.J.; Do, P.T.M.; Lobo, R.F. The Synergy of the Support Acid Function and the Metal Function in the Catalytic Hydrodeoxygenation of *m*-Cresol. *Top. Catal.* **2012**, *55*, 1–11.

14. Ausavasukhi, A.; Huang, Y.; To, A.T.; Sooknoi, T.; Resasco, D.E. Hydrodeoxygenation of m-cresol over gallium-modified beta zeolite catalysts. *J. Catal.* **2012**, *290*, 90–100.

15. Zanuttini, M.; Lago, C.; Querini, C.; Peralta, M. Deoxygenation of *m*-cresol on Pt/γ-Al_2O_3 catalysts. *Catal. Today* **2013**, *213*, 9–17.

16. Nie, L.; de Souza, P.M.; Noronha, F.B.; An, W.; Sooknoi, T.; Resasco, D.E. Selective conversion of *m*-cresol to toluene over bimetallic Ni-Fe catalysts. *J. Mol. Catal. A* **2014**, *388–389*, 47–55.

17. Pham, T.N.; Shi, D.; Resasco, D.E. Evaluating strategies for catalytic upgrading of pyrolysis oil in liquid phase. *Appl. Catal. B* **2014**, *145*, 10–23.

18. To, A.T.; Resasco, D.E. Role of a phenolic pool in the conversion of *m*-cresol to aromatics over HY and HZSM-5 zeolites. *Appl. Catal. A* **2014**, *487*, 62–71.

19. To, A.T.; Resasco, D.E. Hydride transfer between a phenolic surface pool and reactant paraffins in the catalytic cracking of *m*-cresol/hexanes mixtures over an HY zeolite. *J. Catal.* **2015**, *329*, 57–68.

20. Zhu, X.; Nie, L.; Lobban, L.L.; Mallinson, R.G.; Resasco, D.E. Efficient Conversion of *m*-Cresol to Aromatics on a Bifunctional Pt/HBeta Catalyst. *Energy Fuels* **2014**, *28*, 4104–4111.

21. Cheng, Y.T.; Jae, J.; Shi, J.; Fan, W.; Huber, G.W. Production of Renewable Aromatic Compounds by Catalytic Fast Pyrolysis of Lignocellulosic Biomass with Bifunctional Ga/ZSM-5 Catalysts. *Angew. Chem. Int. Ed.* **2012**, *124*, 1416–1419.

22. Vichaphund, S.; Ahtong, D.; Sricharoenchaikul, V.; Atong, D. Production of aromatic compounds from catalytic fast pyrolysis of Jatropha residues using metal/HZSM-5 prepared by ion-exchange and impregnation methods. *Renew. Energy* **2015**, *79*, 28–37.

23. Kim, J.Y.; Lee, J.H.; Park, J.; Kim, J.K.; An, D.; Song, I.K.; Choi, J.W. Catalytic pyrolysis of lignin over HZSM-5 catalysts: Effect of various parameters on the production of aromatic hydrocarbon. *J. Anal. Appl. Pyrolysis* **2015**, *114*, 273–280.

24. Zhang, H.; Cheng, Y.T.; Vispute, T.P.; Xiao, R.; Huber, G.W. Catalytic conversion of biomass-derived feedstocks into olefins and aromatics with ZSM-5: The hydrogen to carbon effective ratio. *Energy Environ. Sci.* **2011**, *4*, 2297–2307.

25. Borgna, A.; Magni, S.; Sepúlveda, J.; Padró, C.L.; Apesteguía, C.R. Side-chain alkylation of toluene with methanol on Cs-exchanged NaY zeolites: Effect of Cs loading. *Catal. Lett.* **2005**, *102*, 15–21.

26. Padro, C. Gas-phase synthesis of hydroxyacetophenones by acylation of phenol with acetic acid. *J. Catal.* **2004**, *226*, 308–320.

27. Padró, C.L.; Rey, E.A.; Peña, L.G.; Apesteguía, C.R. Activity, selectivity and stability of Zn-exchanged NaY and ZSM5 zeolites for the synthesis of *o*-hydroxyacetophenone by phenol acylation. *Microporous Mesoporous Mater.* **2011**, *143*, 236–242.

28. Sad, M.E.; Padró, C.L.; Apesteguía, C.R. Selective synthesis of *p*-cresol by methylation of phenol. *Appl. Catal. A* **2008**, *342*, 40–48.

29. Sad, M.E.; Padró, C.L.; Apesteguía, C.R. Study of the phenol methylation mechanism on zeolites HBEA, HZSM5 and HMCM22. *J. Mol. Catal. A* **2010**, *327*, 63–72.

30. Alabi, W.; Atanda, L.; Jermy, R.; Al-Khattaf, S. Kinetics of toluene alkylation with methanol catalyzed by pure and hybridized HZSM-5 catalysts. *Chem. Eng. J.* **2012**, *195*, 276–288.

31. Crocellà, V.; Cerrato, G.; Magnacca, G.; Morterra, C.; Cavani, F.; Cocchi, S.; Passeri, S.; Scagliarini, D.; Flego, C.; Perego, C. The balance of acid, basic and redox sites in Mg/Me-mixed oxides: The effect on catalytic performance in the gas-phase alkylation of *m*-cresol with methanol. *J. Catal.* **2010**, *270*, 125–135.

32. Gevert, S.; Eriksson, M.; Eriksson, P.; Massoth, F. Direct hydrodeoxygenation and hydrogenation of 2, 6-and 3, 5-dimethylphenol over sulphided CoMo catalyst. *Appl. Catal. A* **1994**, *117*, 151–162.

33. Derrouiche, S.; Bianchi, D. Heats of Adsorption Using Temperature Programmed Adsorption Equilibrium: Application to the Adsorption of CO on Cu/Al_2O_3 and H_2 on Pt/Al_2O_3. *Langmuir* **2004**, *20*, 4489–4497.

34. Odebunmi, E.O.; Ollis, D.F. Catalytic hydrodeoxygenation: I. Conversions of *o*-, *p*-, and *m*-cresols. *J. Catal.* **1983**, *80*, 56–64.

35. Massoth, F.; Politzer, P.; Concha, M.; Murray, J.; Jakowski, J.; Simons, J. Catalytic hydrodeoxygenation of methyl-substituted phenols: Correlations of kinetic parameters with molecular properties. *J. Phys. Chem. B* **2006**, *110*, 14283–14291.

36. Adebajo, M.O.; Howe, R.F.; Long, M.A. Methylation of benzene with methanol over zeolite catalysts in a low pressure flow reactor. *Catal. Today* **2000**, *63*, 471–478.

37. Young, L.; Butter, S.; Kaeding, W. Shape selective reactions with zeolite catalysts: III. Selectivity in xylene isomerization, toluene-methanol alkylation, and toluene disproportionation over ZSM-5 zeolite catalysts. *J. Catal.* **1982**, *76*, 418–432.

38. Baduraig, A.; Odedairo, T.; Al-Khattaf, S. Disproportionation and methylation of toluene with methanol over zeolite catalysts. *Top. Catal.* **2010**, *53*, 1446–1456.

39. Zhu, Z.; Chen, Q.; Xie, Z.; Yang, W.; Li, C. The roles of acidity and structure of zeolite for catalyzing toluene alkylation with methanol to xylene. *Microporous Mesoporous Mater.* **2006**, *88*, 16–21.

40. Conte, M.; Lopez-Sanchez, J.A.; He, Q.; Morgan, D.J.; Ryabenkova, Y.; Bartley, J.K.; Carley, A.F.; Taylor, S.H.; Kiely, C.J.; Khalid, K. Modified zeolite ZSM-5 for the methanol to aromatics reaction. *Catal. Sci. Technol.* **2012**, *2*, 105–112.

41. Nimmanwudipong, T.; Runnebaum, R.C.; Block, D.E.; Gates, B.C. Catalytic conversion of guaiacol catalyzed by platinum supported on alumina: Reaction network including hydrodeoxygenation reactions. *Energy Fuels* **2011**, *25*, 3417–3427.

42. Ilias, S.; Bhan, A. Tuning the selectivity of methanol-to-hydrocarbons conversion on H-ZSM-5 by co-processing olefin or aromatic compounds. *J. Catal.* **2012**, *290*, 186–192.

43. Zhang, X.; Zhang, Q.; Wang, T.; Ma, L.; Yu, Y.; Chen, L. Hydrodeoxygenation of lignin-derived phenolic compounds to hydrocarbons over Ni/ZrO$_2$-SiO$_2$ catalysts. *Bioresour. Technol.* **2013**, *134*, 73–80.

44. Ahn, J.H.; Kolvenbach, R.; Neudeck, C.; Al-Khattaf, S.S.; Jentys, A.; Lercher, J.A. Tailoring mesoscopically structured H-ZSM5 zeolites for toluene methylation. *J. Catal.* **2014**, *311*, 271–280.

45. Yang, Y.; Wang, Q.; Zhang, X.; Wang, L.; Li, G. Hydrotreating of C$_{18}$ fatty acids to hydrocarbons on sulphided NiW/SiO$_2$-Al$_2$O$_3$. *Fuel Process. Technol.* **2013**, *116*, 165–174.

Insights to Achieve a Better Control of Silicon-Aluminum Ratio and ZSM-5 Zeolite Crystal Morphology through the Assistance of Biomass

Alessandra V. Silva, Leandro S. M. Miranda, Márcio Nele, Benoit Louis and Marcelo M. Pereira

Abstract: The present study attempts to provide insights for both the chemical composition (Si/Al) and the crystal morphology of ZSM-5 zeolites while using biomass template compounds in the synthesis. The solution containing biomass-derivative compounds was obtained after treating biomass in a sodium hydroxide aqueous solution under reflux. The latter alkaline solution was used as a solvent for zeolite nuclei ingredients to form the gel phase under hydrothermal conditions (170 °C during 24, 48 or 72 h). This approach allowed for preparing MFI zeolites having a broad range of Si/Al ratio, *i.e.*, from 25 to 150. Likewise, MFI crystals with different morphologies could be obtained, being different from the pristine zeolite formed in the absence of biomass.

Reprinted from *Catalysts*. Cite as: Silva, A.V.; Miranda, L.S.M.; Nele, M.; Louis, B.; Pereira, M.M. Insights to Achieve a Better Control of Silicon-Aluminum Ratio and ZSM-5 Zeolite Crystal Morphology through the Assistance of Biomass. *Catalysts* **2016**, *6*, 30.

1. Introduction

ZSM-5 zeolite, discovered by Mobil Company, is a member of commonly called "big five" zeolites and has been used in several acid-catalyzed processes. For instance, its use can lead to improved gasoline quality in fluid catalytic cracking processes [1]. Ethylene and propylene selectivity can also be improved based on ZSM-5 additives [2]. Likewise this zeolite has been successfully tested in NO_x reduction [3], glycerol conversion [3,4] and in the Methanol-To-Olefins process [5]. Barely reactive σ C–H and σ C–C bonds in saturated hydrocarbons have been activated by means of zeolite catalysts [6]. We have recently demonstrated that the first step in alkane activation occurs through the protolytic cleavage of either σ C–H or σ C–C bonds, followed by bimolecular reactions involving carbocations [7]. The product selectivity is usually strongly affected by the zeolite Si/Al ratio, especially in the chemistry involving hydrocarbons [8]. In addition, the propylene selectivity (in MTO reaction) is largely enhanced in large ZSM-5 crystals and yet by higher SAR [9].

225

It is therefore of prime importance to design ZSM-5 catalysts with tailored SAR, crystal size, morphology and texture since these parameters play an important role in several types of acid-catalyzed reactions. In addition, it is preferrable to control these properties via a one-pot synthesis, thus avoiding sequential post-treatments.

Living organisms such as shells and corals commonly use self-assembly processes involving inorganic and organic moieties [10]. Moreover, the origin of life itself could be related to an interaction between silicon and aluminum compounds and organic molecules [11,12]. As a consequence, research combining biomass-derived compounds in alumino-silicate preparation, particularly the ones involving long synthesis duration and the formation of metastable species has great potential. Indeed, zeolites are metastable materials prepared either with an assistance of organic templates or by seed addition [13]. A clear effect between the template size and the zeolite pore architecture remains rather difficult to demonstrate. Crystal size and morphology are largely affected by the self-assembly process involved during crystal growth. In zeolite synthesis, meta-stable intermediate structures are produced, being themselves affected by the presence of different ingredients and especially the presence of organic co-components, in analogy to the aforementioned natural processes.

The use of sugar cane biomass (bagasse) in the preparation of ZSM-5 zeolite was already reported by our group [14]. This procedure greatly affected the morphology of ZSM-5 crystals, resulting in sizes between 50–100 nm. In contrast, SAR values were not greatly affected and remained in the 20–30 range, similar to the zeolite synthesized in the absence of biomass residues. It is important to highlight that small ZSM-5 crystals increased both alkane cracking activity and selectivity toward propylene and ethylene compared with large crystals [14]. This simple and inexpensive method was also applied to the preparation of γ-alumina and resulted both in surface area and pore volume increases of up to 2.5-fold and 1.5-fold, respectively, when compared to alumina preparation in the absence of biomass [15]. Furthermore, mesopores in the 2-10 nm range were produced. HRTEM measurements showed that the presence of biomass enhanced the the formation of pores by interconnection of (002) and (1-11) planes. However, no improvement in alumina textural properties prepared in the presence of a solution containing biomass-extracted compounds could be observed. These results suggest that biomass or biomass-derived components formed under *in-situ* synthesis conditions induce changes (minor or major) in the self-assembly process. We would like to stress that biomass introduction itself or a solution containing biomass-extracted compounds for preparing an inorganic material is, so far, a case-by-case phenomenon, rather than a general and well established cause-effect principle. Hence, so far catalytic properties can barely be understood '*a priori*'. Herein the effect of biomass derived-compounds, obtained by prior alkaline sugar-cane bagasse hydrolysis, was evaluated in ZSM-5 zeolite preparations.

2. Results and Discussion

In order to investigate whether the alkaline hydrolyzate from sugar-cane bagasse has an influence on the chemical composition or crystal size/morphology of ZSM-5 zeolite when added as a co-component in its synthesis, we performed the following steps:

i   Select a reliable standard preparation based on the International Zeolite Association to obtain a high Al-containing ZSM-5 zeolite [16].

ii   Repeat this procedure in the presence of biomass-derived compounds extracted by alkaline hydrolysis treatment of biomass residues.

iii   Characterize the as-prepared zeolites in terms of textural properties, SAR, microstructure, crystallinity and catalytic activity in the n-hexane cracking reaction.

The Leuven group has recently shown by ^{27}Al-^{1}H REDOR NMR of precursor gel species that TPA^{+} cations were readily incorporated at early stages, in their final configuration, in the MFI aluminosilicate (SAR = 50) [17]. This is in agreement with earlier proposed MFI self-assembly pathways [18–20].

In order to assess any effect related to the biomass presence, we have performed preliminary experiments: zeolite synthesis without fiber addition (reference materials), and analysis of hydrolyzate composition as well as fiber compositions before and after alkaline treatment. Table 1 describes both the yield of mass loss from the bagasse, as well as the yield of the extraction. As expected, the mass loss observed after the more concentrated alkaline treatment (0.5 mol/L) is the highest. Likewise, the yield of extracted products remains the highest too (Table 1). According to former reports [21], an alkaline hydrolysis of sugar cane bagasse yielded mainly molecules from the hydrolysis of lignin, especially cinnamic acid derivatives such as coumaric and ferulic acids. The direct injection of crude hydrolyzate into a mass spectrometer equipped with an electron spray ionization source showed the main signals (M-H^{+}) corresponding to the aforementioned acids. In parallel, the quantification of pentoses and hexoses by chromatography confirmed the absence of these carbohydrates in the hydrolyzate. However, partial hydrolysis or decomposition of these polymers cannot be completely ruled out but a more detailed analysis of this solution is not within the scope of this report.

With this alkaline hydrolyzate in hand, we then proceeded to evaluate its influence in the synthesis of ZSM-5 when used as a solvent. The ZSM-5 zeolite preparation in the absence of hydrolyzate led to a highly crystalline zeolite, as shown in Table 2. The chemical analysis of SAR (including both non- and framework aluminum) showed an increase ranging from 24 to 37 for crystallization times of 24 to 48 h. SAR did not change further after 72 h of synthesis. This slight increase in SAR with ageing time is regularly observed for ZSM-5 synthesis as siliceous species can be continuously incorporated in the framework [20]. However higher intensities in XRD diffraction lines for 48 h were observed compared to the ones detected after a 24 h material synthesis (roughly a 15% raise based on the two theta = 23° reflexion). An optimum synthesis time has therefore to be found to guarantee a high crystallinity.

Table 1. Analysis of alkaline hydrolysis solution of sugar cane bagasse.

Biomass Treatment (mol/L NaOH)	Yield of Bagasse Mass Loss (%) [a]	Extract Yield (%) [b]
0.1	14	8
0.5	52	45

[a] [(initial bagasse mass − recovery bagasse mass)/initial bagasse mass] × 100; [b] [mass extract/(initial bagasse mass − recovery bagasse mass)] × 100.

Table 2. ZSM-5 textural and structural properties.

Catalysts	SAR XRF	Micro $cm^3 \cdot g^{-1}$	Meso $cm^3 \cdot g^{-1}$	Rel. XRD	Particle Size (μm)
Ref. [1]	27	0.15	0.025	100	1-3
Ref. [2]	26	n.d.	n.d.	100	1-3
Ref. [3]	37	n.d.	n.d.	100	1-3
ZSM-5 A(24)	107	0.17	0.018	88	1-3
ZSM-5 B(48)	143	0.16	0.017	98	1-4
ZSM-5 C(72)	153	0.16	0.018	97	6-10
ZSM-5 D(24)	24	0.16	0.017	93	3-7
ZSM-5 E(48)	81/98 *	0.16	0.017	106	7-8
ZSM-5 F(72)	119	0.17	0.017	88	7-11

* These two zeolites were synthesized twice to assess the reproducibility of the protocol.

The textural properties of as-prepared ZSM-5 zeolites as a function of both the type of biomass derivative-solution and ageing time are presented in Table 2. The pristine MFI zeolite pattern, prepared without biomass-derivate compounds (Refs. [1,2]) were included. Firstly, it is important to highlight that all materials exhibited all the characteristic diffraction lines of a unique MFI structure (Figure 1). All zeolites were compared with (Ref. [2]) to estimate the relative crystallinity and exhibited high crystallinities, above 87%. Additional DRX experiment of ZSM-5 E (48) and a commercial ZSM-5 sample (results not shown) were performed and both zeolites showed similar crystallinity, therefore we can assure that the relative crystallinity values detailed in Table 2 for the zeolite samples prepared with and without the biomass-derivatives assistance are comparable to those of conventional prepared zeolites.

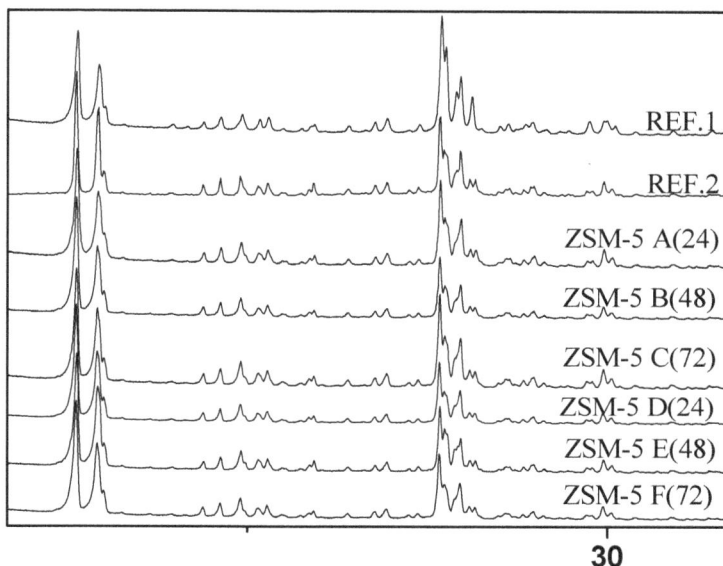

Figure 1. Diffraction patterns for the different zeolites: Ref. [1] series and Ref. [2]. series Zeolite synthesized for 24 h; NaOH 0.1 mol/L (ZSM-5 A(24)); for 48 h; NaOH 0.1 mol/L (ZSM-5 B(48)); for 72 h NaOH 0.1 mol/L (ZSM-5 C(72)); for 24 h; NaOH 0.5 mol/L (ZSM-5 D(24)); for 48 h; NaOH 0.5 mol/L (ZSM-5 E(48)); for 72 h; NaOH 0.5 mol/L (ZSM-5 F(72)).

Secondly, ZSM-5 properties were discussed in terms of SAR, textural property, crystal size, and morphology (Table 2). As observed for ZSM-5 reference materials (prepared in the absence of biomass), longer ageing times led to higher SAR values. However, a broad range of SAR was observed and yet the extract obtained at 0.1 mol/L of NaOH resulted in higher values compared to those reached at 0.5 mol/L of NaOH.

The SAR based on the bulk composition was remarkably affected and values in the range of 25 to 150 were measured. Microporosities and mesoporosities are rather similar for all zeolites in line with literature data [22,23]. For the sake of clarity, a parent ZSM-5 was characterized by the nitrogen adsorption/desorption technique and similar microporous volumes were measured. ZSM-5 prepared in the presence of biomass derived-compounds exhibited a slightly lower mesoporous volume compared to its parent ZSM-5 counterpart. The ZSM-5 structure is orthorhombic

(Pnma space group) when high thermal treatment is conducted [24,25], usually resulting in prismatic crystal formation [26] as observed for reference zeolites (Figure 2). The standard method used for preparing ZSM-5 zeolite resulted in 1-3 µm range crystal size with no changes in morphology with ageing time. In contrast, both morphology and particle size were affected by the presence of biomass derived-compounds. The extract obtained at 0.1 mol/L led to regular (size and type) ZSM-5 crystals at an ageing time of 24 h. A further raise in ageing time to 48 h led to the formation of crystals having the same size, but extremely different in terms of morphology that was constructed by an assembly of elongated French-fries having 100 nm in width. At longer ageing times, larger crystals were formed, exhibiting again a conventional morphology. In contrast to ZSM-5 A(24) sample, ZSM-5 C(72) sample exhibits sharper crystal outer rim and appears to be more faceted. While studying the influence of alkali cation nature and gel ionic strength, Petrik *et al.* were able to synthesize MFI zeolites with significant changes in morphology and drawn the conclusion that minor alterations of the composition may lead to major changes in the crystal growth [27].

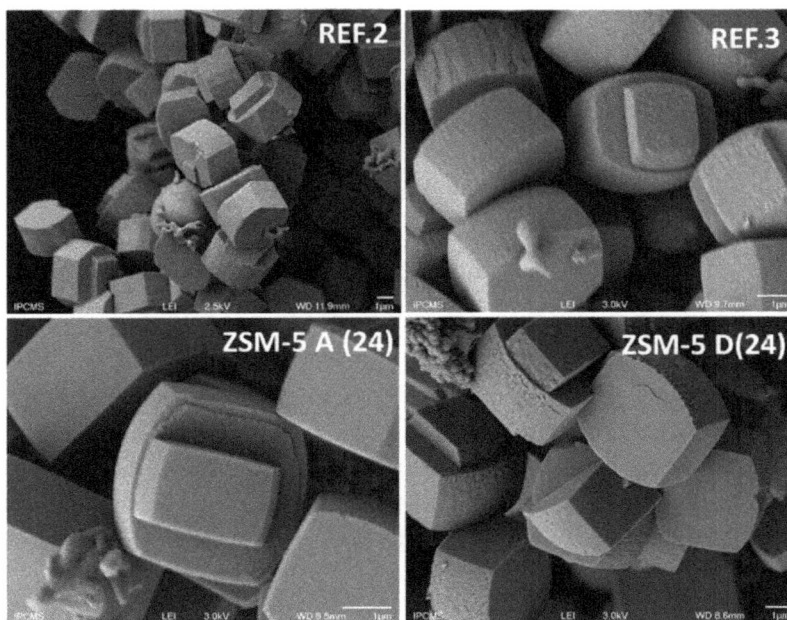

Figure 2. *Cont.*

Figure 2. SEM images (1 μm scale-bar) of zeolites synthesized for different durations.

In the ZSM-5 zeolite synthesis with biomass extract (obtained by hydrolysis in NaOH 0.5 mol/L, aged during 24 h, ZSM-5 D), conventional MFI coffin-shape crystals were formed, being larger than those observed without biomass-derived compounds (Refs. [1,2]). After extending the ageing time to 48 h (ZSM-5 E(48)) a further increase in the crystal size was observed, but the common shape for ZSM-5 crystals was still maintained. It seems however that the surface roughness diminished for ZSM-5 E(48) which may indicate a higher crystallinity. This assumption is in line with XRD data (Table 2). In contrast, after an ageing time of 72 h, ZSM-5 F(72) exhibits a similar morphology to ZSM-5 D(24) but with a high level of intergrowth. A peculiar crystal aggregation mechanism seems to occur, resulting in extensive twinning observed for longer synthesis durations. This might be due to dissolution/recrystallization phenomena, involving peculiar self-assembly processes during the crystal growth. Since zeolites are metastable materials, it is possible to shift from one structure to another one [28], or to crystallize in various morphologies by a simple change in the synthesis duration. This phenomenon was already observed for SiC substrate self-reconstruction into zeolites [29,30].

The cracking of *n*-hexane or so-called alpha test [31] is an useful tool for ranking the catalytic activity of acidic zeolites since catalytic activity is related to both the number and type of acid sites and also to the morphology and crystal organization. The *n*-hexane activity is indeed lower compared to the commercial ZSM-5 [22]

231

zeolite and remains similar to deactivated ZSM-5 [32], including the reference ZSM-5 (Ref. [1]). The ZSM-5 zeolite prepared with an extract of biomass (0.1 mol/L), samples A, B and C, exhibited similar relative crystallinity and textural properties. The rate of hexane consumption for the zeolite obtained after an ageing time of 72 h, was two-fold higher than the one achieved after 24 h ageing (Table 3). In contrast, SAR was 1.5-fold higher in the former, *i.e.* probably resulting in a lower quantity of Brønsted acid sites. ZSM-5 C(72) exhibits a high degree of twinning which led to higher exposure of straight channel pore openings. Hence, this may render easier the access to active sites and therefore speed up the cracking rate. In contrast, only minor changes could be observed in the selectivity toward propylene for all samples. Since the aim of this manuscript was to evaluate the acid properties of those zeolites the lower conversion regime was adopted therefore sequential reactions by re-conversion of primary and secondary products of *n*-hexane were avoided. As a consequence a lower effect on selectivity was observed and yet heavier compounds than n-C_6 were severely suppressed. For instance similar propane/propene ratios (as presented in Table 2) varied between 0.34–0.39 and yet the coke amount (not shown) varies in the range 0.3–0.4 wt % in spent ZSM-5.

Table 3. Data obtained in the *n*-hexane catalytic cracking reaction.

Catalyst	Conversion	Rate nC6 *	C3=	Propane/Propylene
Ref. [1]	2.1	0.78	0.26	0.35
A	1.8	0.38	0.26	0.35
B	2.7	0.57	0.28	0.39
C	3.5	0.71	0.32	0.37
D	2.9	0.60	0.31	0.36
E	3.4	0.71	0.33	0.36
F	2.1	0.43	0.28	0.34

* $\text{mmol} \cdot \text{g} \cdot \text{cat}^{-1} \cdot \text{min}^{-1}$.

Regarding, ZSM-5 D(24) to F(72) catalysts, it appears that the highest rate of *n*-hexane cracking was achieved for ZSM-5 E(48), that exhibits the highest crystallinity. In contrast, the alkane cracking rate was the lowest for ZSM-5 F(72). The latter zeolite seems to be built by different aggregation mechanism(s), since its surface remains rather heterogeneous. The self-assembly of smaller building blocks (like slabs) may generate a higher surface permeability in those smaller sub-units [33,34]. According to Karger *et al.*, these surface barriers may induce a higher resistance for the molecules to enter inside the pores in smaller crystals [33], thus reducing the reaction rate.

These results clearly demonstrate that the ZSM-5 properties (SAR, crystal size, crystallization, and morphology) are remarkably affected by the presence biomass-derived compounds that were obtained by alkaline hydrolysis of sugar-cane bagasse. These effects could be rationalized in terms of a direct interaction of the compounds present in the hydrolyzate and aluminate/silicate species. Besides, the crystal nucleus surface affects the self-assembly process and therefore the crystal growth. Moreover, an interaction between biomass derived-compounds and inorganic precursors is occurring during the ageing time (after the crystals are readily formed), probably via a continuous process of redissolution/recrystallization at longer synthesis times [35]. We highlight that this process is barely observed during ZSM-5 synthesis in the absence if biomass-derived compounds and only a moderate increase in SAR was achieved during ageing time. As aluminate species were already dissolved in the gel, the incorporation of silicates is favored by a sequential process. It is therefore expected that longer ageing times result in higher SAR values, as observed from Refs. [1–3]. The process of raising SAR can be related to both an introduction of silicates and alternatively to aluminum removal (as extra-framework aluminum species).

The molecular and supramolecular events taking place during zeolite synthesis are very complex and not yet fully understood. By taking as model compounds the acids detected by ESI-MS/MS, we can hypothesize that these phenolic acids, as well as other components present in the hydrolyzate such as carbohydrates (probably polysaccharides), may possibly complex aluminum and/or silicon. Indeed, many complexes of phenolates, cinnamic acid derivatives with aluminum were readily reported in literature [36,37]. In addition, those complexes may further react with carbohydrates. In the frame of zeolite synthesis, these complexes may interfere in the crystal organization synthesis modulating the incorporation and/or removal of T-elements (T: Al or Si) from the framework. Finally, we believe that after consolidation and understanding of all events involved in zeolite preparation with the assistance of biomass-derivate, this approach could work as a powerful tool for tailoring zeolite properties.

3. Experimental Section

Untreated sugar cane bagasse (20 g) was ground and particles in the 20–80 mesh size range were selected. The bagasse was washed using de-ionized water at 50 °C, filtered and dried in a Buchner funnel. This procedure was repeated eight times. Then, the bagasse was used in alkaline hydrolysis conditions, suspended in a 20 mL NaOH solution (either 0.1 or 0.5 mol/L). These alkaline treatments of biomass residues were conducted for 1 h under reflux. Afterwards, the suspension was filtered to remove unreacted fibers and diluted to 100 mL. Thirty mL of this solution was used for the zeolite synthesis. Prior to the ZSM-5 zeolite preparation, the pH of the biomass derived-compound solution obtained after hydrolysis with 0.1 and 0.5 mol/L of NaOH was adjusted to 12 by adding 0.1 and 0.5 mol/L HCl solution, respectively. Then 0.08 g NaAlO$_2$ and 8 mL TPAOH (20 wt % in water) were added to this solution under vigorous stirring. Likewise, 0.76 g NaCl were added, prior to a dilution in 30 mL distilled water. Finally, 6 mL TEOS were added dropwise. The solution was allowed to age for 1h at room temperature and then autoclaved at 170 °C during 24, 48 or 72 h. 1.2–1.3 g of ZSM-5 zeolites were obtained after Millipore membrane filtration. The zeolite was dried at 120 °C overnight and then calcined at 600 °C.

X-Ray diffraction (XRD) was performed using an Ultima IV diffractometer (Rigaku, Tokyo, Japan, Cu K$_\alpha$ = 0.1542 nm) at a scanning rate of 0.02 s^{-1} in 2θ ranges from 5 to 80°. A fixed power source was used (40 kV, 20 mA). The relative crystallinity of ZSM-5 phase was determined by the area of the peaks at 2θ 23.1° to 24.1°. The crystallinity was arbitrarily set to 100% for pristine reference zeolites (1 and 2) exhibiting the same crystallinity. Scanning electron microscopy (SEM) micrographs were acquired on a FEG 6700F microscope (JEOL, Tokyo, Japan) working at a 9 kV accelerating voltage. X-Ray fluorescence was carried out to estimate the SAR using a Magic X instrument (Phillips, Almalo, The Netherlands). The textural properties were evaluated using n-hexane adsorption as presented elsewhere [38]. ZSM-5 samples were placed in an 85 µL crucible and analyzed by an Iris TG 209 F1 thermal analyzer (Netzsch, Selb, Germany). Firstly, the sample was pre-treated to remove water (this step comprises heating in N$_2$ flow (60 mL· min^{-1}) from room temperature to 600 °C at 10°· min^{-1} plus 30 min kept at final temperature. Secondly, the sample was cooled to 160 °C. Finally, n-hexane adsorption was carried out continuously, followed by weight variation, during cooling from 160 °C to 40 °C at 4°· min^{-1} under a flow of 8% of n-hexane in N$_2$. Silicon-to-aluminum ratios were determined by XRF.

The catalytic cracking of n-hexane (99.89% pure, water 0.005 wt % no volatile residue 0.001% wt/wt from (VETEC/ Sigma-Aldrich Chemistry Ltda., St. Louis, MO, USA) was conducted in a high-throughput unit. Eight reactors in parallel (187 mm in length and 6 mm of internal diameter) made of quartz, with the catalyst localized 90 mm below the top of the reactor, were placed in the middle part of an oven (400 mm external diameter). The temperature was measured by means of a thermocouple located in the center of each reactor and differed less than two degrees between the reactors. All experiments were performed at atmospheric pressure in a continuous down flow fixed-bed micro-reactor. Prior to catalyst evaluation, all the materials were simultaneously heated in nitrogen from room temperature to the reaction temperature (500 or 600 °C) at 10°/min. Then, nitrogen was fed through a saturator containing n-hexane at 20 °C to the reactors. This procedure ensured a 60 mL/min of 11% v/v n-hexane flow in nitrogen. Since the catalysts were evaluated sequentially, the activation time (at the final temperature) is different for each reactor. This difference did not affect activity and selectivity results. A typical run was carried out with 0.1 to 0.03 g of catalyst (the amounts of catalyst were adjusted in order to provide iso conversion conditions). Reaction products were analyzed on-line after three different times on stream (3, 17, and 32 min) by gas chromatography using a GC-2010 apparatus (Shimadzu, Tokyo, Japan) equipped with a Chrompack KCl/Al_2O_3 column operated under isothermal conditions (303 K) and nitrogen flow (2 mL/min). Catalytic activity and selectivity were estimated as an average of results obtained after 17 and 32 min on-stream. These values differed by less than 10% and 2% for activity and selectivity, respectively (each catalyst was analyzed three times). Propylene selectivity was estimated as percentage in *wt/wt* divided by the total conversion in *wt/wt* and the alkene/alkane ration was estimated as the total alkenes formation in *wt/wt* % referred to total alkane formation in *wt/wt* %. The catalysts showed a moderated to low deactivation (the catalytic activity at 3 min on time on stream differs less than 15%, when compared to the one measured after 32 min).

4. Conclusions

Pure ZSM-5 zeolite phases were successfully synthesized in the presence of alkaline solutions containing biomass-derived compounds. The presence of these compounds remarkably affected the silicon/aluminum ratio varying from 25 to 150, compared with SAR obtained in the absence of biomass-derived compounds (from 27 to 37). Moreover ZSM-5 crystals with different crystal sizes as well as peculiar morphologies were obtained.

The solutions containing biomass-derived compounds were obtained by alkaline hydrolysis treatment under reflux of second-generation biomass; therefore this procedure can be easily implemented in regular ZSM-5 synthesis and further extended to other zeolite preparation. Similar cracking activities and selectivities were achieved over as-prepared biomass-derived zeolites with respect to parent ZSM-5 zeolites.

Acknowledgments: Alessandra V. Silva is grateful to the Programa de Pós Graduação em Tecnologia de Processos Químicos e Bioquímicos da Escola de Química.

Author Contributions: MMP and BL have developed this research area. BL proposed the concept and initiated the zeolite syntheses at the University of Strasbourg. MMP shared his experience in biomass valorization and wrote the main part of the paper. AVS performed the ZSM-5 zeolite preparations and most of the characterizations; LSMM and MN performed the thorough characterization of biomass components by LC-MS-MS techniques.

Conflicts of Interest: The authors declare no conflict of interest.

References

1.   Donnelly, S.P.; Mizrahi, S.; Sparrell, P.T.; Huss, A.J.; Schipper, P.H.; Herbst, J.A. How ZSM-5 works in FCC. In Proceedings of ACS Meeting, New Orleans, LA, USA, 30 August–4 September 1987; 1987; pp. 621–626.

2. Buchanan, J.S. The chemistry of olefins production by ZSM-5 addition to catalytic cracking units. *Catal. Today* **2000**, *55*, 207–212.

3. Burch, R.; Scire, S. Selective catalytic reduction of nitric oxide with ethane and methane on some metal exchanged ZSM-5 zeolites. *Appl. Catal. B* **1994**, *3*, 295–318.

4. Zakaria, Z.Y.; Linnekoski, J.; Amin, N.A.S. Catalyst screening for conversion of glycerol to light olefins. *Chem. Eng. J.* **2012**, *207–208*, 803–813.

5. Chang, C.D. Methanol conversion to light olefins. *Catal. Rev.* **1984**, *26*, 323–345.

6. Sassi, A.; Sommer, J. Isomerization and cracking reactions of branched octanes on sulfated zirconia. *Appl. Catal. A* **1999**, *188*, 155–162.

7. Louis, B.; Pereira, M.M.; Santos, F.M.; Esteves, P.M.; Sommer, J. Alkane activation over acidic zeolites: The first step. *Chem. Eur. J.* **2010**, *16*, 573–576.

8. Olah, G.A.; Molnar, A. *Hydrocarbon Chemistry*, 2nd ed.; Wiley Interscience: New York, NY, USA, 2003.

9. Lønstad Bleken, F.; Chavan, S.; Olsbye, U.; Boltz, M.; Ocampo, F.; Louis, B. Conversion of Methanol into light olefins over ZSM-5 zeolite: Strategy to enhance propene selectivity. *Appl. Catal. A* **2012**, *447–448*, 178–185.

10. Mann, S. The chemistry of form. *Angew. Chem. Int. Ed.* **2000**, *39*, 3392–3406.

11. Feuillie, C.; Sverjensky, D.A.; Hazen, R.M. Attachment of ribonucleotides on α-alumina as a function of pH, ionic strength, and surface loading. *Langmuir* **2015**, *31*, 240–248.

12. Fraser, D.G.; Fitz, D.; Jakschitz, T.; Rode, B.M. Selective adsorption and chiral amplification of amino acids in vermiculite clay-implications for the origin of biochirality. *Phys. Chem. Chem. Phys.* **2011**, *13*, 831–838.

13. Cundy, C.S.; Cox, P.A. The hydrothermal synthesis of zeolites: Precursors, intermediates and reaction mechanism. *Microporous Mesoporous Mater.* **2005**, *82*, 1–78.

14. Ocampo, F.; Cunha, J.A.; de Lima Santos, M.R.; Tessonnier, J.P.; Pereira, M.M.; Louis, B. Synthesis of zeolite crystals with unusual morphology: Application in acid catalysis. *Appl. Catal. A* **2010**, *390*, 102–109.

15. Cardoso, C.S.; Licea, Y.E.; Huang, X.; Willinger, M.; Louis, B.; Pereira, M.M. Improving textural properties of γ-alumina by using second generation biomass in conventional hydrothermal method. *Microporous Mesoporous Mater.* **2015**, *207*, 134–141.

16. Chou, Y.H.; Cundy, C.S.; Garforth, A.A.; Zholobenko, V.L. Mesoporous ZSM-5 catalysts: Preparation, characterisation and catalytic properties. Part I: Comparison of different synthesis routes. *Microporous Mesoporous Mater.* **2006**, *89*, 78–87.

17. Magusin, P.C.M.M.; Zorin, V.E.; Aerts, A.; Houssin, C.J.Y.; Yakovlev, A.L.; Kirschhock, C.E.A.; Martens, J.A.; van Santen, R.A. Template-aluminosilicate structures at the early stages of zeolite ZSM-5 formation. A combined preparative, solid-state NMR, and computational study. *J. Phys. Chem. B* **2005**, *109*, 22767–22774.

18. Burkett, S.L.; Davis, M.E. Mechanism of structure direction in the synthesis of Si-ZSM-5: An investigation by intermolecular ^{1}H-^{29}Si CP MAS NMR. *J. Phys. Chem.* **1994**, *98*, 4647–4653.

19. Larsen, S.C. Nanocrystalline zeolites and zeolite structures: Synthesis, characterization, and applications. *J. Phys. Chem. C* **2007**, *111*, 18464–18474.

20. Cundy, C.S.; Cox, P.A. The hydrothermal synthesis of zeolites: History and development from the earliest days to the present time. *Chem. Rev.* **2003**, *103*, 663–702.

21. Ou, S.Y.; Luo, Y.L.; Huang, C.H.; Jackson, M. Production of coumaric acid from sugarcane bagasse. *Innovative Food Sci. Emerg. Technol.* **2009**, *10*, 253–259.

22. Maia, A.J.; Louis, B.; Lam, Y.L.; Pereira, M.M. Ni-ZSM-5 catalysts: detailed characterization of metal sites for proper catalyst design. *J. Catal.* **2010**, *269*, 103–109.

23. Xue, T.; Chen, L.; Wang, Y.M.; He, M.-Y. Seed-induced synthesis of mesoporous ZSM-5 aggregates using tetrapropylammonium hydroxide as single template. *Microporous Mesoporous Mater.* **2012**, *156*, 97–105.

24. Brandenberger, S.; Kröcher, O.; Casapu, M.; Tissler, A.; Althoff, R. Hydrothermal deactivation of Fe-ZSM-5 catalysts for the selective catalytic reduction of no with NH_3. *Appl. Catal. B* **2011**, *101*, 649–659.

25. Le van Mao, R.; Le, S.T.; Ohayon, D.; Caillibot, F.; Gelebart, L.; Denes, G. Modification of the micropore characteristics of the desilicated ZSM-5 zeolite by thermal treatment. *Zeolites* **1997**, *19*, 270–278.

26. Louis, B.; Rocha, C.C.; Balanqueux, A.; Boltz, M.; Losch, P.; Bernardon, C.; Bénéteau, V.; Pale, P.; Pereira, M.M. Unraveling the importance of zeolite crystal morphology. *L'Actualité Chim.* **2015**, *393–394*, 108–111.

27. Petrik, L. The influence of cation, anion and water content on the rate of formation and pore size distribution of zeolite ZSM-5. *S. Afr. J. Sci.* **2009**, *105*, 251–257.

28. Ocampo, F.; Yun, H.; Pereira, M.M.; Tessonnier, J.P.; Louis, B. Design of MFI zeolite-based composites with hierarchical pore structure: a new generation of structured catalysts. *Cryst. Growth Des.* **2009**, *9*, 3721–3729.

29. Ivanova, S.; Louis, B.; Ledoux, M.J.; Pham-Huu, C. Auto-Assembly of Nanofibrous Zeolite Crystals via Silicon Carbide Substrate Self-Transformation. *J. Am. Chem. Soc.* **2007**, *129*, 3383–3391.

30. Losch, P.; Boltz, M.; Soukup, K.; Song, I.-H.; Yun, H.S.; Louis, B. Binderless zeolite coatings on macroporous alpha-SiC foams. *Microporous Mesoporous Mater.* **2014**, *188*, 99–107.

31. Miale, J.N.; Chen, N.Y.; Weisz, P.B. Catalysis by crystalline aluminosilicates: Attainable catalytic cracking rate constants, and superactivity. *J. Catal.* **1966**, *6*, 278–287.

32. Wallenstein, D.; Kanz, B.; Haas, A. Influence of coke deactivation and vanadium and nickel contamination on the performance of low ZSM-5 levels in fcc catalysts. *Appl. Catal. A* **2000**, *192*, 105–123.

33. Tzoulaki, D.; Heinke, L.; Lim, H.; Li, J.; Olson, D.; Caro, J.; Krishna, R.; Chmelik, C.; Karger, J. Assessing Surface Permeabilities from Transient Guest Profiles in Nanoporous Host Materials. *Angew. Chem. Int. Ed.* **2009**, *48*, 3525–3528.

34. Karwacki, L.; Kox, M.H.F.; de Winter, D.A.M.; Drury, M.R.; Meeldijk, J.D.; Stavitski, E.; Schmidt, W.; Mertens, M.; Cubillas, P.; John, N.; *et al.* Morphology-dependent zeolite intergrowth structures leading to distinct internal and outer-surface molecular diffusion barriers. *Nature Mater.* **2009**, *8*, 959–965.

35. Louis, B.; Laugel, G.; Pale, P.; Pereira, M.M. Rational design of microporous and mesoporous solids for catalysis: From the molecule to the reactor. *ChemCatChem* **2011**, *3*, 1263–1272.

36. Shaw, A.J.; Tsao, G.T. Isomerization of D-glucose with sodium aluminate: mechanism of the reaction. *Carbohydr. Res.* **1978**, *60*, 327–325.

37. Lambert, J.B.; Lu, G.; Singer, S.R.; Kolb, V.M. Silicate complexes of sugars in aqueous solution. *J. Am. Chem. Soc.* **2004**, *126*, 9611–9625.

38. Lowen, J.; Jonker, R. Method of analyzing microporous material. US Patent N 60/261920, 15 January 2001.

Adsorption and Diffusion of Xylene Isomers on Mesoporous Beta Zeolite

Aixia Song, Jinghong Ma, Duo Xu and Ruifeng Li

Abstract: A systematic and detailed analysis of adsorption and diffusion properties of xylene isomers over Beta zeolites with different mesoporosity was conducted. Adsorption isotherms of xylene isomers over microporous and mesoporous Beta zeolites through gravimetric methods were applied to investigate the impact of mesopores inside Beta zeolites on the adsorption properties of xylene isomers in the pressure range of lower 20 mbar. It is seen that the adsorption isotherms of three xylene isomers over microporous and mesoporous Beta zeolites could be successfully described by the single-site Toth model and the dual-site Toth model, respectively. The enhanced adsorption capacities and decreased Henry's constants (K_H) and the initial heats of adsorption (Q_{st}) for the all xylene isomers are observed after the introduction of mesopores in the zeolites. For three xylene isomers, the order of Henry's constant is o-xylene > m-xylene > p-xylene, whereas the adsorption capacities of Beta zeolite samples for xylene isomers execute the following order of o-xylene > p-xylene > m-xylene, due to the comprehensive effects from the molecular configuration and electrostatic interaction. At the same time, the diffusion properties of xylene isomers in the mesoporous Beta zeolites were also studied through the desorption curves measured by the zero length column (ZLC) method at 333–373 K. It turned out that the effective diffusion time constant (D_{eff}/R^2) is a growing trend with the increasing mesoporosity, whereas the tendency of the activation energy is just the reverse, indicating the contribution of mesopores to facilitate molecule diffusion by shortening diffusion paths and reducing diffusion resistances. Moreover, the diffusivities of three xylene isomers in all Beta zeolites follow an order of p-xylene > m-xylene > o-xylene as opposed to K_H, conforming the significant effects of adsorbate-adsorbent interaction on the diffusion.

Reprinted from *Catalysts*. Cite as: Song, A.; Ma, J.; Xu, D.; Li, R. Adsorption and Diffusion of Xylene Isomers on Mesoporous Beta Zeolite. *Catalysts* **2015**, *5*, 2098–2114.

1. Introduction

Zeolites are typical microporous crystalline materials with narrow pore size distribution ranging from 0.3 to 1.5 nm, having a widespread application in adsorption, separation, and catalysis with its unique properties, such as developed ordered porous structure, high specific surface area, and strong acidity, high hydrothermal and thermal stability [1–3]. Adsorption and diffusion of molecules in zeolites are very significant performances for understanding these processes of zeolite applications, especially the catalytic performance of chemical reaction over zeolites. However, adsorption and diffusion limitations are often observed when the processes involve large molecules [4]. To alleviate this, hierarchical porous materials have attracted a wide spread attention with the advantages of enhancing the mass transfer and improving the accessibility of the molecules in intracrystalline. There are different approaches to synthesize hierarchical zeolites. Among which, introducing mesopores into the original structure of microporous zeolites is one of the most efficient methods [5–7]. Such mesostructured zeolite materials have not only the advantages of micropores, which ensure acidity and shape selectivity, but also the advantages of mesopores, which promote the adsorption of molecules [8–10] and diffusion rate for mass transfer within zeolites [8,11–13].

As zeolite Beta has large window diameter and strong acidic sites, it has become a member of zeolitic materials of great substantial industrial importance, displaying unique catalytic reactivity in chemical reactions of hydrocarbons. Beta zeolite belongs to a complex intergrowth family, which consists of two (polymorph A and B) or more polymorphs [14]. It is a high-silica zeolite with three-dimension channel system and consists of straight ring channels of a free aperture of 0.76×0.64 nm and zig-zag ring channels of 0.55×0.55 nm which are similar to other large-pore molecular sieves such as FAU and EMT zeolites. Even Beta zeolite can be classified as a large pore zeolite with a 12-ring structure, it still suffers from strong steric and diffusion limitations from crystal sizes in the micrometer range in a number of catalytic reactions with participation of bulky molecules [15]. Thus, it is of prime importance to develop mesoporous Beta zeolite which can improve the catalytic activities substantially in various organic reactions involving large molecules compared with microporous Beta zeolites.

Three xylene isomers, namely *o*-xylene, *m*-xylene and *p*-xylene, are very important basic chemicals and find wide and various applications in chemical industry. Mixed xylenes are produced from petroleum by various catalytic reactions. However, as a result of important application to individual xylene isomers, it is necessary that mixed xylene is separated out through adsorption separation and isomerization reaction processes to obtain desired xylene isomers. In fact, the remarkable advances in *p*-xylene production from *m*-xylene isomerization using zeolite catalyst have been achieved over the last several decades [16–20]; meanwhile, the separation of xylene isomers through zeolite adsorbents has also gained considerable interest. The great success of zeolites as catalysts and adsorbents is exactly due to their unique shape selective properties from pore structure [21,22]. Shape-selective conversion and separation of xylenes over zeolites can be attributed the differences in the adsorptions and diffusivities of the three xylene isomers in the zeolite structures.

In preceding articles, we have used an organo-silylated nano-SiO$_2$ as silicon source to synthesize mesoporous Beta zeolite with intracrystalline mesopores. In consideration of the essentiality for studying the adsorption and mass transport of molecules in the pores of nanoporous materials, where sorption and reaction sites are located, to understand catalytic and separation processes of organic molecules, in the present study, three xylene isomers were employed as probe molecules to explore the adsorption and diffusion behaviors of xylene isomers molecules on the mesoporous Beta zeolites by their adsorption isotherms and ZLC desorption curves. The impacts of mesopore in Beta zeolites on the adsorption and diffusion of xylene isomers with different molecular configuration will also be investigated.

2. Results and Discussion

2.1. Characterization

Figure 1 presents the X-ray powder diffraction (XRD) patterns of the four Beta zeolite samples. It is seen that the samples exhibit the diffraction peaks around 7.8[101] and 22.4[302], which are the characteristic peaks of BEA topology structure [1]. Compared to zeolite Beta-0, a slight decrease in the intensity of most peaks and broader reflections were observed for the mesoporous Beta zeolite samples. Nitrogen adsorption/desorption isotherms at 77 K on the four Beta zeolite samples and corresponding density functional theory (DFT) pore size distributions are displayed in Figure 2. Beta-0 sample exhibits a typical type I isotherm according to the International Union of Pure and Applied Chemistry (IUPAC) classification [23], reflecting solely microporous structure. However, the Beta-1, -2, -3 samples represent a steep uptake caused by the micropores filled at the low relative pressure region of $p/p_0 < 0.01$, followed by the gradual and continual rising curves and the hysteresis loops, indicating distinctly the presence of both micropores and mesopores. The pore sizes of the three mesoporous Beta zeolite samples mainly center in the range of 2–6 nm from the DFT pore size distribution curves. Table 1 listed the pore structural parameters of four Beta zeolite samples based on N$_2$ adsorption/desorption isotherms. Mesoporous Beta-1, Beta-2, and Beta-3 possess an increasing sequence for mesopore volumes and external surface areas. Further, the typical scanning electron microscopy (SEM) and transmission electron microscope (TEM) images of the mesoporous Beta-3 zeolite sample are displayed in Figure 3. It is seen that the zeolite consists of particles with dimensions in the range of 400–600 nm made up of nanocrystals with sizes smaller than 100 nm. These images also show mesoporous voids between the primary particles. Moreover, as showed in the TEM images, intracrystalline mesopores and lattice fringes of zeolite crystalline are well resolved in large scale through the zeolite particles, indicating the preservation of microporous structures and the presence of additional porosity within the crystals.

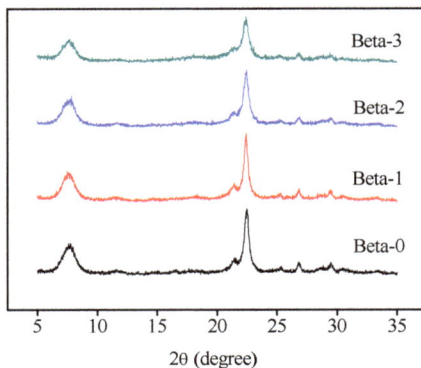

Figure 1. XRD patterns of Beta zeolite samples.

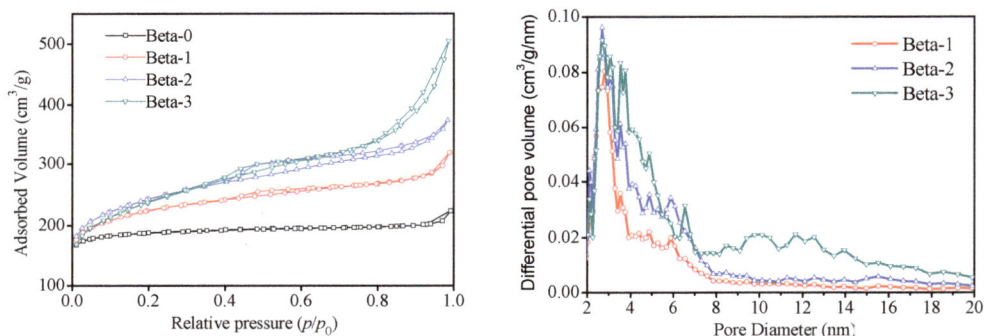

Figure 2. Nitrogen adsorption/desorption isotherms at 77 K and corresponding DFT pore size distributions of Beta zeolite samples.

Table 1. Pore structure parameters of Beta zeolite samples from N_2 adsorption isotherms.

Sample	S_{BET} (m²/g)	S_{mic}[b] (m²/g)	S_{ext}[b] (m²/g)	V_{mic}[b] (cm³/g)	V_{meso}[c] (cm³/g)
Beta-0	743	681	62	0.27	0.06
Beta-1	814	591	223	0.25	0.18
Beta-2	879	539	340	0.23	0.29
Beta-3	851	392	459	0.17	0.51

[a] Determined by the BET method. [b] Obtained from the t-plot method, S_{mic} is micropore surface area; S_{ext} is external surface area, V_{mic} is micropore volume, [c] V_{meso} is mesopore volume (=V_{total}(total pore volume) -V_{mic}).

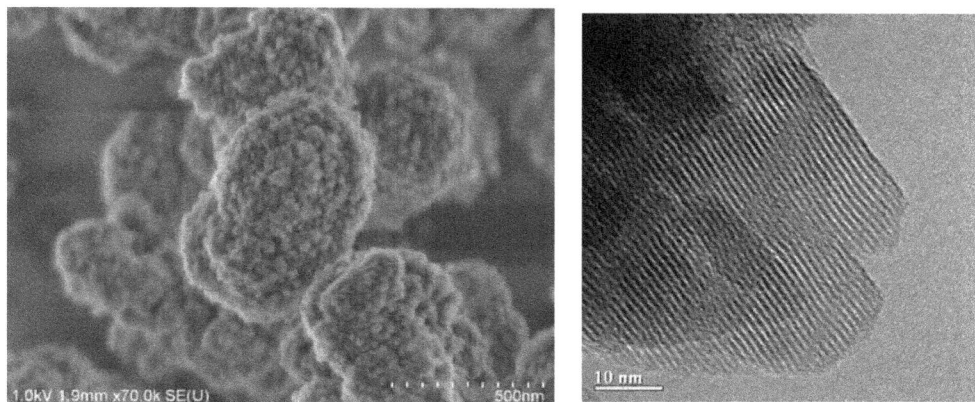

Figure 3. SEM/TEM of mesoporous Beta-3 zeolite sample.

2.2. Adsorption Results

2.2.1. Adsorption Equilibrium Isotherms

Representative adsorption equilibrium isotherms of three xylene isomers on the microporous Beta-0 and the mesoporous Beta-1,-2,-3 zeolite samples at 308 K are shown in Figure 4. It is observed that the isotherms of three xylene isomers on the Beta-0 display a sharp increase of adsorption capacities at low pressure region corresponding to micropore filling and then turn out to be a plateau at relative high pressures, which represent the typical characteristics of type-I adsorption isotherm. Compared to Beta-0, the adsorption isotherms of three xylene isomers on Beta-1,-2,-3 zeolites show a similar shape at first, however, the adsorption amount has a continual increase instead of a plateau as the pressure increases to relative higher pressure. This observation consists with the results of the N_2 adsorption/desorption isotherms, indicating again the presence of mesopores in Beta-1,-2,-3 zeolites. Moreover, the increase of adsorption capacities with pressure on four Beta zeolite samples follows, obviously, the order of Beta-0 < Beta-1 < Beta-2 < Beta-3, which can be reasonably ascribed to the increase of the mesopore volume (V_{meso}) caused by newly created mesopores.

2.2.2. Modeling of Equilibrium Isotherms

The single-site Toth model and the dual-site Toth model were used to fit the adsorption equilibrium data for Beta-0 and Beta-1,-2,-3 zeolites, respectively. As illustrated in Figure 4, there is a good agreement between the symbols of experimental data and the solid lines obtained from the fitting models. All the fitting parameters of three xylene isomers on the four Beta zeolites are summarized in Tables 2 and 3.

243

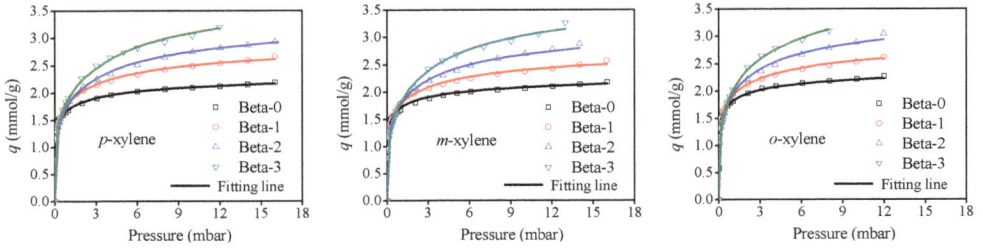

Figure 4. Adsorption isotherms of xylene isomers on Beta zeolite samples at 308 K.

Table 2. Toth fitting parameters of xylene isomers on Beta-0.

T (K)	q_m (mmol·g^{-1})	b (mbar^{-1}·10^3)	n	R^2
		m-xylene		
308	2.74	4.93	0.25	0.9945
323	2.60	2.59	0.25	0.9815
338	2.40	1.20	0.27	0.9782
		p-xylene		
308	2.79	2.20	0.26	0.9781
323	2.64	1.88	0.26	0.9913
338	2.44	0.99	0.27	0.9828
		o-xylene		
308	2.90	6.83	0.24	0.9894
323	2.75	2.95	0.25	0.9880
338	2.59	1.94	0.25	0.9888

The single-site Toth model can be expressed as follows [24]:

$$q = \frac{q_m b p}{\left(1 + (bp)^n\right)^{1/n}} \tag{1}$$

where q is the equilibrium adsorbed amount, q_m is the representation of the saturation adsorption capacity, b-value reflects the interaction between adsorbate and adsorbent, n characterizes the surface heterogeneity. The farther n is from 1, more heterogeneous is the system. When n is equal to 1, the Toth model becomes the Langmuir model. Due to the change of adsorption environment for absorbate on mesoporous zeolites, the Toth isotherm model does not fit any more. In order to distinguish two categories of adsorption sites in the mesoporous Beta zeolite, the dual-site Toth model can be employed as follows:

$$q = \frac{q_{m_1} b_1 p}{\left(1 + (b_1 p)^{n_1}\right)^{1/n_1}} + \frac{q_{m_2} b_2 p}{\left(1 + (b_2 p)^{n_2}\right)^{1/n_2}} \tag{2}$$

244

where the meaning of q_m, b, n is the same with that in above Toth model. The q_{m1}, b_1, and n_1 characterize parameters of adsorption site I, which represent the adsorption sites located inside the micropores, while q_{m2}, b_2, n_2 characterize parameters of adsorption site II which represent the adsorption sites located inside mesopores.

Table 3. Dual-site Toth fitting parameters of xylene isomers on Beta-1,-2,-3.

Sample	Sorbate	T (K)	q_{m1} (mmol·g^{-1})	b_1 (mbar^{-1}·10^3)	n_1	q_{m2} (mmol·g^{-1})	b_2 (mbar^{-1})	n_2	R^2
Beta-1	m-xylene	308	1.98	4.53	0.35	1.13	1.37	0.49	0.9865
		323	1.80	2.24	0.31	1.01	0.92	0.53	0.9960
		338	1.67	0.98	0.32	0.90	0.36	0.63	0.9949
	p-xylene	308	2.02	1.98	0.30	1.21	0.91	0.58	0.9948
		323	1.87	1.65	0.32	1.10	0.42	0.55	0.9979
		338	1.76	0.75	0.31	0.98	0.17	0.76	0.9976
	o-xylene	308	2.13	6.52	0.35	1.34	1.68	0.41	0.9969
		323	1.91	2.49	0.30	1.25	1.32	0.44	0.9982
		338	1.80	1.61	0.31	1.05	0.66	0.49	0.9969
Beta-2	m-xylene	308	1.64	4.21	0.32	1.78	0.54	0.70	0.9979
		323	1.49	1.97	0.38	1.66	0.22	0.82	0.9973
		338	1.36	0.54	0.38	1.51	0.18	0.73	0.9964
	p-xylene	308	1.69	1.67	0.44	1.85	0.50	0.67	0.9992
		323	1.55	1.18	0.39	1.67	0.19	0.86	0.9981
		338	1.44	0.38	0.38	1.52	0.12	0.75	0.9973
	o-xylene	308	1.74	4.70	0.38	1.97	0.88	0.59	0.9951
		323	1.63	2.14	0.32	1.83	0.42	0.64	0.9971
		338	1.51	1.21	0.35	1.73	0.27	0.59	0.9974
Beta-3	m-xylene	308	1.35	3.64	0.35	2.53	0.45	0.78	0.9980
		323	1.20	1.36	0.50	2.40	0.19	0.99	0.9983
		338	1.05	0.27	0.53	2.25	0.12	0.99	0.9992
	p-xylene	308	1.41	0.77	0.49	2.63	0.40	0.73	0.9985
		323	1.26	0.46	0.51	2.50	0.16	0.81	0.9982
		338	1.13	0.19	0.58	2.36	0.09	0.91	0.9984
	o-xylene	308	1.51	3.92	0.34	2.95	0.68	0.59	0.9989
		323	1.40	1.76	0.37	2.77	0.25	0.72	0.9992
		338	1.27	0.63	0.40	2.63	0.16	0.68	0.9990

Table 2 shows that the saturation adsorption capacities q_m of microporous sample Beta-0 at each temperature follow an order of m-xylene < p-xylene < o-xylene. The order of b-values is p-xylene < m-xylene < o-xylene. Besides, the values of n, which characterize the system heterogeneity, are comparable. For the mesoporous Beta-n zeolite samples, as listed in Table 3, q_{m1}-values of three adsorbates decrease with the reduction of microporous volume in the order: Beta-1 > Beta-2 > Beta-3 while q_{m2}-values have a growing trend with the increase of mesoporosity as expected. In addition, the total adsorption amounts of our samples at relative higher pressure are as follows: Beta-0 < Beta-1 < Beta-2 < Beta-3, revealing a fact that the presence of mesopores provides more space for adsorption and it is helpful for the increase of adsorption capacity. It is easy to be seen that the values of b_1, b_2 derived from the dual-site Toth model for Beta-1,-2,-3 are smaller than b derived from the Toth model for Beta-0 and both b_1 and b_2 decrease with the increase of mesoporosity. Moreover, the b_1-values are much greater than b_2, which is similar to the result of mesoporous ZSM-5 [8], reflecting a weaker adsorbate-adsorbent interaction in mesopores than in micropores.

With regard to the saturate adsorption capacities (q_m and q_{m1}, q_{m2}) of three xylene isomers, the order is: o-xylene > p-xylene > m-xylene. The reason why o-xylene has a greater adsorption amount can be divided into two characters. On the one hand, the packing configuration of o-xylene molecules, called "face-to-face" packing orientation, can pack much more closely in the channel of Beta zeolite samples than p-xylene molecules which have a "shoulder to shoulder" packing orientation [25,26]. This may contribute to the higher saturation capacities of o-xylene than p-xylene. On the other hand, the higher ortho adsorption also corresponds to the bigger b-values, which reflect the interaction between adsorbate and adsorbent. As listed in Tables 2 and 3, the values of b for Beta-0 and b_1, b_2 for Beta-1,-2,-3 follow the same order of o-xylene > m-xylene > p-xylene. The result shows that o-xylene with a dipole moment interacts more strongly with the electric field of Beta zeolite than p-xylene without dipole moment. Thus, the ortho adsorption is further enhanced by electrostatic effects [27]. However, as to m-xylene, the electrostatic interaction seems to be only of secondary importance in the adsorption. The smallest adsorption capacity of m-xylene, which is inconsistent with the order of interactions, indicates that molecular configuration and size become the dominant factors in the adsorption of m-xylene on Beta zeolites. Apparently, there is only a slight difference in adsorption amounts between p-xylene and m-xylene, reflecting the combined effects of molecular configuration and electrostatic interaction.

2.2.3. Henry's Constants and Initial Heats of Adsorption

Henry's constants (K_H) and the initial heats of adsorption, which are the characterizations of interaction force between the adsorbate and adsorbent at very low adsorption pressure, were calculated from the isotherm data in this work. According to the Virial equation:

$$\ln(p/q) = -\ln(K_H) + A_1 q + A_2 q^2 + A_3 q^3 + \dots \qquad (3)$$

where $A_1, A_2, A_3, \dots$ are the virial coefficients. The plot of $\ln(p/q)$ as a function of q should approach a line while the adsorption amount q approximates to zero. The Henry's constant (K_H) could be directly calculated from the intercept $-\ln(K_H)$ [28,29].

Moreover, the heat of adsorption at zero coverage Q_{st} were obtained from 1/T dependence of $\ln(K_H)$ in terms of van't Hoff expression [30,31]:

$$\ln K_H = -\frac{Q_{st}}{RT} + C \qquad (4)$$

The results are listed in Table 4. It is seen that Henry's constants and the initial heats of adsorption for three xylene isomers in the four samples follow the same order of Beta-3 < Beta-2 < Beta-1 < Beta-0, indicating the weaker interaction between the adsorbate and adsorbent in mesoporous Beta zeolites. The difference in values of K_H between microporous Beta-0 and mesoporous Beta-1,-2,-3 seems to be remarkable, corresponding to the variation of b-values and b_1-values discussed above.

The initial heats of adsorption for three xylene isomers in Beta-0, ranging from 55.5 kJ/moL for p-xylene to 69.5 kJ/moL for o-xylene, were higher than the values of 53.6 kJ/moL for 2,2-dimethylbutane and 63.4 kJ/moL for n-hexane in Beta zeolite [32]. The values of Henry's constants and the initial heats of adsorption for three xylene isomers go in order of p-xylene < m-xylene < o-xylene, which is accordance with the results of b and b_1. This can be explained by the dipole moment (D) of p-xylene, m-xylene and o-xylene, which are 0, 0.33 and 0.62, respectively. The non-polar and the smallest molecule kinetic diameter of p-xylene result in the smallest Henry's constant value among three xylene isomers.

2.3. Diffusion Results

ZLC diffusion experiments of xylene isomers (p-xylene, m-xylene and o-xylene) on the four Beta zeolites with different mesoporosity were carried out at the high flow rate of 80 mL/min and temperatures from 333 K to 373 K. The partial pressure of three xylene isomers was maintained at 0.04 mbar, which are low enough to ensure the experiments were carried out within the Henry's law region. The fact that the system is controlled by kinetics was verified by measuring ZLC desorption curves of xylene isomers at two different flow rates under the same conditions. It turned out that two desorption curves have similar long time slopes, *i.e.*, they are nearly parallel to each other in the tails [33]. The experimental desorption plots were fitted using Equations (5)–(7) to obtain the fitting parameters, L, β and D_{eff}/R^2. Representative experimental and theoretical ZLC curves of xylene isomers are showed in Figure 5. A good agreement is observed between the symbols represented experimental data and the solid lines obtained from fitting ZLC theoretical model with the goodness of fit $R^2 > 0.98$.

The correctness and effectiveness of the method was demonstrated by the above points.

Table 4. Henry's constants (K_H) and heats of adsorption at zero coverage (Q_{st}) for xylene isomers on Beta zeolite samples.

Sample	Sorbates	T (K)	K_H (mmol/g mbar·10^3)	Q_{st} (kJ/moL)
Beta-0	p-xylene	308	10.9	55.5
		323	4.35	
		338	1.53	
	m-xylene	308	13.4	66.1
		323	4.38	
		338	1.68	
	o-xylene	308	17.8	69.5
		323	4.94	
		338	1.77	
Beta-1	p-xylene	308	0.84	43.5
		323	0.41	
		338	0.19	
	m-xylene	308	1.28	55.0
		323	0.47	
		338	0.21	
	o-xylene	308	1.37	56.1
		323	0.51	
		338	0.22	
Beta-2	p-xylene	308	0.11	35.1
		323	0.06	
		338	0.04	
	m-xylene	308	0.19	42.8
		323	0.09	
		338	0.05	
	o-xylene	308	0.31	43.6
		323	0.15	
		338	0.07	
Beta-3	p-xylene	308	0.04	25.5
		323	0.03	
		338	0.02	
	m-xylene	308	0.08	31.0
		323	0.05	
		338	0.03	
	o-xylene	308	0.15	33.2
		323	0.07	
		338	0.05	

From the desorption curves displayed in Figure 5, it is seen, obviously, that, under the same experimental conditions, the time spent for desorption to a certain c/c_0 reduce with the increase of mesoporosity in Beta zeolites, reflecting the enhancement of the speed of desorption by introduction of mesopores inside zeolites. For three different xylene isomers adsorbates, the desorption rates are in the order of p-xylene > m-xylene > o-xylene. A summary of the effective diffusion time constants (D_{eff}/R^2) with parameters L and β extracted from the ZLC model fittings corresponding to Figure 5 are listed in Table 5. It is seen that the fitted L values, which were used as a criterion for the ZLC model to guarantee that the desorption process is controlled by the kinetics [34,35], are all greater than 5. The effective diffusivity values (D_{eff}/R^2) of all sorbates are of the same magnitude around $10^{-4} \cdot s^{-1}$, which is in reasonably good agreement with the reported results [36,37]. Figure 6 reveals the variation of the effective diffusion time constants for p-xylene, m-xylene and o-xylene with the mesopore volume at 333 K. It can be seen that there is a similar trend for all adsorbates that the values of D_{eff}/R^2 approximate linearly increase with mesopore volumes of the Beta zeolites. The values of effective diffusion time constants D_{eff}/R^2 for three xylene isomers increase with the mesoporosities of the Beta zeolites at the measured temperature. This confirms that the molecular transport in mesoporous Beta zeolites is obviously enhanced. In addition, the values of D_{eff}/R^2 for the three xylene isomers in the four Beta zeolite samples at 373 K is in the following order: p-xylene > m-xylene > o-xylene, which shows the same trend as expected from their molecular kinetic diameter and Henry's constants. The favorable molecular configuration of adsorbate and weak interaction of adsorbate-adsorbent will be beneficial on the diffusion rate of adsorbates. Moreover, the relation of corresponding activation energy values calculated according the relationship between the temperature and diffusivity to mesopore volume are displayed in Figure 6. It is seen for the opposite trend with diffusivity, *i.e.*, the E_a values decrease with the increase of the mesopores in four samples, which is similar to the previous reported results [8]. In addition, a significant downtrend can be seen for all adsorbates from Beta-0 to Beta-3 (e.g., from 24.7 kJ/moL to 15.6 kJ/moL for p-xylene, from 30.3 kJ/moL to 20.9 kJ/moL for m-xylene, from 31.1 kJ/moL to 21.9 kJ/moL for o-xylene, respectively). Such an observation reveals a general pattern that the activation energy involved in diffusion in a mesoporous zeolite would be relatively lower than that in a microporous zeolite.

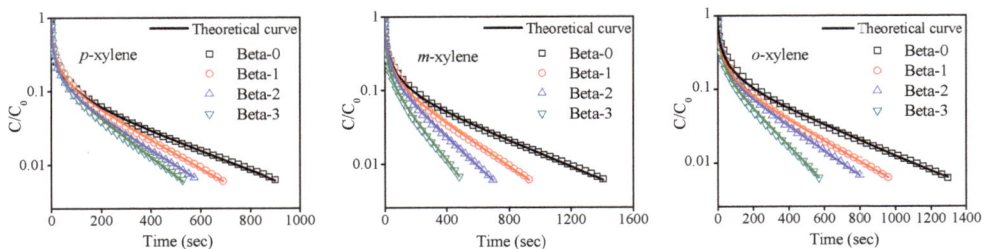

Figure 5. Experimental data (symbols) and theoretical ZLC curves (lines) for xylene isomers in Beta zeolite samples at 373 K.

Table 5. Zero length column (ZLC) fitting data for xylene isomers on Beta zeolite samples at 373 K.

Sample	L	β	$D_{eff}/R^2 (s^{-1} \cdot 10^{-4})$
		p-xylene	
Beta-0	17.7	2.97	3.74
Beta-1	15.6	2.94	5.18
Beta-2	15.9	2.95	6.06
Beta-3	13.1	2.91	7.48
		m-xylene	
Beta-0	13.2	2.91	2.84
Beta-1	12.9	2.90	4.26
Beta-2	15.8	2.95	5.01
Beta-3	16.9	2.96	6.85
		o-xylene	
Beta-0	15.4	2.94	2.73
Beta-1	13.4	2.91	3.97
Beta-2	13.4	2.91	4.64
Beta-3	14.3	2.93	6.35

The above results provide objective evidence that the presence of mesopores in zeolite crystals can facilitate the transport of molecules by more paths, shortened diffusion length and the less diffusion resistance, attributed to the introduction of mesoporous into Beta zeolite.

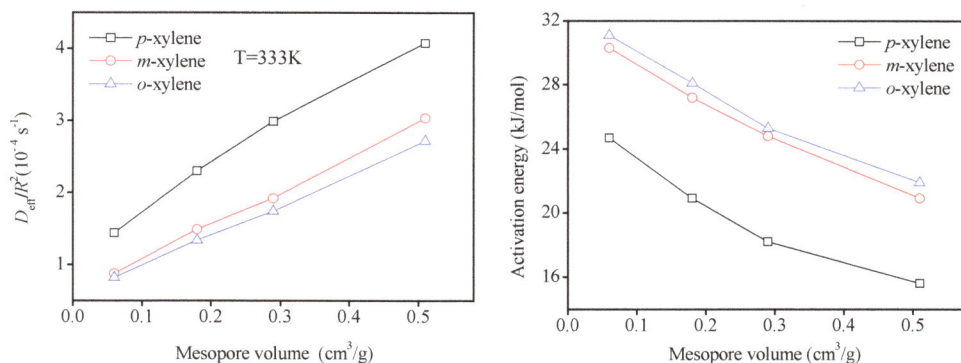

Figure 6. Effective diffusivities at different temperatures and activation energy of xylene isomers as the function of mesopore volume in Beta zeolite samples.

3. Experimental Section

3.1. Materials and Characterization

The mesoporous Beta zeolite samples used in this work were hydrothermally synthesized by employing organfunctioned nano-silica as silicon source, which was described in the previous paper [38]. Firstly, fumed silica (Aerosil 200, Degussa, Marl, NRW, Germany) was organic functioned with phenylaminopropyl-trimethoxysilane (PHAPTMS, Aldrich, Saint Louis, MO, USA), then obtained organo-silylated nano-SiO_2 was used as silicon source, according to the molar composition of Al_2O_3: 38 org-SiO_2: 1.5 Na_2O: 10 TEAOH: 532 H_2O, three mesoporous Beta zeolite samples with different mesoporosities were synthesized by adjusting silanization degree of nano-silica, and coded as Beta-n (n = 1, 2, 3). The corresponding microporous Beta zeolite was synthesized by applying nano-silica without silanization under the same composition and conditions, and denoted Beta-0 as a reference.

X-ray powder diffraction (XRD) patterns were examined using a SHIMADZU XRD-6000 diffractometer (Shimadzu, Nakagyo-ku, Kyoto, Japan) operating at 40 kV and 30 mA, which employed Ni-filtered and Cu-Kα radiation with the step of 0.02° and counting time of 10 s. Nitrogen adsorption/desorption isotherms of the samples at 77 K were gained on a Quantachrome Quadrasorb SI gas adsorption analyzer (Quantachrome Instruments, Boynton Beach, FL, USA). The total surface area was obtained by using the BET (Brunauer-Emmett-Teller) equation, whereas the external surface area and micropore volume were calculated by the t-plot method [39]. The total pore volume was evaluated at p/p = 0.95. The mesopore size distribution was derived from the adsorption branch of the isotherms using the density functional theory (DFT) method [40]. Scanning electron microscopy (SEM) and transmission electron microscope (TEM) images were obtained on a Hitachi S-4800II FE-SEM

(Hitachi HighTechnolo-gies, Tokyo, Japan) and JEOL JEM-1011 microscope (Jeol, Tokyo, Japan), respectively.

3.2. Adsorption Isotherms Measurements

The adsorption isotherms measurements were carried out by an intelligent gravimetric analyzer (IGA-002, Hiden, Warrington, UK) with three xylene isomers as the representative probe adsorbates and four Beta zeolite samples as absorbents. The key component of this apparatus is a sensitive microbalance with the accuracy of ± 0.1 µg, which was used to record the changes in weight of the samples under the computer control. Before the sorption measurement, the zeolite sample about 50 mg was activated under vacuum below 10^{-7} mbar and maintained at 673 K for 300 min to remove the impurities and moisture until the sample weight keeping constant. In the process of measurement, the vapor of the sorbate was gradually introduced into the sample chamber until the set pressure value was obtained. On reaching the adsorption equilibrium under the set point, the weight change was recorded by the computer software and the vapor pressure was increased to the next designed pressure. Finally, a complete adsorption isotherm was drawn at a certain pressure range. The measurements of adsorption isotherms for xylene isomers on Beta zeolites in this work were conducted in the pressure range of 0–20 mbar and temperature range of 308–338 K, respectively.

3.3. ZLC Measurement and Theory

The ZLC method was first proposed by Eic and Ruthven in the late 1980s [41]. Since then, it had become one of the most common methods to measure the intracrystalline diffusion coefficients [42–45]. This method has the advantage of eliminating the intrusion of extraneous heat and mass transfer resistances. The ZLC system was detailedly depicted in [41]. The experimental apparatus is similar to conventional gas chromatography; however, the packed column in gas chromatograph (GC) is replaced by a ZLC column, which is loaded with a very thin layer of sample sandwiched between two porous sintered metal disks. In accordance with the ZLC theory, the sorbate concentration was controlled under the very low relative partial pressures to guarantee the experiments were carried out within the linear region of the adsorption isotherms, i.e., within the Henry's law region. The effluent stream concentration of ZLC was monitored online by the flame ionization detector.

Assuming the adsorption occurring in a very thin layer of zeolite spherical crystals within the linear region of adsorption equilibrium at the very low adsorbate partial pressure, as well as neglecting the hold-up in the ZLC bed and the perfect mixing throughout the ZLC system, the relationship between relative effluent concentration c/c_0 and time t is given by references [41,46]:

$$\frac{c}{c_0} = 2L \sum_{n=1}^{\infty} \frac{\exp\left(-\frac{\beta_n^2 D_{\text{eff}} t}{R^2}\right)}{\left[\beta_n^2 + L(L-1)\right]} \tag{5}$$

where D_{eff}/R^2 is the effective diffusion time constant, R is the particle radius, L is a dimensionless parameter and the eigenvalues β_n are given by the roots of the auxiliary equation:

$$\beta_n \cot\beta_n + L - 1 = 0 \tag{6}$$

and

$$L = \frac{1}{3}\frac{FR^2}{KV_s D_{\text{eff}}} \tag{7}$$

where F is purge flow rate, V_s is the volume of adsorbent and K is the dimensionless Henry's law constant.

Based on the above relationships, D_{eff}/R^2 and L values can be extracted by fitting Equations (5)–(7) to the experimental ZLC data applying the full range method by matlab software [47].

Additionally, the relationship between the temperature and diffusivity is described by the Arrhenius form [48]:

$$D = D_0 \exp\left(-E_a/R_g T\right) \tag{8}$$

From measurements at three different temperatures, estimates for the activation energy E_a of diffusion may be calculated by

$$E_a = R_g T^2 \left(\frac{\partial \ln D_{\text{eff}}/R^2}{\partial T}\right) \tag{9}$$

where R_g is the ideal gas constant, the activation energy for diffusion E_a can be calculated from the slope of line which is made according to $\ln D_{\text{eff}}/R^2$ and $1/T$ [49].

4. Conclusions

The adsorption and diffusion properties of three xylene isomers on Beta zeolites with different mesoporosities were discussed to investigate the effects of mesopores in zeolites for molecules with different kinetic diameter, molecular configuration and dipole. The adsorption isotherms of three xylene isomers on mesoporous Beta zeolites presented the shapes of type IV, reflecting the introduction of secondary mesoporosity with together N_2 adsorption. The enhanced adsorption capacities, reduced values of Henry's constants K_H and initial heats of adsorption Q_{st} with increasing mesoporosity are observed. The successful fittings of adsorption isotherms of three xylene isomers on mesoporous Beta zeolites by the dual-site Toth model reflect the differences of adsorption for xylene isomers on microporous and mesoporous surfaces, a much weaker sorbent-sorbate interaction over the surface of mesopore than that over micropore surface are found. K_H and Q_{st} follow the order of p-xylene $<$ m-xylene $< o$-xylene, non-polar and smaller kinetic diameter could account for the smallest K_H and Q_{st} value of p-xylene. The ZLC desorption curves of three xylene isomers on the Beta zeolites, obtained at same experimental condition, present the enhanced desorption speed with increasing mesoporosity inside zeolite crystals. The effective diffusion time constants (D_{eff}/R^2) derived from desorption curves display the increased trend with the increase of mesopore volume, whereas activation energies decrease with the increase of mesoporosity. The results reflect the improved effect of mesopore on mass transfer in zeolite, due to more paths, shortened diffusion length and the less diffusion resistance in mesoporous zeolite relative to microporous zeolite. The diffusivity of three xylene isomers follow the order of o-xylene $< m$-xylene $< p$-xylene, which is ascribed to the smaller molecular kinetic diameter and weaker interaction of adsorbate-adsorbent for p-xylene relative to m-xylene and o-xylene.

Acknowledgments: This work was supported by the National Natural Science Foundation of China (Grant No. 51272169).

Author Contributions: Jinghong Ma and Ruifeng Li designed the experiments. Aixia Song and Duo Xu conducted the experiments. Duo Xu, Aixia Song, Jinghong Ma and Ruifeng Li analyzed the data. Jinghong Ma wrote the first draft of the manuscript, which was then revised by all other authors.

Conflicts of Interest: The authors declare no conflict of interest.

References

1. Baerlocher, C.; McCusker, L.B.; Olson, D.H. *Atlas of Zeolite Framework Types*; Elsevier Science: Amsterdam, The Netherlands, 2007; p. 73.

2. Ackley, M.W.; Rege, S.U.; Saxena, H. Application of natural zeolites in the purification and separation of gases. *Microporous Mesoporous Mater.* **2003**, *61*, 25–42.

3. Degnan, T.F., Jr. Applications of zeolites in petroleum refining. *Top. Catal.* **2000**, *13*, 349–356.

4. Christensen, C.H.; Johannsen, K.; Törnqvist, E.; Schmidt, I.; Topsøe, H. Mesoporous zeolite single crystal catalysts: Diffusion and catalysis in hierarchical zeolites. *Catal. Today* **2007**, *128*, 117–122.

5. Verboekend, D.; Pérez-Ramírez, J. Design of hierarchical zeolite catalysts by desilication. *Catal. Sci. Technol.* **2011**, *1*, 879–890.

6. Choi, M.; Cho, H.S.; Srivastava, R.; Venkatesan, C.; Choi, D.H.; Ryoo, R. Amphiphilic organosilane-directed synthesis of crystalline zeolite with tunable mesoporosity. *Nat. Mater.* **2006**, *5*, 718–723.

7. Tao, Y.; Kanoh, H.; Abrams, L.; Kaneko, K. Mesopore-modified zeolites: Preparation, characterization, and applications. *Chem. Rev.* **2006**, *106*, 896–910.

8. Zhao, H.; Ma, J.; Zhang, Q.; Liu, Z.; Li, R. Adsorption and diffusion of n-heptane and toluene over mesoporous ZSM-5 zeolites. *Ind. Eng. Chem. Res.* **2014**, *53*, 13810–13819.

9. Lee, S.; Kim, H.; Choi, M. Controlled decationization of X zeolite: Mesopore generation within zeolite crystallites for bulky molecular adsorption and transformation. *J. Mater. Chem. A* **2013**, *1*, 12096–12102.

10. Gu, F.N.; Wei, F.; Yang, J.Y.; Lin, N.; Lin, W.G.; Wang, Y.; Zhu, J.H. New strategy to synthesis of hierarchical mesoporous zeolites. *Chem. Mater.* **2010**, *22*, 2442–2450.

11. Bonilla, M.R.; Valiullin, R.; Kärger, J.; Bhatia, S.K. Understanding adsorption and transport of light gases in hierarchical materials using molecular simulation and effective medium theory. *J. Phys. Chem. C* **2014**, *118*, 14355–14370.

12. Groen, J.C.; Zhu, W.; Brouwer, S.; Huynink, S.J.; Kapteijn, F.; Moulijn, J.A.; Pérez-Ramírez, J. Direct demonstration of enhanced diffusion in mesoporous ZSM-5 zeolite obtained via controlled desilication. *J. Am. Chem. Soc.* **2007**, *129*, 355–360.

13. Zhao, L.; Shen, B.; Gao, J.; Xu, C. Investigation on the mechanism of diffusion in mesopore structured ZSM-5 and improved heavy oil conversion. *J. Catal.* **2008**, *258*, 228–234.

14. Newsam, J.M.; Treacy, M.M.; Koetsier, W.T.; de Gruyter, C.B. Structural characterization of zeolite beta. *Proc. R. Soc. Lond. A* **1988**, *420*, 375–405.

15. Brito, A.; Borges, M.E.; Otero, N. Zeolite Y as a heterogeneous catalyst in biodiesel fuel production from used vegetable oil. *Energy Fuels* **2007**, *21*, 3280–3283.

16. Perez-Pariente, J.; Sastre, E.; Fornes, V.; Martens, J.A.; Jacobs, P.A.; Corma, A. Isomerization and disproportionation of m-xylene over zeolite β. *Appl. Catal.* **1991**, *69*, 125–137.

17. Ratnasamy, P.; Bhat, R.N.; Pokhriyal, S.K.; Hegde, S.G.; Kumar, R. Reactions of aromatic hydrocarbons over zeolite β. *J. Catal.* **1989**, *119*, 65–70.

18. Llopis, F.J.; Sastre, G.; Corma, A. Xylene isomerization and aromatic alkylation in zeolites NU-87, SSZ-33, β, and ZSM-5: Molecular dynamics and catalytic studies. *J. Catal.* **2004**, *227*, 227–241.

19. Byun, Y.; Jo, D.; Shin, D.N.; Hong, S.B. Theoretical investigation of the isomerization and disproportionation of m-Xylene over medium-pore zeolites with different framework topologies. *ACS Catal.* **2014**, *4*, 1764–1776.

20. Fernandez, C.; Stan, I.; Gilson, J.P.; Thomas, K.; Vicente, A.; Bonilla, A.; Pérez-Ramírez, J. hierarchical ZSM-5 zeolites in shape-selective xylene isomerization: Role of mesoporosity and acid site speciation. *Chem. A Eur. J.* **2010**, *16*, 6224–6233.

21. Dehkordi, A.M.; Khademi, M. Adsorption of xylene isomers on Na-BETA zeolite: Equilibrium in batch adsorber. *Microporous Mesoporous Mater.* **2013**, *172*, 136–140.

22. Rasouli, M.; Yaghobi, N.; Allahgholipour, F.; Atashi, H. Para-xylene adsorption separation process using nano-zeolite Ba-X. *Chem. Eng. Res. Des.* **2014**, *92*, 1192–1199.

23. Sing, K.S. Reporting physisorption data for gas/solid systems with special reference to the determination of surface area and porosity (Recommendations 1984). *Pure Appl. Chem.* **1985**, *57*, 603–619.

24. Do, D.D. *Adsorption Analysis: Equilibria and Kinetics*; Imperial College Press: London, UK, 1998; p. 64.

25. Cavalcante, C.L., Jr.; Azevêdo, D.C.; Souza, I.G.; Silva, A.C.M.; Alsina, O.L.; Lima, V.E.; Araujo, A.S. Sorption and diffusion of p-xylene and o-xylene in aluminophosphate molecular sieve AlPO$_{4-11}$. *Adsorption* **2000**, *6*, 53–59.

26. Chiang, A.S.; Lee, C.K.; Chang, Z.H. Adsorption and diffusion of aromatics in AlPO$_{4-5}$. *Zeolites* **1991**, *11*, 380–386.

27. Peralta, D.; Barthelet, K.; Pérez-Pellitero, J.; Chizallet, C.; Chaplais, G.; Simon-Masseron, A.; Pirngruber, G.D. Adsorption and separation of xylene isomers: CPO-27-Ni vs HKUST-1 vs NaY. *J. Phys. Chem. C* **2012**, *116*, 21844–21855.

28. Ruthven, D.M.; Kaul, B.K. Adsorption of n-hexane and intermediate molecular weight aromatic hydrocarbons on LaY zeolite. *Ind. Eng. Chem. Res.* **1996**, *35*, 2060–2064.

29. Silva, J.A.; Rodrigues, A.E. Sorption and diffusion of n-pentane in pellets of 5A zeolite. *Ind. Eng. Chem. Res.* **1997**, *36*, 493–500.

30. Vavlitis, A.P.; Ruthven, D.M.; Loughlin, K.F. Sorption of n-pentane, n-octane, and n-decane in 5A zeolite crystals. *J. Colloid Interface Sci.* **1981**, *84*, 526–531.

31. Saha, D.; Deng, S. Adsorption equilibrium, kinetics, and enthalpy of N$_2$O on zeolite 4A and 13X. *J. Chem. Eng. Data* **2010**, *55*, 3312–3317.

32. Bárcia, P.S.; Silva, J.A.; Rodrigues, A.E. Adsorption equilibrium and kinetics of branched hexane isomers in pellets of Beta zeolite. *Microporous Mesoporous Mater.* **2005**, *79*, 145–163.

33. Liu, Z.; Fan, W.; Xue, Z.; Ma, J.; Li, R. Diffusion of n-alkanes in mesoporous 5A zeolites by ZLC method. *Adsorption* **2013**, *19*, 201–208.

34. Brandani, S. Effects of nonlinear equilibrium on zero length column experiments. *Chem. Eng. Sci.* **1998**, *53*, 2791–2798.

35. Brandani, S.; Jama, M.A.; Ruthven, D.M. ZLC measurements under non-linear conditions. *Chem. Eng. Sci.* **2000**, *55*, 1205–1212.

36. Ruthven, D.M.; Eic, M.; Richard, E. Diffusion of C$_8$ aromatic hydrocarbons in silicalite. *Zeolites* **1991**, *11*, 647–653.

37. Brandani, S.; Jama, M.; Ruthven, D. Diffusion, self-diffusion and counter-diffusion of benzene and p-xylene in silicalite. *Microporous Mesoporous Mater.* **2000**, *35*, 283–300.

38. Zhang, Q.; Ming, W.; Ma, J.; Zhang, J.; Wang, P.; Li, R. De novo assembly of mesoporous beta zeolite with intracrystalline channels and its catalytic performance for biodiesel production. *J. Mater. Chem. A* **2014**, *2*, 8712–8718.

39. Rouquerol, F.; Rouquerol, J.; Sing, K. *Adsorption by Powders and Porous Solids*; Academic Press: San Diego, CA, USA, 1999; p. 267.

40. Landers, J.; Gor, G.Y.; Neimark, A.V. Density functional theory methods for characterization of porous materials. *Colloids Surf. A* **2013**, *437*, 3–32.

41. Eic, M.; Ruthven, D.M. A new experimental technique for measurement of intracrystalline diffusivity. *Zeolites* **1988**, *8*, 40–45.

42. Gunadi, A.; Brandani, S. Diffusion of linear paraffins in NaCaA studied by the ZLC method. *Microporous Mesoporous Mater.* **2006**, *90*, 278–283.

43. Ruthven, D.M. Diffusion of aromatic hydrocarbons in silicalite/HZSM-5. *Adsorption* **2007**, *13*, 225–230.

44. Thang, H.V.; Malekian, A.; Eić, M.; On, D.T.; Kaliaguine, S. Diffusive characterization of large pore mesoporous materials with semi-crystalline zeolitic framework. *Stud. Surf. Sci. Catal.* **2003**, *146*, 145–148.

45. Malekian, A.; Vinh-Thang, H.; Huang, Q.; Eic, M.; Kaliaguine, S. Evaluation of the main diffusion path in novel micro-mesoporous zeolitic materials with the zero length column method. *Ind. Eng. Chem. Res.* **2007**, *46*, 5067–5073.

46. Hufton, J.R.; Ruthven, D.M. Diffusion of light alkanes in silicalite studied by the zero length column method. *Ind. Eng. Chem. Res.* **1993**, *32*, 2379–2386.

47. Han, M.; Yin, X.; In, Y.; Chen, S. Diffusion of aromatic hydrocarbon in ZSM-5 studied by the improved zero length column method. *Ind. Eng. Chem. Res.* **1999**, *38*, 3172–3175.

48. Cavalcante, C.L., Jr.; Silva, N.M.; Souza-Aguiar, E.F.; Sobrinho, E.V. Diffusion of paraffins in dealuminated Y mesoporous molecular sieve. *Adsorption* **2003**, *9*, 205–212.

49. Gobin, O.C.; Huang, Q.; Vinh-Thang, H.; Kleitz, F.; Eic, M.; Kaliaguine, S. Mesostructured silica SBA-16 with tailored intrawall porosity part 2: Diffusion. *J. Phys. Chem. C* **2007**, *111*, 3059–3065.